SOLID STATE PHYSICS

VOLUME 27

Contributors to This Volume

A. Corciovei

G. Costache

P. H. Dederichs

R. P. Huebener

D. Vamanu

J. Zak

SOLID STATE PHYSICS

Advances in Research and Applications

Editors

HENRY EHRENREICH

Division of Engineering and Applied Physics
Harvard University, Cambridge, Massachusetts

FREDERICK SEITZ

The Rockefeller University, New York, New York

DAVID TURNBULL

Division of Engineering and Applied Physics
Harvard University, Cambridge, Massachusetts

VOLUME 27

1972

ACADEMIC PRESS • NEW YORK AND LONDON

ACADEMIC PRESS, INC.
111 Fifth Avenue, New York, New York 10003

United Kingdom Edition published by
ACADEMIC PRESS, INC. (LONDON) LTD.
24/28 Oval Road, London NW1

LIBRARY OF CONGRESS CATALOG CARD NUMBER: 55-12200

PRINTED IN THE UNITED STATES OF AMERICA

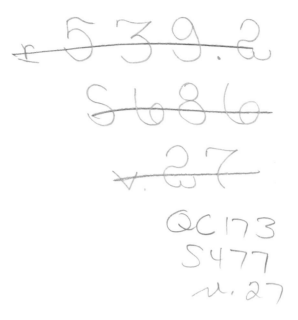

Contents

The *kq*-Representation in the Dynamics of Electrons in Solids

J. Zak

Thermoelectricity in Metals and Alloys

R. P. Huebener

Dynamical Diffraction Theory by Optical Potential Methods

P. H. Dederichs

Ferromagnetic Thin Films

A. Corciovei, G. Costache, and D. Vamanu

Contributors to Volume 27

Numbers in parentheses indicate the pages on which the authors' contributions begin.

A. CORCIOVEI, *Institute for Atomic Physics, Bucharest, Romania* (237)

G. COSTACHE, *Institute for Atomic Physics, Bucharest, Romania* (237)

P. H. DEDERICHS, *Institut für Festkörperforschung der Kernforschungsanlage Jülich, Jülich, West Germany* (135)

R. P. HUEBENER, *Argonne National Laboratory, Argonne, Illinois* (63)

D. VAMANU, *Institute for Atomic Physics, Bucharest, Romania* (237)

J. ZAK, *Department of Physics, Technion—Israel Institute of Technology, Haifa, Israel* (1)

Preface

This volume begins with an article by Zak in which he discusses a quantum mechanical representation, the *kq*-representation, whose properties he has investigated extensively in recent years. In addition to many other features of interest, this representation is particularly useful in providing deeper understanding of the motion of Bloch electrons in external electric and magnetic fields, which has been a controversial subject of long standing. This paper complements discussions of related topics that have appeared earlier in this serial publication, most notably the article on Formalisms in Band Theory by Blount (Volume 13) and that concerning aspects of Group Theory in Electron Dynamics by Brown (Volume 22). In the second article, Huebener gives a comprehensive survey of thermoelectric phenomena in metals and alloys and their theoretical interpretation.

Although multiple scattering theories have been familiar since the early part of this century, their general usefulness for treating a wide variety of physical problems has been recognized only relatively more recently. The review by Dederichs summarizes the applications of dynamical diffraction theory to some of the important areas of interest: electrons, X-rays, and neutrons in perfect and perturbed periodic lattices. This chapter should be of immediate interest to those concerned, for example, with low-energy electron diffraction, channeling of high-energy electrons, and dynamical scattering of Mössbauer quanta. While the article is self-contained it emphasizes developments which have occurred since an earlier review of the dynamical theory of X-ray diffraction by R. W. James in Volume 15. In the final article, Corciovei, Costache, and Vamanu review the unique features of the magnetic behavior of ferromagnetic thin films and their interpretation in terms of the phenomenological theories.

HENRY EHRENREICH
FREDERICK SEITZ
DAVID TURNBULL

Supplementary Monographs

Supplement 1: T. P. Das and E. L. Hahn
Nuclear Quadrupole Resonance Spectroscopy, 1958

Supplement 2: William Low
Paramagnetic Resonance in Solids, 1960

Supplement 3: A. A. Maradudin, E. W. Montroll, G. H. Weiss, and I. P. Ipatova, Theory of Lattice Dynamics in the Harmonic Approximation, 1971 (Second Edition)

Supplement 4: Albert C. Beer
Galvanomagnetic Effects in Semiconductors, 1963

Supplement 5: R. S. Knox
Theory of Excitons, 1963

Supplement 6: S. Amelinckx
The Direct Observation of Dislocations, 1964

Supplement 7: J. W. Corbett
Electron Radiation Damage in Semiconductors and Metals, 1966

Supplement 8: Jordan J. Markham
F-Centers in Alkali Halides, 1966

Supplement 9: Esther M. Conwell
High Field Transport in Semiconductors, 1967

Supplement 10: C. B. Duke
Tunneling in Solids, 1969

Supplement 11: Manuel Cardona
Optical Modulation Spectroscopy of Solids, 1969

Supplement 12: A. A. Abrikosov
An Introduction to the Theory of Normal Metals, 1971

Supplement 13: P. M. Platzman and P. A. Wolff
Waves and Interactions in Solid State Plasmas (In Preparation)

Articles to Appear Shortly

F. C. BROWN	Ultraviolet Spectroscopy of Solids Using Synchrotron Radiation
J. I. BUDNICK	Hyperfine Interactions and Spin Densities in Magnetically Ordered Metals
G. BURNS — B. D. SILVERMAN	Ferroelectricity
J. E. ENDERBY	Electronic Properties of Liquid Metals
Y. V. GULYAEV	Acoustoelectric Phenomena in Semiconductors
J. F. HAMILTON — L. SLIFKIN	The Photographic Process
J. R. HARDY — A. M. KARO	The Lattice Dynamics and Statics of Alkali Halide Crystals
N. D. LANG	Density Functional Formalism and Its Application to the Study of Metal Surfaces
G. D. MAHAN	Many Body Theory of X-ray Emission in Solids
S. C. MOSS	Structure of Amorphous Solids
T. NAGAMIYA	Helical Spin Ordering — 2
P. O. NILSSON	Optical Properties of Metals and Alloys
R. O. SIMMONS	Noble Gas Crystals
G. SLACK — C. T. WALKER	Thermal Conductivity in Nonmetals

The *kq*-Representation in the Dynamics of Electrons in Solids*

J. Zak

Department of Physics, Technion—Israel Institute of Technology, Haifa, Israel

I. Introduction

Many phenomena in solids are connected with their electronic motion. In general, this motion can be very complicated because of the complexity of the forces in a solid. It turns out, however, that for many purposes a good model is the idealized Bloch model,[1] which is based on the motion of a charged particle in a periodic potential. The Bloch model forms a basis for most of solid state theory and it is therefore of great importance and interest to study the motion of a single electron in a periodic potential.

* This work has been sponsored partly by the European Research Office of the U.S. Army under contract No. DAJA37-70-C-0542.

[1] F. Bloch, *Ann. Phys.* **52**, 555 (1928).

The motion of an electron in a periodic structure has some fundamental features that makes it similar to the motion of a free particle. On a scale much larger than the lattice constant a crystal is as good a homogeneous medium as the free space is. One should therefore expect some gross features of the motion in a crystalline world to be similar to the motion in free space. This similarity was first established for the properties of an electron gas in a metal.[2] Later, it was shown[3] that the motion of electrons in perturbed crystals (in an impurity potential or a magnetic field with the condition that the perturbation varies on a scale much larger than the lattice constant) is very much like the motion in free space with the mass of the electron replaced by its effective mass.

Further and more precise theoretical developments[4–6] have shown that the fundamental laws of motion in a periodic potential resemble to a great extent the laws in free space and are very simple in their structure. Thus, in the lowest approximation of perturbation theory the effective Hamiltonian for a Bloch electron in a magnetic field $\mathbf{H}$ is $\epsilon_n(\mathbf{k} + (e/2c)\mathbf{H} \times i\, \partial/\partial\mathbf{k})$ where $\epsilon_n(\mathbf{k})$ is the dispersion law for the n's energy band. This Hamiltonian is very similar in structure to the Hamiltonian for a free electron (the term "free electron" will be used in the absence of the crystal) in a magnetic field. In the latter, the band index is absent and the dispersion law is quadratic. The effective Hamiltonian for a Bloch electron in a magnetic field is so simple in structure that one would expect its derivation also to be simple.[4] It is known, however, that the conventional derivations of this result[4–6] are much more complicated than may be expected from the result itself. The same is true for other results in the fundamental dynamics of electrons in solids. A particularly striking example is the acceleration theorem of a Bloch electron in an electric field $\mathbf{E}$: $\dot{\mathbf{k}} = -e\mathbf{E}$. An exact derivation of this theorem is not simple at all,[7] while the theorem itself is extremely simple.

These examples show therefore that on one hand the fundamental laws of the dynamics of electrons in solids are very simple in structure while on the other hand, their derivation is rather complicated. One is

[2] A. Sommerfeld and H. Bethe, Elektronentheorie der Metalle, "Handbuch der Physik," pp. 333–622. Springer, Berlin 24/2, 1933.

[3] C. Kittel and A. H. Mitchell, *Phys. Rev.* **96**, 1488 (1954); J. M. Luttinger and W. Kohn, *ibid.* **97**, 869 (1955).

[4] W. Kohn, *Phys. Rev.* **115**, 1460 (1959).

[5] E. I. Blount, *Phys. Rev.* **126**, 1636 (1962).

[6] L. M. Roth, *J. Phys. Chem. Solids* **23**, 433 (1962).

[7] L. D. Landau and E. M. Lifshitz, "Quantum Mechanics." Pergamon Press, London, 1958.

led to ask the question whether a more suitable way of describing the motion of Bloch electrons does not exist?

It is well known in physics that the description and solution of some problems can often be significantly simplified by a suitable choice of coordinates. Usually the choice of the best coordinates is dictated by the symmetry of the problem. In a solid the most fundamental symmetry is its invariance under translations. A question could be asked, what are the coordinates that reflect best the symmetry of a crystal?

In the Bloch theory of conduction electrons in solids the concept of quasimomentum k is of great importance because it is a conserved quantity and in addition it carries the significant information about the momentum of the electron. One should expect k to be a suitable coordinate in describing the motion of an electron in a periodic structure. It can however be seen that k does not carry complete quantum-mechanical information about the motion. From the point of view of elementary quantum mechanics, it is possible along with k to specify also to some extent the coordinate of the electron. This is so because k carries only partial information about the linear momentum p of the electron and when k is given, p is defined only up to an additive constant $2\pi\hbar/a$ where a is the lattice constant. It can easily be seen (and will be shown in Section II) that along with k quantum mechanics allows also to specify the quasicoordinate q of the electron[8] which gives the position x up to the additive constant a. The coordinate q carries the most relevant information about the location of the electron in a periodic structure. Indeed, the periodic potential of a crystal varies only inside a unit cell and is therefore a function of the quasicoordinate q only. From the point of view of the periodic structure it is completely irrelevant in which of the unit cells the electron is located. What counts is its position inside a unit cell and this is given by q. Since k carries the most relevant information about the momentum and similarly q about the coordinate and since they also reflect the symmetry of the periodic structure it is to be expected that these coordinates will be suitable for describing the motion.

In order to clarify the significance of the coordinates k and q it is instructive to compare them with the spherical coordinates in describing motion in a central field. In the latter problem the potential energy depends only on r, the absolute value of the radius vector, while the constants of motion, the angular momentum, are functions of the angles θ and φ. In the Bloch problem, the periodic potential depends only on q while the constant of motion, the quasimomentum, is k itself. This

[8] J. Zak, *Phys. Rev. Lett.* **19**, 1385 (1967).

analogy shows that k and q reflect the symmetry of the periodic problem in a way similar to the way the spherical coordinates reflect the symmetry of the central field problem.

Since k and q carry most relevant information about the motion of an electron in a crystal it is to be expected that their use will significantly simplify the description of many problems in solids.

In this article we will demonstrate the usefulness of the coordinates k and q in the following examples: The impurity problem in semiconductors, the dynamics in a constant electric field and the effective Hamiltonian for a Bloch electron in a magnetic field.

II. The kq-Representation

The most often used representations in elementary quantum mechanics are the coordinate representation (x-representation) and the momentum representation (p-representation). It is well known that for a spinless particle with one degree of freedom the coordinate x and the momentum p each separately form a complete set of commuting operators.[9] This means, for example, that x specifies a complete set of functions $\delta(x - x_0)$ such that

$$x\, \delta(x - x_0) = x_0\, \delta(x - x_0), \tag{2.1}$$

where $\delta(x - x_0)$ is the Dirac δ-function and x_0 is the eigenvalue of x in the state $\delta(x - x_0)$. Similarly, the momentum p also specifies a complete set of functions. The claim that x forms a "complete set of commuting operators" means that the only operator that commutes with x is necessarily a function of x. No function of the momentum operator p commutes with x. It is said that x forms a quantum-mechanical representation, the x-representation. The momentum p has similar properties and it also defines a quantum-mechanical representation. A representation in quantum mechanics gives a scheme (or a language) for describing the motion. Thus, in the x-representation, any operator is a function of x and derivatives with respect to x (the latter comes from the momentum, $p = -i\hbar\, \partial/\partial x$) and the wave function ψ depends on x.

Let us now describe a representation that is based on the quasi-momentum k and quasicoordinate q. The operator that defines k is well known in solid state physics and is the translation operator (in what follows we will assume $\hbar = 1$)

$$T(a) = \exp(ipa), \tag{2.2}$$

[9] P. A. M. Dirac, "The Principles of Quantum Mechanics." Oxford Univ. Press, London and New York, 1958.

where p is the linear momentum and a is the lattice constant. The eigenvalues of $T(a)$ are $\exp(ika)$ and this is usually the way the quasimomentum k is defined.[10] k assumes values from 0 to $2\pi/a$ (or from $-\pi/a$ to π/a) and is used for specifying Bloch functions

$$T(a)\,\psi_k(x) = \exp(ika)\,\psi_k(x). \tag{2.3}$$

The operator $T(a)$ by itself does not form a complete set of commuting operators (the way x or p does) because as will be shown below another operator can be found that commutes with (2.2) and is not a function of it. It is for this reason that the function $\psi_k(x)$ in (2.3) is not defined completely by the requirement that it is an eigenfunction of $T(a)$. Physically this means that along with k one can specify the motion of the electron by another coordinate. In order to define $\psi_k(x)$ completely one has to require that $\psi_k(x)$ is also an eigenfunction of other operators that commute with $T(a)$ and that are independent of the latter. As can be easily seen an operator that commutes with $T(a)$ is[8]

$$\tau(2\pi/a) = \exp(ix2\pi/a). \tag{2.4}$$

The eigenvalues of this operator are $\exp(iq2\pi/a)$ where q varies from 0 to a. It is clear that the operator (2.4) defines the quasicoordinate q that was mentioned in the Introduction. The structure of the operators $T(a)$ and $\tau(2\pi/a)$ is very similar. The first of them represents a translation by a in regular space while the second is a translation operator in p-space ($x = i\,\partial/\partial p$). The importance of the two operators $T(a)$ and $\tau(2\pi/a)$ is that together they form a complete set of commuting operators or a quantum mechanical representation.[11] This means that the operators $T(a)$ and $\tau(2\pi/a)$ commute and also that any operator that commutes with both of them is necessarily a function of them. If we require that the function $\psi_k(x)$ in (2.3) is also an eigenfunction of $\tau(2\pi/a)$, this will define $\psi_k(x)$ completely. It can be checked that the requirement

$$\tau(2\pi/a)\,\psi_{kq}(x) = \exp(iq2\pi/a)\,\psi_{kq}(x) \tag{2.5}$$

leads to the function[11]

$$\psi_{kq}(x) = \left(\frac{2\pi}{a}\right)^{1/2} \sum_n \exp(ikna)\,\delta(x - q - na), \tag{2.6}$$

where k and q are the quasimomentum and the quasicoordinate respec-

[10] C. Kittel, "Quantum Theory of Solids." Wiley, New York, 1963.
[11] J. Zak, *Phys. Rev.* **168**, 686 (1968).

tively. Being eigenfunctions of a complete set of commuting operators $T(a)$ and $\tau(2\pi/a)$, the functions $\psi_{kq}(x)$ form a complete system of functions. These functions satisfy the following relations[11]:

$$\int \psi_{kq}^*(x)\, \psi_{k'q'}(x)\, dx = \sum_m \delta(k - k' - (2\pi/a)m) \sum_n \delta(q - q' - na) \tag{2.6a}$$

$$\int \psi_{kq}^*(x)\, \psi_{kq}(x')\, dk\, dq = \delta(x - x'). \tag{2.6b}$$

Relation (2.6a) is the orthogonality condition, while (2.6b) expresses the completeness of the functions (2.6). In relation (2.6a), sums on δ-functions appear to assure periodicity in k and in q. The labels k and q in (2.6) define the function $\psi_{kq}(x)$ completely (up to a phase factor), meaning that the coordinates k and q define a quantum-mechanical representation.[9] This representation is called the kq-representation.

The functions (2.6) are clearly Bloch-like functions because they satisfy Eq. (2.3). Bloch functions $\psi_{nk}(x)$ satisfy in addition to relation (2.3) also Schrödinger's equation (this being the origin of the band index n). The eigenfunctions (2.6) of $T(a)$ and $\tau(2\pi/a)$ satisfy Eqs. (2.3) and (2.5). It is for this reason that in the functions (2.6) the quasicoordinate q appears instead of the band index in the Bloch functions. It is interesting to compare the functions $\psi_{kq}(x)$ in (2.6) with the expression for Bloch functions $\psi_{nk}(x)$ by means of Wannier functions[12] $a_n(x - la)$

$$\psi_{nk} = \sum_l \exp(ikna)\, a_n(x - la) \tag{2.7}$$

Comparing (2.7) with (2.6) it is seen that the $\psi_{kq}(x)$ functions are Bloch-type functions that correspond to infinitely localized Wannier functions.

In order to be able to write down Schrödinger's equation in the kq-representation one needs the expressions for the operators x and p in the coordinates k and q. For this one has to calculate the matrices $\langle kq|x|k'q'\rangle$ and $\langle kq|p|k'q'\rangle$. For example,

$$\begin{aligned}\langle kq|p|k'q'\rangle &= \int dp''\, dp' \langle kq|p''\rangle\langle p''|p|p'\rangle\langle p'|k'q'\rangle \\ &= \int dp''\, p'' \langle kq|p''\rangle\langle p''|k'q'\rangle \\ &= -i(\partial/\partial q')\, \delta(q' - q)\, \delta(k' - k).\end{aligned} \tag{2.8}$$

[12] E. I. Blount, *Solid State Phys.* **13** (1961).

Similarly, one finds also the matrix $\langle kq|x|k'q'\rangle$ and from these expressions, it follows that[11]

$$p = -i\,\partial/\partial q \tag{2.9}$$

$$x = i\,\partial/\partial k + q \tag{2.10}$$

It should be noticed that the asymmetry in the expressions for x and p is connected with the special choice of phase[11] for the functions (2.6).

All the results until now were written for one dimension. Their generalization to three dimensions is straightforward. The translations in three dimensions in direct and reciprocal spaces are

$$T(\mathbf{a}_m) = \exp(i\mathbf{p}\cdot\mathbf{a}_m), \tag{2.11}$$

$$\tau(\mathbf{b}_n) = \exp(i\mathbf{r}\cdot\mathbf{b}_n), \tag{2.12}$$

where $\mathbf{a}_m$ ($m = 1, 2, 3$) are the unit cell vectors of the Bravais lattice and $\mathbf{b}_n$ ($n = 1, 2, 3$) are the unit cell vectors of the reciprocal lattice, $\mathbf{a}_m \cdot \mathbf{b}_n = 2\pi\delta_{mn}$. The eigenfunctions of $T(\mathbf{a}_m)$ and $\tau(\mathbf{b}_n)$ are

$$\psi_{kq}(\mathbf{r}) = \left(\frac{V_0}{(2\pi)^3}\right)^{1/2} \sum_{R_n} \exp(i\mathbf{k}\cdot\mathbf{R}_n)\,\delta(\mathbf{r}-\mathbf{q}-\mathbf{R}_n), \tag{2.13}$$

where V_0 is the volume of a unit cell in the Bravais lattice, $\mathbf{R}_n$ is a Bravais lattice vector and $\mathbf{k}$ and $\mathbf{q}$ are the quasimomentum and quasicoordinate in three dimensions. They vary in unit cells of the reciprocal and direct space correspondingly. Finally, the operators $\mathbf{p}$ and $\mathbf{r}$ are

$$\mathbf{p} = -i\,\partial/\partial\mathbf{q}, \tag{2.14}$$

$$\mathbf{r} = i\,\partial/\partial\mathbf{k} + \mathbf{q}. \tag{2.15}$$

The main feature of the *kq*-representation is in the simultaneous use of partial information about both the coordinate and the momentum. By measuring $\mathbf{k}$ and $\mathbf{q}$ one can tell where in the unit cell of k-space and q-space the values of the operators are but not in which of the cells they are. This is a consequence of the meaning of the quasicoordinates $\mathbf{k}$ and $\mathbf{q}$: $\mathbf{k}$ and $\mathbf{k} + \mathbf{K}_m$ ($\mathbf{K}_m$ is a vector of the reciprocal lattice) define the same eigenvalue of $\exp(i\mathbf{p}\cdot\mathbf{R}_n)$ and similarly $\mathbf{q}$ and $\mathbf{q} + \mathbf{R}_n$ define the same eigenvalue of $\exp(i\mathbf{r}\cdot\mathbf{K}_m)$. The knowledge of $\mathbf{k}$ and $\mathbf{q}$ leads only to partial information of the momentum and the coordinate and there is no violation of the uncertainty principle.

From the definition of the coordinates k and q it should be clear that the *kq*-representation can be defined on any pair of conjugate coordinates that satisfy the commutation relation $[p, x] = -i$. The crystal is just a natural medium for the quasicoordinates k and q. In general, if there are two operators A and B (for example, the angular momentum and the angle) that satisfy the commutation relation $[A, B] = -i$ one can always choose an arbitrary constant a and define the operators (2.2) and (2.4).[13] For the angular momentum-angle operators this constant is no longer arbitrary and is given by the very nature of the angular coordinate. From the meaning of the angle α it is clear that α and $\alpha + 2\pi$ should give the same result. Therefore for the angular coordinate the constant a should be equal to 2π.

It is of interest to point out the difference between using the quasicoordinate description in a crystal and in the angular coordinate problem. In the former, the use of quasicoordinates is only a convenience and in some problems the x-representation will be more convenient while in others one will prefer the kq-representation. For the angular coordinate this is not so. The angular coordinate is by its nature a quasicoordinate and the only way to describe it is the framework of the kq-representation.[13a]

Having established the kq-representation it is possible to write Schrödinger's equation in this representation. In the conventional r-representation (coordinate representation) Schrödinger's equation for an electron in a periodic potential $V(\mathbf{r})$, in a constant magnetic field $\mathbf{H}$ and an additional electric potential $v(\mathbf{r})$ (the latter can be caused, for example, by an impurity or a constant electric field) is

$$\Big[\Big\{\Big(\mathbf{p} + \frac{e}{2c}\mathbf{H} \times \mathbf{r}\Big)^2 \Big/ 2m\Big\} + V(\mathbf{r}) + v(\mathbf{r})\Big]\,\psi(\mathbf{r}) = \epsilon\psi(\mathbf{r}), \tag{2.16}$$

where a special gauge $\mathbf{A} = \frac{1}{2}(\mathbf{H} \times \mathbf{r})$ was chosen for the vector potential $\mathbf{A}$. By using expressions (2.14) and (2.15), this equation can be written in the kq-representation

$$\Big\{\Big\{\Big[-i\frac{\partial}{\partial\mathbf{q}} + \frac{e}{2c}\mathbf{H} \times \Big(\mathrm{i}\frac{\partial}{\partial\mathbf{k}} + \mathbf{q}\Big)\Big]^2 \Big/ 2m\Big\} + V(\mathbf{q}) + v(i\,\partial/\partial\mathbf{k} + \mathbf{q})\Big\}\; C(\mathbf{kq})$$
$$= \epsilon C(\mathbf{kq}), \tag{2.17}$$

where $C(\mathbf{kq})$ is the wave function in the kq-representation

$$\psi(\mathbf{r}) = \int d\mathbf{k}\,d\mathbf{q}\;C(\mathbf{kq})\,\psi_{kq}(\mathbf{r}). \tag{2.18}$$

[13] Y. Aharanov, H. Pendleton, and A. Peterson, *Int. J. Theoret. Phys.* **2**, 213 (1969).
[13a] J. Zak, *Phys. Lett.* **29A**, 393 (1969); *Phys. Rev.* **187**, 1803 (1969).

The eigenfunctions $\psi_{kq}(\mathbf{r})$ of the translation operators (2.11) and (2.12) are given by relation (2.13). A very important feature of Eq. (2.17) is the appearance of the periodic potential as a function of $\mathbf{q}$ only. This feature of the *kq*-representation was already mentioned before on the basis of qualitative considerations. Now this can be shown quantitatively. Being a periodic function, $V(\mathbf{r})$ can be expanded in a series

$$V(\mathbf{r}) = \sum_{\mathbf{K}_m} V(\mathbf{K}_m) \exp(i\mathbf{K}_m \cdot \mathbf{r}), \tag{2.19}$$

where $\mathbf{K}_m$ are vectors of the reciprocal lattice. From (2.18) it follows that

$$C(\mathbf{kq}) = \int d\mathbf{r}\, \psi(\mathbf{r})\, \psi_{kq}^*(\mathbf{r}) = \sum_{\mathbf{R}_n} \exp(-i\mathbf{k} \cdot \mathbf{R}_n)\, \psi(\mathbf{q} + \mathbf{R}_n) \tag{2.20}$$

and therefore $C(\mathbf{kq})$ satisfies the following boundary conditions

$$C(\mathbf{k} + \mathbf{K}_m, \mathbf{q}) = C(\mathbf{k}, \mathbf{q}), \tag{2.21}$$

$$C(\mathbf{k}, \mathbf{q} + \mathbf{R}_n) = \exp(i\mathbf{k} \cdot \mathbf{R}_n)\, C(\mathbf{k}, \mathbf{q}), \tag{2.22}$$

which are the same boundary conditions satisfied by a Bloch function $\psi_{nk}(\mathbf{q})$. It can now be seen that

$$V(i\, \partial/\partial\mathbf{k} + \mathbf{q})\, C(\mathbf{kq}) = V(\mathbf{q})\, C(\mathbf{kq}), \tag{2.23}$$

which follows at once from the expansion (2.19) and the boundary condition (2.21). The periodic potential V depends therefore in Eq. (2.17) on $\mathbf{q}$ only. This dependence of the periodic potential on $\mathbf{q}$ only is of great importance because, as we will see later, it will lead to a separation of variables in Schrödinger's equation.

The wave function $C(\mathbf{kq})$ in the *kq*-representation is a Bloch-type function satisfying the same boundary conditions (2.21), (2.22) as a Bloch function. This feature of $C(\mathbf{kq})$ leads to the following expansion in Bloch functions

$$C(\mathbf{kq}) = \sum_{n} B_n(\mathbf{k})\, \psi_{nk}(\mathbf{q}) \tag{2.24}$$

where $B_n(\mathbf{k})$ are periodic in $\mathbf{k}$ and the summation is on the band indices. A formal derivation of this expansion will be given in the next section. Expansion (2.24) will turn out to be very important in applying the *kq*-representation to the dynamics of electrons in solids. The connections between the wave functions in different representations are given in Appendix A.

III. Bloch Electron in the *kq*-Representation

Let us first discuss the motion of an electron in a periodic potential in the absence of any perturbation. Equation (2.17) then will become

$$[(1/2m)(-i\,\partial/\partial\mathbf{q})^2 + V(\mathbf{q})]\, C(\mathbf{kq}) = \epsilon C(\mathbf{kq}) \tag{3.1}$$

The dependence of this equation on the quasi momentum $\mathbf{k}$ appears only through the boundary conditions (2.21) and (2.22) on the wavefunction. The Bloch solutions of Eq. (3.1) are (Bloch functions in the *kq*-representation; See Eq. (2.20)).

$$C_{nk_B}(\mathbf{kq}) = \psi_{nk}(\mathbf{q}) \sum_{\mathbf{K}_m} \delta(\mathbf{k} - \mathbf{k}_B - \mathbf{K}_m). \tag{3.2}$$

In (3.2) a distinction is made between the quasimomentum $\mathbf{k}$ as a coordinate of the wavefunction and the Bloch quasimomentum $\mathbf{k}_B$ as an eigenvalue of the translation operator (2.11)

$$T(\mathbf{R}_n)\, C_{nk_B}(\mathbf{kq}) = \exp(i\mathbf{k}_B \cdot \mathbf{R}_n)\, C_{nk_B}(\mathbf{kq}). \tag{3.3}$$

The physical meaning of the Bloch functions (3.2) in the *kq*-representation is very simple: $C_{nk_B}(\mathbf{kq})$ does not vanish only when $\mathbf{k} = \mathbf{k}_B$ up to a vector of a reciprocal lattice $\mathbf{K}_m$. Equation (3.3) gives the Bloch theorem written in the *kq*-representation. It is of some interest to discuss this theorem in light of the operators $T(\mathbf{R}_n)$ and $\tau(\mathbf{K}_m)$. The Bloch theorem is a consequence of the fact that $T(\mathbf{R}_n)$ commutes with the periodic potential $V(\mathbf{r})$ in (2.19), the latter being a function of the operators $\tau(\mathbf{K}_m)$. It is therefore seen that the operators $T(\mathbf{R}_n)$ and $\tau(\mathbf{K}_m)$ on which the *kq*-representation is built play a fundamental role in the Bloch theorem.

Having the Bloch functions (3.2) in the *kq*-representation we can now prove the validity of expansion (2.24). The Bloch functions form a complete set of functions and therefore any function $C(\mathbf{kq})$ can be expanded in terms of them

$$C(\mathbf{kq}) = \sum_n \int d\mathbf{k}_B\, B_n(\mathbf{k}_B)\, C_{nk_B}(\mathbf{kq}).$$

By using the explicit form of the Bloch functions $C_{nk_B}(\mathbf{kq})$ one comes to expansion (2.24).

In further discussion of the dynamics of Bloch electrons the following remark will be quite useful. In the *kq*-representation one of the coordinates namely $\mathbf{q}$ is straightforwardly connected with the Bravais lattice of the crystal. The other coordinate, the quasimomentum $\mathbf{k}$, which has been known since the beginning of solid state theory is a much more abstract concept and is connected with the reciprocal space. For a qualitative discussion and sometimes also for a quantitative theory it is more convenient to replace $\mathbf{k}$ by the Bravais lattice vector. This can be achieved by using Eq. (2.20) which connects $C(\mathbf{kq})$ with $\psi(\mathbf{q} + \mathbf{R}_n)$. From (2.20) and (3.1) it follows that for a Bloch electron $\psi(\mathbf{q} + \mathbf{R}_n)$ satisfies the following Schrödinger equation

$$[(1/2m)(-i\,\partial/\partial\mathbf{q})^2 + V(\mathbf{q})]\,\psi(\mathbf{q} + \mathbf{R}_n) = \epsilon\psi(\mathbf{q} + \mathbf{R}_n) \tag{3.4}$$

The last equation looks very much like Schrödinger's equation in the ordinary r-representation, with $\mathbf{r}$ replaced by $\mathbf{q} + \mathbf{R}_n$. However, the introduction of the quasicoordinate leads to a very significant difference: The Hamiltonian depends only on $\mathbf{q}$. The independence of the Hamiltonian of $\mathbf{R}_n$ is to be expected and is a consequence of the translational symmetry of the crystal. What this means is that the Hamiltonian for an unperturbed Bloch electron is the same no matter which unit cell we choose. Since the Hamiltonian does not depend on $\mathbf{R}$ (in what follows $\mathbf{R}$ will be used instead of $\mathbf{R}_n$) the coordinate conjugate to $\mathbf{R}$, which is the quasimomentum $\mathbf{k}$, is a constant of motion. That $\mathbf{R}$ and $\mathbf{k}$ are conjugate coordinates can be seen from the two alternative expressions for $\mathbf{r}$: in the kq-representation $\mathbf{r} = i\,\partial/\partial\mathbf{k} + \mathbf{q}$ (Eq. (2.15)); compare it with $\mathbf{r} = \mathbf{R} + \mathbf{q}$. It follows that $\mathbf{R} = i\,\partial/\partial\mathbf{k}$. The commutator of $\mathbf{R}$ and $\mathbf{k}$ therefore equals i and this means that $\mathbf{R}$ and $\mathbf{k}$ are conjugate coordinates. In a free space all points are identical and the conserved quantity is the linear momentum. In a crystalline space the identical points lie on a Bravais lattice, and the conserved quantity is the quasimomentum. The analogy between $\mathbf{r}$ and $\mathbf{p}$ for a free electron and $\mathbf{R}$ and $\mathbf{k}$ for a Bloch electron goes much further. To see it let us expand $\psi(\mathbf{R} + \mathbf{q})$ in Bloch functions. By inverting relation (2.20), by using the expansion of $C(\mathbf{kq})$ in Bloch functions (Eq. (2.24)) and remembering that Bloch functions are periodic in $\mathbf{k}$ one finds

$$\psi(\mathbf{R} + \mathbf{q}) = \sum_n \psi_{nk}(\mathbf{q})\,F_n(\mathbf{R}), \tag{3.5}$$

where $\psi_{nk}(\mathbf{q})$ is an operator with $\mathbf{k}$ replaced by $-i\,\partial/\partial\mathbf{R}$ (here again $\mathbf{R}$ and $\mathbf{k}$ appear as conjugate coordinates) and $F_n(\mathbf{R})$ is the Fourier transform of $B_n(\mathbf{k})$ in (2.24):

$$F_n(\mathbf{R}) = \int \exp(i\mathbf{k}\cdot\mathbf{R})\,B_n(\mathbf{k}).$$

The operator $\psi_{nk}(\mathbf{q})$ in (3.5) was already used previously[6] and can be defined as follows

$$\psi_{nk}(\mathbf{q}) = \sum_{\mathbf{R}_n} \exp(\mathbf{R}_m\cdot\partial/\partial\mathbf{R})\,a_n(\mathbf{q} - \mathbf{R}_m),$$

where $a_n(\mathbf{q})$ are the Wannier functions. A detailed derivation of expansion (3.5) is given in Appendix A. By using expansion (3.5) in Eq. (3.1) for $\mathbf{H} = 0$, $v = 0$, one finds the equation that $F_m(\mathbf{R})$ satisfies for an unperturbed Bloch electron

$$\epsilon_m(\mathbf{k}) F_m(\mathbf{R}) = \epsilon F_m(\mathbf{R}) \tag{3.6}$$

Here again $\mathbf{k} = -i\,\partial/\partial\mathbf{R}$. The solution of (3.6) corresponding to $\epsilon = \epsilon_n(\mathbf{k})$ ($\mathbf{k}$ is the quasimomentum) is

$$F_m(\mathbf{R}) = \exp(i\mathbf{k}\cdot\mathbf{R})\,\delta_{nm} \tag{3.7}$$

The result (3.7) is of great interest. It shows that in the $\mathbf{R}$ coordinate the Bloch electron behaves like a free particle. This is to be expected because all the points in a Bravais lattice are identical. We see again that for a Bloch electron $\mathbf{k}$ and $\mathbf{R}$ take the place of $\mathbf{p}$ and $\mathbf{r}$ of a free electron. The Hamiltonian of the unperturbed Bloch electron in the space of the Bravais lattice is given by $\epsilon_m(\mathbf{k})$ with $\mathbf{k} = -i\,\partial/\partial\mathbf{R}$.

If a perturbation is present that varies slowly on the scale of a lattice constant then to a good approximation this perturbation will be a function of $\mathbf{R}$ only. Since in the absence of the perturbation the Hamiltonian is $\epsilon_m(-i\,\partial/\partial\mathbf{R})$ one would expect that when a magnetic field $\mathbf{H}$ and an electric perturbation v are present the effective Hamiltonian will be

$$\epsilon_m(-i\,\partial/\partial\mathbf{R} + (e/2c)\,\mathbf{H}\times\mathbf{R}) + v(\mathbf{R}). \tag{3.8}$$

This simple result follows in a qualitative way from the kq-representation with $\mathbf{k}$ replaced by its conjugate coordinate $\mathbf{R}$. As will be shown in the following sections the whole dynamics of the electrons in solids will be significantly simplified by the use of the kq-representation.

In conclusion of this section let us make the following remark. Although Eq. (3.1) and (3.4) are identical in their form it turns out that sometimes it is more convenient to use Eq. (3.4) in order to arrive at simple qualitative results (as was done in the former paragraph). Formally, we could clearly arrive at the same results by using Eq. (3.1) in the kq-representation. Qualitatively, however, the coordinate $\mathbf{R}$ has a much simpler meaning in a crystal than $\mathbf{k}$ and it is therefore much easier to visualise the results when expressed in $\mathbf{R}$. There is also another reason that makes the treatment in $\mathbf{R}$ sometimes more convenient than in $\mathbf{k}$. This is the appearance of $\mathbf{k}$ in the boundary condition (2.22) on $C(\mathbf{kq})$. It is because of this condition that the conjugate of $\mathbf{k}$ (which is $\mathbf{R}$) is not a constant of motion although $\mathbf{k}$ itself does not appear in the Hamiltonian of Eq. (3.1). In what follows sometimes the kq-representation and sometimes the Rq-representation will be used. The wave functions of these representations are connected by Relation (2.20) and it is a matter of convenience which of them to use.

IV. The Impurity Problem in Semiconductors

The impurity problem is one of the examples demonstrating the simplicity of the dynamics of electrons in solids in the *kq*-representation. This is a case when the perturbation caused by the impurity can often be assumed to be slowly varying on a scale of atomic distances.[3] Such a perturbation turns out to be a function of **R** only (**R** a Bravais lattice vector) and from the discussion in Section III one should expect that to some approximation result (3.8) with $H = 0$ will represent the Hamiltonian of the problem in the R-coordinate. In order to show this result and also to develop a general theory we start with Schrödinger's equation for the impurity problem (Eq. (2.17) with $\mathbf{H} = 0$) in the *kq*-representation (the derivations in this section follow very closely Zak[14])

$$[(1/2m)(-i\,\partial/\partial\mathbf{q})^2 + V(\mathbf{q}) + v(i\,\partial/\partial\mathbf{k} + \mathbf{q})]\ C(\mathbf{kq}) = \epsilon C(\mathbf{kq}) \qquad (4.1)$$

From here, by using expansion (2.20), Eq. (4.1) can be rewritten in the **R** and **q** coordinates

$$[(1/2m)(-i\,\partial/\partial\mathbf{q})^2 + V(\mathbf{q}) + v(\mathbf{R} + \mathbf{q})]\,\psi(\mathbf{q} + \mathbf{R}) = \epsilon\psi(\mathbf{q} + \mathbf{R}). \qquad (4.2)$$

There is a very interesting feature of Eq. (4.2) that can be noticed at once. If the perturbation potential does not change appreciably on a distance of a unit cell, then one can neglect the dependence of the perturbation on **q** and v will depend only on **R**. The variables **R** and **q** separate in the Hamiltonian and one should therefore expect that in the **R** coordinate the electron will move as if there were no crystal. As we have already mentioned, qualitatively one should expect that the Hamiltonian for the impurity problem will be

$$\epsilon_m(\mathbf{k}) + v(\mathbf{R}), \qquad (4.3)$$

where $\epsilon_m(\mathbf{k})$ is the same operator as has appeared in Eq. (3.6).

Let us now turn to the quantitative description of the problem. For this we substitute expansion (3.5) into Eq. (4.2)

$$[(1/2m)(-i\,\partial/\partial\mathbf{q})^2 + V(\mathbf{q}) + v(\mathbf{R} + \mathbf{q})] \sum_n \psi_{nk}(\mathbf{q}) F_n(\mathbf{R})$$
$$= \epsilon \sum_n \psi_{nk}(\mathbf{q}) F_n(\mathbf{R}). \qquad (4.3a)$$

[14] J. Zak, *Phys. Rev.* **2B**, 384 (1970).

In addition to the operator $\psi_{nk}(\mathbf{q})$ we define an operator $\psi^{+}_{nk}(\mathbf{q})$,

$$\psi^{+}_{nk}(\mathbf{q}) = \sum_{\mathbf{R}_m} \exp(-R_m\, \partial/\partial\mathbf{R})\, a_n{}^*(\mathbf{q} - \mathbf{R}_m),$$

by which Eq. (4.3a) is multiplied from the left and then integrated over $\mathbf{q}$. The following equation is obtained for $F_m(\mathbf{R})$:

$$\epsilon_m(\mathbf{k})\, F_m(\mathbf{R}) + \sum_n v_{mn}(\mathbf{R}, \mathbf{k})\, F_n(\mathbf{R}) = \epsilon F_m(\mathbf{R}), \tag{4.4}$$

where $\mathbf{k} = -i\, \partial/\partial\mathbf{R}$ and the operator v_{mn} is

$$\begin{aligned} v_{mn}(\mathbf{R}, \mathbf{k}) &= \int \psi^{\dagger}_{mk}(\mathbf{q})\, v(\mathbf{R} + \mathbf{q})\, \psi_{nk}(\mathbf{q})\, d^3q \\ &= \int u^{\dagger}_{mk}(\mathbf{q})\, v(\mathbf{R})\, u_{nk}(\mathbf{q})\, d^3q, \end{aligned} \tag{4.5}$$

where $u^{\dagger}_{mk}(\mathbf{q})$ is an operator obtained by replacing $\mathbf{k}$ by $-i\, \partial/\partial\mathbf{R}$ in the function $u^{*}_{nk}(\mathbf{q})$, the periodic part of the Bloch function. Equation (4.4) is an exact multiband difference equation for the wave function $F_m(\mathbf{R})$ in the mR-representation (m is the band index and $\mathbf{R}$ is a Bravais lattice vector).

In the case when $v(\mathbf{R})$ is a slowly varying function one can consider $\mathbf{R}$ a continuous variable. It is then convenient to interchange the order of two terms in (4.5)

$$v(\mathbf{R})\, u_{nk}(\mathbf{q}) = u_{nk}(\mathbf{q})\, v(\mathbf{R}) + [v(\mathbf{R}), u_{nk}(\mathbf{q})] \tag{4.6}$$

where the brackets in (4.6) stand for a commutator. Equation (4.5) will become

$$v_{mn}(\mathbf{R}, \mathbf{k}) = v(\mathbf{R})\, \delta_{mn} + \int u^{\dagger}_{mk}(\mathbf{q})[v(\mathbf{R}), u_{nk}(\mathbf{q})]\, d^3q \tag{4.7}$$

The first term in (4.7) when substituted in Eq. (4.4) leads to the conventional effective mass equation[3] with the Hamiltonian of Eq. (4.3):

$$[\epsilon_m(\mathbf{k}) + v(\mathbf{R})]F_m(\mathbf{R}) = \epsilon F_m(\mathbf{R}). \tag{4.8}$$

For a slowly varying $v(\mathbf{R})$ it is now easy to obtain higher order terms in the effective Hamiltonian from Eq. (4.7). To do this let us write $u_{nk}(\mathbf{q})$ in powers of $\mathbf{k}$ about $\mathbf{k}_0$. Usually, $\mathbf{k}_0$ is chosen as an extremum of the energy function[3] $\epsilon_m(\mathbf{k})$. For simplicity we assume $\mathbf{k}_0 = 0$ and no degeneracy. Then

$$u_{nk}(\mathbf{q}) = u_{n0}(\mathbf{q}) + (\partial u_{nk}/\partial\mathbf{k})_0\, \mathbf{k} + \tfrac{1}{2}(\partial^2 u_{nk}/\partial\mathbf{k}\, \partial\mathbf{k})_0\, \mathbf{k}\mathbf{k} + \cdots . \tag{4.9}$$

This expansion can easily be generalized for any $\mathbf{k}_0$ and for a degenerate case.[15] The first term in (4.9) does not contain derivatives with respect to $\mathbf{R}$ and it gives a vanishing commutator in (4.7). The second term in (4.9) will lead to the following one-band first-order correction[15a] in (4.7):

$$(\partial v(R)/\partial \mathbf{R}) \int u_{m0}^*(\mathbf{q})(\partial u_{mk}(\mathbf{q})/\partial \mathbf{k})_0 \, d^3q \tag{4.10}$$

In general, this is a nonvanishing correction. However, most crystals of interest possess inversion symmetry and for them the integral in (4.10) vanishes.[3] We are left with the second-order one-band correction in (4.7)

$$A_{ij}^{(m)} \frac{\partial^2 v(\mathbf{R})}{\partial R_i \, \partial R_j} + B_{ij}^{(m)} \frac{\partial v(\mathbf{R})}{\partial R_i} \frac{\partial}{\partial R_j} \tag{4.11}$$

where

$$A_{ij}^{(m)} = \frac{1}{2} \int u_{m0}^*(\mathbf{q}) \left(\frac{\partial^2 u_{mk}(\mathbf{q})}{\partial k_i \, \partial k_j}\right)_0 d^3q + \int \left(\frac{\partial u_{mk}^*(\mathbf{q})}{\partial k_i}\right)_0 \left(\frac{\partial u_{mk}(\mathbf{q})}{\partial k_j}\right)_0 d^3q \tag{4.12}$$

$$B_{ij}^{(m)} = \int u_{m0}^*(\mathbf{q}) \left(\frac{\partial^2 u_{mk}(\mathbf{q})}{\partial k_i \, \partial k_j}\right)_0 d^3q + \int \left(\frac{\partial u_{mk}(\mathbf{q})}{\partial k_i}\right)_0 \left(\frac{\partial u_{mk}^*(\mathbf{q})}{\partial k_j}\right)_0 d^3q. \tag{4.13}$$

For a known band structure (known $u_{mk}(\mathbf{q})$), the coefficients $A_{ij}^{(m)}$ and $B_{ij}^{(m)}$ can easily be evaluated. With the definitions (4.12) and (4.13), the one-band effective mass equation to second order terms in the variation of the potential will become

$$\left[\epsilon_m(\mathbf{k}) + v(\mathbf{R}) + A_{ij}^{(m)} \frac{\partial^2 v(\mathbf{R})}{\partial R_i \, \partial R_j} + B_{ij}^{(m)} \frac{\partial v(\mathbf{R})}{\partial R_i} \frac{\partial}{\partial R_j}\right] F_m(\mathbf{R}) = \epsilon F_m(\mathbf{R}). \tag{4.14}$$

This equation takes into account corrections that come from one particular band only.[16]

In a similar way one can find corrections to the effective mass equation that follows from other bands. For this purpose, Eq. (4.7) is used with $n \neq m$. The first order correction will be (we will mention here only the lowest order nonvanishing terms)

$$\frac{\partial v(\mathbf{R})}{\partial \mathbf{R}} \int u_{m0}^*(\mathbf{q}) \left(\frac{\partial u_{nk}}{\partial \mathbf{k}}\right)_0 d^3q. \tag{4.15}$$

[15] G. Dresselhaus, A. F. Kip, and C. Kittel, *Phys. Rev.* **98**, 368 (1955).

[15a] J. O. Dimmock, "Semiconductors and Semimetals," Vol. 3. Academic Press, New York, 1967. In this reference the first order term was derived by using the **k · p** approximation.

[16] The evaluation of the second order terms in Ref. 14 contains an algebraic error and the correction term is highly overestimated.

Unlike expression (4.10) the contribution of (4.15) is nonvanishing even when the crystal possesses inversion symmetry. Accounting for terms (4.15) in the effective mass equation is however more complicated because the terms (4.15) are nondiagonal in the band index. The relative importance of the terms (4.12), (4.13), and (4.15) with respect to their contribution to the effective mass equation will certainly depend on the particular band structure of the solid, and their evaluation can be carried out for each specific case. For example, in the tight binding approximation the expressions (4.12), (4.13), and (4.15) can be simplified. The periodic part of the Bloch function that enters these expressions assumes in the tight binding approximation the following form

$$u_{mk}(\mathbf{q}) = \sum_{\mathbf{R}} \exp\{-i(\mathbf{q} - \mathbf{R}) \cdot \mathbf{k}\}\, a_m(\mathbf{q} - \mathbf{R}), \tag{4.16}$$

where $a_m(\mathbf{q} - \mathbf{R})$ are atomic functions and for simplicity we assume that there is no degeneracy and that a_m are chosen to be real. Then, the expressions (4.12), (4.13), and the expression that multiplies $\partial v(\mathbf{R})/\partial \mathbf{R}$ in (4.15) become

$$A_{ij}^{(m)} = \frac{1}{2} \sum_{\mathbf{R}} \int x_i x_j a_m(\mathbf{r} + \mathbf{R})\, a_m(\mathbf{r})\, dV \tag{4.12a}$$

$$B_{ij}^{(m)} = 0 \tag{4.13a}$$

$$\sum_{\mathbf{R}} \int \mathbf{r} a_m(\mathbf{r} + \mathbf{R})\, a_n(\mathbf{r})\, dV. \tag{4.15a}$$

The integration in expressions (4.12a) and (4.15a) is on the whole space.

In conclusion of this section we would like to point out that the *kq*-representation is well suited for evaluating the contribution to the effective mass approximation of terms that vary slowly on a range of a unit cell. The simplicity is achieved in the *kq*-representation by splitting the radius vector **r** into two parts **R** (or **k**) and **q**. With respect to the **R**-coordinate the crystal is a homogeneous medium and the Bloch part (unperturbed part) of Eq. (4.2) does not depend on **R**. On the other hand, the impurity potential, being a slowly varying function depends mainly on **R**. The **q** coordinate can be neglected in the lowest approximation and is also very convenient for taking care of higher order effects. The correction terms (4.12), (4.13), and (4.15) to the effective mass approximation are in some sense obtained as an expansion of the impurity potential $v(\mathbf{R} + \mathbf{q})$ in powers of **q**. The splitting of the radius vector **r** into **R** and **q** (or **k** and **q**) that is achieved by the *kq*-representation leads to both a simple calculational procedure and a simple qualitative picture for the impurity problem.

V. Bloch Electrons in a Constant Electric Field

One of the simplest and most useful theorems in the theory of solids is the acceleration theorem for a Bloch electron in an electric field **E**:

$$\dot{\mathbf{k}} = -e\mathbf{E} \tag{5.1}$$

where $\dot{\mathbf{k}}$ is the rate of change of the quasimomentum. Because of its importance many different proofs of this theorem have been presented in the literature.[7,10,17,18] Theorem (5.1) looks very simple. Its meaning and derivation, however, are very far from being simple. It is instructive to compare this theorem with the analogous theorem for an electron moving in free space in an electric field

$$\dot{\mathbf{p}} = -e\mathbf{E} \tag{5.2}$$

where $\mathbf{p}$ is the linear momentum of the electron. The quantum-mechanical derivation of (5.2) is very simple and follows from the simple rule for the rate of change of any operator A

$$A = i[H, A] \tag{5.3}$$

where H is the Hamiltonian of the problem under consideration. For an electron in free space in an electric field **E** the Hamiltonian is $p^2/2m + e\mathbf{E} \cdot \mathbf{r}$ and theorem (5.2) follows at once from Relation (5.3). In order to apply the same method for deriving Eq. (5.1) one should know how to write the Hamiltonian for a Bloch electron in an electric field as a function of **k**. In the usual Bloch theory this is achieved by working in the nk-representation,[7] where n is the band index and **k** is the quasimomentum. Such an approach requires the expansion of the wave function $\psi(\mathbf{r})$ in Bloch functions[7] $\psi_{nk}(\mathbf{r})$

$$\psi(\mathbf{r}) = \sum_n a_{nk}\psi_{nk}(\mathbf{r})\, d^3k \tag{5.4}$$

which subsequently leads to a multiband representation of the Hamiltonian. The theorem (5.1) is then obtained by a matrix equation

[17] A. H. Wilson, "The Theory of Metals." Cambridge Univ. Press, London and New York, 1965.

[18] G. H. Wannier, "Elements of Solid State Theory." Cambridge Univ. Press, London and New York, 1959.

corresponding to (5.3). We shall not follow here this approach in detail (the interested reader should consult Landau and Lifschitz[7]) because it involves a rather complicated algebra. Instead, we will show that the *kq*-representation is most suitable for the derivation of the acceleration theorem (5.1) because this representation contains the quasimomentum explicitly. In the *kq*-representation, Schrödinger's equation for a Bloch electron in an electric field is (Eq. (2.17) with $\mathbf{H} = 0$ and $v(\mathbf{r}) = e\mathbf{E} \cdot \mathbf{r}$)

$$\left[-\frac{1}{2m}\frac{\partial^2}{\partial \mathbf{q}^2} + V(\mathbf{q}) + e\mathbf{E} \cdot \left(i\frac{\partial}{\partial \mathbf{k}} + \mathbf{q}\right)\right] C(\mathbf{kq}) = \epsilon C(\mathbf{kq}). \tag{5.5}$$

From here and the quantum mechanical rule (5.3) it follows at once

$$\dot{\mathbf{k}} = i[H, \mathbf{k}] = -e\mathbf{E}. \tag{5.6}$$

It is therefore seen that in the *kq*-representation the derivation of the acceleration theorem follows straightforwardly from Schrödinger's equation (5.5).

Having such a simple derivation of the acceleration theorem (5.6) it is very easy to understand its meaning. Namely, when the qausimementum $\mathbf{k}$ is used for describing the motion of a Bloch electron in an electric field, the derivative of $\mathbf{k}$ with respect to time equals the electric force $-e\mathbf{E}$. From the derivation of (5.6) it is clear that in this meaning the acceleration theorem is completely exact. This is also the meaning given to the acceleration theorem in Landau and Lifschitz.[7] In the original derivation (see Wilson[17]) the approach was different as was the interpretation. In this latter approach the derivation was based on finding the probability for a Bloch electron in an electric field to remain in a given energy band. This probability calculation has predicted that for short times (the shortness of the time wasn't specified) the electron will remain in the same band but the $\mathbf{k}$ vector of the Bloch state (before the electric field was turned on) will vary with time according to relation (5.6). It should be pointed out that in this interpretation the acceleration theorem (5.6) is approximate and holds only for short times.[17] It is probably because of these two different interpretations that the acceleration theorem has led, in some cases, to not well-defined results. The reference here is to the problem of the oscillating Bloch electron in an electric field (Kittel,[10] p. 198) and, in connection with it, to the existence of a Stark ladder.[19] The remainder of this section deals with this problem.

It is generally believed that an electron in a crystal with well-separated bands will oscillate in space when exposed to a constant electric field.[10,18]

[19] J. Zak, *Phys. Rev. Lett.* **20**, 1477 (1968).

This belief is based on the interpretation of the acceleration theorem and was actually never proven. By using the *kq*-representation, a general time-dependent solution of a Bloch electron in an electric field is found and it is shown that this belief is unjustified and that no such oscillations follow from the present theory.

In the absence of an electric field an electron in a crystal is described by a Bloch function $\psi_{nk}(\mathbf{r})$ corresponding to the energy $\epsilon_n(\mathbf{k})$. When an electric field $\mathbf{E}$ is turned on the quasimomentum varies with time according to the following rule (see the acceleration theorem (5.1)):

$$\mathbf{k} = \mathbf{k}_0 - e\mathbf{E}t, \tag{5.7}$$

where k_0 is the value of $\mathbf{k}$ at time zero. It is a very tempting idea to assume that to some approximation the function $\psi_{nk}(\mathbf{r})$ with $\mathbf{k}$ replaced by expression (5.7) will be a solution of the time-dependent equation for a Bloch electron in an electric field. The first attempt to use this idea was made by Houston[20] who defined the following function (known as the Houston function)

$$\psi(\mathbf{r}t) = \exp\left\{-i\int_0^t \epsilon_n(\mathbf{k}_0 - e\mathbf{E}t')\,dt'\right\}\psi_{nk_0-eEt}(\mathbf{r}). \tag{5.8}$$

It was claimed by Houston[20] that the function (5.8) is an approximate solution of the time dependent equation for a Bloch electron in an electric field

$$i\,\partial\psi/\partial t = [p^2/2m + V(\mathbf{r}) + e\mathbf{E}\cdot\mathbf{r}]\psi. \tag{5.9}$$

In order to check this claim, (5.8) is substituted into (5.9) and one arrives at the following equation

$$[p^2/2m + V(\mathbf{r}) + e\mathbf{E}(i\,\partial/\partial\mathbf{k} + \mathbf{r})]\,\psi_{nk}(\mathbf{r}) = \epsilon_n(\mathbf{k})\,\psi_{nk}(\mathbf{r}), \tag{5.10}$$

where $\mathbf{k}$ is given by (5.7). Now if one can assume that the term

$$e\mathbf{E}\cdot(i\,\partial/\partial\mathbf{k} + \mathbf{r})\,\psi_{nk}(\mathbf{r}) = \exp[i\mathbf{k}\cdot\mathbf{r}]\,e\mathbf{E}\cdot i(\partial/\partial\mathbf{k})\,u_{nk}(\mathbf{r}) \tag{5.11}$$

in Eq. (5.10) can be neglected ($u_{nk}(\mathbf{r})$ is the periodic part of the Bloch function $\psi_{nk}(\mathbf{r})$) then it is clear that $\psi(\mathbf{r}t)$ in (5.8) will be a solution of the time-dependent Eq. (5.9). This was Houston's argument.[20]

Before discussing the correctness of solution (5.8), let us first consider

[20] W. V. Houston, *Phys. Rev.* **57**, 184 (1940).

its consequences (for simplicity a one-dimensional case is treated). By calculating the average velocity in state (5.8) one finds[10]

$$\langle v \rangle = \partial \epsilon_n(k)/\partial k, \tag{5.12}$$

where again k is given by (5.7). This means that the average velocity of a Bloch electron in an electric field is a periodic function of time. Relation (5.12) can be integrated to give the coordinate of the electron as a function of time (Kittel,[10] p. 198 and Wannier,[18] p. 191),

$$x(t) - x(0) = -(1/eE)(\epsilon_n(k_0 - eEt) - \epsilon_n(k_0)], \tag{5.13}$$

which is also periodic with the period $2\pi/eEa$, a being the lattice constant. Relation (5.13), if correct, means that a Bloch electron in an electric field will oscillate in real space. These oscillations also have led to the belief that a Stark ladder exists for a Bloch electron in an electric field.[18]

Let us now turn to a discussion of Houston's function (5.8). It first should be remarked that the function in (5.8) is not even uniquely defined. The reason for this is that a Bloch function is defined only up to a k-dependent phase factor. The usual requirement on a Bloch function $\psi_{nk}(x)$ is the periodicity with respect to k and an arbitrary factor of the form $e^{if(k)}$ is therefore permissable ($f(k)$ should be periodic). However, such a factor will lead to a completely different time-dependence in (5.8) and to additional terms which can be arbitrarily large in Eq. (5.10). Houston's function in the form (5.8) is therefore not well defined. It turns out that a well defined Houston's function can be constructed by using the kq-representation. As will be seen later the use of the kq-representation will also lead to a much better understanding of the whole problem.

By means of the kq-representation the following theorem can be stated: any solution of Schrödinger's equation for a Bloch electron in an electric field, Eq. (5.5), can be used to construct an exact time dependent solution $\psi(\mathbf{r}t)$ of Eq. (5.9)

$$\psi(\mathbf{r}t) = e^{-i\epsilon t} C_\epsilon(\mathbf{k} - e\mathbf{E}t, \mathbf{r}), \tag{5.14}$$

where C_ϵ is a solution of (5.5) corresponding to the energy ϵ. A similar function was given by Wannier[21] and it was checked that it satisfies Eq. (5.9). It can now be shown that Houston's function (with a well defined phase) can be obtained from (5.14) by solving Eq. (5.5) in a one-band approximation (for simplicity we consider the one-dimensional

[21] G. H. Wannier, *Rev. Mod. Phys.* **34**, 645 (1962).

case). To do so $C(kq)$ is expanded in Bloch functions[11] (see expansion (2.24))

$$C(kq) = \sum_n B_n(k)\, \psi_{nk}(q). \tag{5.15}$$

By substituting (5.15) in (5.5) one finds the usual equation for the function $B_n(k)$[22,23]:

$$i\, \partial B_n(k)/\partial k + \epsilon_n(k)\, B_n(k) + eE \sum_m X_{nm}(k)\, B_m(k) = \epsilon B_n(k), \tag{5.16}$$

where

$$X_{nm}(k) = i \int u^*_{nk}(q)(\partial u_{nk}(q)/\partial k)\, dq, \tag{5.17}$$

where $u_{nk}(q)$ is the periodic part of the Bloch function and the integration in (5.17) is on a unit cell of the Bravais lattice. Equation (5.16) can be easily solved if interband terms with $X_{nm}(k)$ are neglected. The result is as follows[19,22,23]:

$$B_n(k) = \exp\left\{-(i/eE) \int_0^k [\epsilon_\nu - \epsilon_n(k') - eEX_{nn}(k')]\, dk'\right\} \tag{5.18}$$

with

$$\epsilon_\nu = (a/2\pi) \int_0^{2\pi/a} [\epsilon_n(k') + eEX_{nn}(k')]\, dk' + eEa\nu \tag{5.19}$$

and ν assuming any integer value from $-\infty$ to $+\infty$. By substituting solution (5.18) into expansion (5.15) (for one-band) and by using theorem (5.14) the following time-dependent solution for a Bloch electron in an electric field is obtained

$$\psi(xt) = \exp\left\{-i \int_0^t [\epsilon_n(k - eEt') + eEX_{nn}(k - eEt')]\, dt'\right\} \psi_{nk-eEt}(x). \tag{5.20}$$

In writing down solution (5.20) a k-dependent phase factor,

$$\exp\left\{-(i/eE) \int_0^k [\epsilon_\nu - \epsilon_n(k') - eEX_{nn}(k')]\, dk'\right\},$$

was neglected because it is irrelevant for the time-dependent solution. The obtained function (5.20) differs from Houston's solution (5.8) only by a phase factor which is, however, of very great importance in making the function well defined. Now if the Bloch function in (5.20) is miltiplied

[22] E. O. Kane, *J. Phys. Chem. Solids* **12**, 181 (1959).
[23] J. Callaway, *Phys. Rev.* **130**, 549 (1963).

by a k-dependent phase factor, $e^{if(k)}$, no change is introduced in $\psi(xt)$ because there is a corresponding change in $X_{nn}(k)$ (see definition (5.17)) that absorbs the above factor. It is therefore seen that a one-band solution of Eq. (5.5) leads to a well-defined Houston function.[24] However this function is based on the one-band solution (5.18) with the spectrum (5.19) which is clearly contradictory. The simplest way to see this contradiction physically is to compare the one-band energy spectrum in the absence of the electric field $\epsilon_n(k)$ with the energy spectrum of the one-band approximation in the presence of the electric field which extends over energies from $-\infty$ to $+\infty$. It does not seem physical that one-band states $\psi_{nk}(x)$ corresponding to energies in the range $\epsilon_n(k)$ could be used for expanding states ranging in energies from $-\infty$ to $+\infty$. The reason for such a contradiction is in neglecting the interband terms in (5.16) with respect to the intraband ones. The intraband terms that were kept in Eq. (5.16) in order to arrive at the results (5.13) and (5.19) led to a spectrum with a fine structure (Stark ladder) of a spacing eEa. At the same time the neglected interband terms $eEX_{nm}(k)$ are of the same order[25] eEa as the spacing in the Stark ladder and it is therefore not very meaningful to solve the eigenvalue equation (5.16) in the one-band approximation.

The conclusion of the last paragraph is that it is impossible to construct a Houston function (5.20) from the theorem (5.14) by using a one-band solution of the eigenvalue equation (5.16) (a one-band solution of (5.16) is contradictory). One could, however, think differently and assume that Houston's function (5.20) is given (we know that now it is correctly defined) and then try to check whether this function satisfies Eq. (5.19). One might hope that the requirement for the function (5.20) to satisfy Eq. (5.19) does not necessarily lead to the contradictory solution (5.18) and (5.19).

In what follows it will be shown that if one assumes that Houston's function (5.20) is a solution of Eq. (5.19) for times equal or longer than the period of oscillations $T = 2\pi/eEa$, then this necessarily leads to the contradictory solution (5.18) and (5.19).

Let us assume that the time dependent function (5.20) is a solution of Eq. (5.19). This leads to an equation (compare with function (5.8) and Eq. (5.10))

$$\begin{aligned}[-(1/2m)(\partial^2/\partial x^2) + V(x) + eE(i\,\partial/\partial k + x)]\,\psi_{nk}(x)\\ = [\epsilon_n(k) + eEX_{nn}(k)]\,\psi_{nk}(x)\end{aligned}$$

[24] This function (Function (5.20)) was already given before [L. Fritsche, *Phys. Status Solidi* **13**, 487 (1966)].

[25] The interband elements $X_{nm}(k)$ are in general of the order of the lattice constant.

Since $\psi_{nk}(x)$ is a Bloch function, the last equation will be satisfied *only* if

$$eE(i\,\partial/\partial k + x)\,\psi_{nk}(x) = eEX_{nn}(k)\,\psi_{nk}(x). \tag{5.21}$$

Now, having relation (5.21), one obtains at once that Eq. (5.5) with expansion (5.15) lead to Eq. (5.16) with no interband terms and therefore to the contradictory solution (5.18) and (5.19). This proves that Houston's function (5.20) cannot be a solution for times greater than the period of oscillations $2\pi/eEa$.

The still open question is whether Houston's function (5.20) can be a solution of the time dependent equation (5.9) for times shorter than the period $2\pi/eEa$. The answer to this question is in general affirmative. One comes to no contradiction by assuming that the function (5.20) is a solution of Eq. (5.19) for times smaller than $2\pi/eEa$. Such an assumption will lead to a one-band equation (5.16) only for a part of the Brillouin zone and therefore to no Stark ladder (the latter is a consequence of the periodicity requirement on the function[19] $B_n(k)$). There is clearly also no oscillating Bloch electron in an electric field because Houston's function can satisfy Eq. (5.9) only for times shorter than the period[25a] $2\pi/eEa$.

The general solution (5.14) can also be used to clarify another aspect of the dynamics in an electric field. There are claims in literature[10,21,26] that in the presence of a uniform electric field Bloch bands can be defined for an electron in a crystal. These claims are usually based on the fact that it is possible to assign a wave vector k to the time-dependent solution.[10,21] Having constructed a general time-dependent solution (Eq. (5.14)) it is easy to analyze the possibility of assigning to it a band index. Theorem (5.14) claims that for any eigenvalue solution of Eq. (55) a time dependent solution (5.14) can be constructed. Since the eigenvalues of Eq. (55) are continuous,[27] the general time-dependent solution (5.14) can therefore be assigned a continuous index ϵ.

[25a] In this section it is shown that the existing proofs of an oscillating Bloch electron in an electric field are not satisfactory. We do not exclude the possibility that other ways to prove the existence of the mentioned oscillations or of a Stark ladder could exist. Recently the author had an opportunity to discuss this problem with Professor Rudolf Peierls. Professor Peierls is of the opinion that there should be a way to prove the existence of an oscillating Bloch electron.

[26] E. N. Adams, *Phys. Rev.* **107**, 698 (1957).

[27] The exact eigenvalue solutions of Eq. (5.5) have clearly a continuous spectrum. This can better be seen by writing the Hamiltonian in the coordinate representation (the Hamiltonian on the right hand side of Eq. (5.9)). For large coordinates r the periodic potential can be neglected and the spectrum is therefore continuous [see, for example, Landau and Lifshitz[7]].

It is interesting to point out that there is a very simple connection between solution (5.14) and the general time-dependent solution of Ref. 10 [Eq. (84) on page 191]:

$$\varphi_k(\mathbf{r}, \mathbf{E}, t) = \exp[i\mathbf{k} \cdot \mathbf{r}]\, u_k(\mathbf{r}t), \tag{5.22}$$

where $u_k(\mathbf{r}t)$ is periodic in $\mathbf{r}$. According to Kittel,[10] function (5.22) satisfies the time-dependent equation in the following gauge

$$i\, \partial\varphi_k/\partial t = [(p - e\mathbf{E}t)^2/2m + V(\mathbf{r})]\, \varphi_k\,. \tag{5.23}$$

It is clear that by using the gauge transformation

$$\psi_k = \exp(-ie\mathbf{E} \cdot \mathbf{r}t)\, \varphi_k = \exp[i(\mathbf{k} - e\mathbf{E}t) \cdot \mathbf{r}]\, u_k \tag{5.24}$$

the function ψ_k will satisfy the time-dependent equation (5.9) which is also satisfied by function (5.14). From the structure of (5.24) [u_k is periodic in $\mathbf{r}$] it is obvious that it can be written in the form (5.14). One comes therefore to the same conclusion that the time-dependent solution (5.24) can be assigned a continuous index ϵ. We see therefore no justification to assign a band index to the time dependent solution φ_k of Kittel[10] (or of Wannier[21]).

It should also be remarked that although the function C_ϵ in (64) is periodic in time, it does not lead to any oscillations in space because the velocity theorem (5.12) does not hold for the solution (5.14). It is clear that for each ϵ, solution (5.14) leads to a set of levels

$$\epsilon + eEa\nu, \tag{5.25}$$

with ν assuming integer values from $-\infty$ to $+\infty$. However since ϵ is continuous no Stark ladder[19] follows from solution (5.14).

Having clarified to what extent a Houston function (5.20) describes the motion of a Bloch electron in an electric field, let us now go back to the meaning of the acceleration theorem (5.1). Two possible interpretations of this theorem were mentioned before: In the first interpretation the question is asked how does the quasimomentum $\mathbf{k}$ of a Bloch electron in an electric field vary with time? No mention is made here of any energy band. The answer to this question is given by the exact relation (5.1), which states that the derivative of $\mathbf{k}$ with respect to time equals the electric force $-eE$. In the second interpretation (Ref. 17) one is interested in the probability for a Bloch electron to remain in a given band after an electric field is turned on. As was shown in Wilson,[17] for short times such a probability is a function of the argument $\mathbf{k} - e\mathbf{E}t$ and the interpretation of this was that the Bloch electron remains in the same band but its **k** vector varies according to the rule (5.1). It can be shown that a very close connection exists between this (second) interpretation and the description of the motion by means of a Houston (5.20). To do this, let us develop a time-dependent perturbation theory for a Bloch electron in an electric field.

In the kq-representation the time-dependent Schrödinger equation for a Bloch electron in an electric fields is (again, for simplicity the one-dimensional case is treated):

$$i\,\partial C(kqt)/\partial t = [-(1/2m)(\partial^2/\partial q^2) + V(q) + eE(i\,\partial/\partial k + q)]\,C(kqt). \quad (5.26)$$

In order to write down this equation in the Bloch representation (nk-representation with n being the band index and k the quasimomentum), expansion (5.15) is used with $B_n(kt)$ depending on time. One has

$$i\,\partial B_n(kt)/\partial t = \epsilon_n(k)\,B_n(kt) + eEi(\partial B_n(kt)/\partial k) + eE\sum_m X_{nm}(k)\,B_m(kt), \quad (5.27)$$

where the matrix elements $X_{nm}(k)$ are given by relation (5.17). Equation (5.27) is the same equation as derived in Wilson,[17] and the coefficients $B_n(kt)$ (the wave function in the nk-representation) have the meaning of the time-dependent probability for the electron to be in band n with the quasi momentum k. If one assumes that the interband elements $X_{nm}(k)$ with m not equal n in (5.27) can be neglected, the latter equation has the following solution

$$B_m(kt) = \delta_{mn}\sum_p \delta(k - k_0 + eEt - p2\pi/a)$$
$$\times \exp\left\{-i\int_0^t [\epsilon(k_0 - eEt') + eEX_{nn}(k_0 - eEt')]\,dt'\right\}, \quad (5.28)$$

where the sum is over p assuming all possible integer values from $-\infty$ to $+\infty$. Solution (5.28) contains the initial condition that at time $t = 0$ the electron is in the Bloch state given by the band n and vector k_0. To see it, (5.28) for $t = 0$ is substituted into expansion (5.15) resulting in the function

$$C(kq) = \psi_{nk}(q)\sum \delta(k - k_0 - p2\pi/a), \quad (5.29)$$

which is just a Bloch function in the kq-representation for the band n and wave vector k_0 (compare (5.29) with Eq. (2.26)). Before analyzing solution (5.28) it should be pointed out that it is an exact solution of Eq. (5.27) without the interband terms $X_{nm}(k)$.

By assigning to the square of the absolute value of $B_m(kt)$ in (5.28) the meaning of a probability, one obtains the result of Wilson,[17] namely, that if the electron was initially in the Bloch state (5.29) it will remain in the same band n but its wave vector k will vary with time according to (5.7). In Wilson,[17] it was claimed that this statement holds only for short times. Having solution (5.28) it can be shown that the probability interpretation cannot possibly be correct for times equal or bigger than one period of oscillations $T = 2\pi/eEa$. To see this, solution (5.28) is substituted into expansion (5.15) and the resulting function $C(kqt)$ is used in relation (2.18) in order to obtain the wave function $\psi(xt)$ in the regular coordinate representation. As one should expect, the function $\psi(xt)$ turns out to be Houston's function (5.20). However, as was previously proven, Honston's function can be a correct solution only for times shorter than $T = 2\pi/eEa$ and the same is therefore true with respect to solution (5.28). It is therefore seen that the probability interpretation of the acceleration theorem is very closely connected with Houston's function.

In conclusion it should be stated that if the acceleration theorem is used with the reference that the electron remains in the same band it can be correct only for times shorter than the period $T = 2\pi/eEa$. If, however, no restriction is made with respect to a particular band the acceleration theorem as given by (5.1) is exact.

VI. Bloch Electrons in a Magnetic Field

In the Introduction of this article it was pointed out that most of the results for the dynamics of electrons in solids are very simple in structure while their derivation is usually not simple at all. The best demonstration of this remark is the effective Hamiltonian for a Bloch electron in a magnetic field. The final result of the effective Hamiltonian in the lowest order of the magnetic field is very simple in structure: in the energy spectrum of a given band, $\epsilon_n(\mathbf{k})$, one has to replace $\mathbf{k}$ by $\mathbf{k} + (e/2c)\mathbf{H} \times i\,\partial/\partial\mathbf{k}$. The operator obtained in such a way is the effective Hamiltonian. The derivation of this result was called by Kohn[4] "shockingly complicated." In this section the kq-representation is used for describing the motion of electrons in a crystal in a constant magnetic field. It will be shown that the use of the kq-representation simplifies the derivation of effective Hamiltonians.

A Bloch electron in a magnetic field is described in the kq-representation by Schrödinger's equation (2.17) with no electric perturbation ($v = 0$).

$$\left\{\left[-i\frac{\partial}{\partial\mathbf{q}} + \frac{e}{2c}\mathbf{H} \times \left(i\frac{\partial}{\partial\mathbf{k}} + \mathbf{q}\right)\right]^2\Big/2m + V(\mathbf{q})\right\} C(\mathbf{kq}) = \epsilon C(\mathbf{kq}) \qquad (6.1)$$

For the magnetic field problem it is more convenient to work with the periodic part $U(\mathbf{kq})$ of the function $C(\mathbf{kq})$. These functions are connected by the relation

$$C(\mathbf{kq}) = \exp(i\mathbf{k}\cdot\mathbf{q})\, U(\mathbf{kq}). \tag{6.2}$$

From the boundary conditions (2.21) and (2.22) on $C(\mathbf{kq})$ it follows

$$U(\mathbf{k}+\mathbf{K}_m, \mathbf{q}) = \exp(-i\mathbf{q}\cdot\mathbf{K}_m)\, U(\mathbf{kq}), \tag{6.3}$$

$$U(\mathbf{kq}+\mathbf{R}_n) = U(\mathbf{kq}). \tag{6.4}$$

The U-function satisfies therefore the same boundary conditions that the periodic part $u_{nk}(\mathbf{q})$ of a Bloch function does (relation (6.2) holds also for a Bloch function $\psi_{nk}(\mathbf{q}) = \exp(i\mathbf{k}\cdot\mathbf{q})\, u_{nk}(\mathbf{q})$). For the function $U(\mathbf{kq})$, Schrödinger's equation (6.1) will become

$$\left\{\left[-i\frac{\partial}{\partial\mathbf{q}} + \mathbf{k} + \frac{e}{2c}\mathbf{H}\times i\frac{\partial}{\partial\mathbf{k}}\right]^2\Big/2m + V(\mathbf{q})\right\} U(\mathbf{kq}) = \epsilon U(\mathbf{kq}). \tag{6.5}$$

It is interesting to compare this equation with the equation that the U-function satisfies in the absence of the magnetic field

$$\left[\left(-i\frac{\partial}{\partial\mathbf{q}} + \mathbf{k}\right)^2\Big/2m + V(\mathbf{q})\right] U(\mathbf{kq}) = \epsilon U(\mathbf{kq}). \tag{6.6}$$

As is seen the magnetic field can be introduced into Schrödinger's equation by the replacement

$$\mathbf{k} \to \mathbf{k} + (e/2c)\,\mathbf{H}\times i\,\partial/\partial\mathbf{k}. \tag{6.7}$$

This is the same rule that is used in quantum mechanics[7] when one introduces the magnetic field into Schrödinger's equation. The rule is to replace the momentum $\mathbf{p}$ by the kinetic momentum $\mathbf{p} + (e/2c)\mathbf{H}\times i\,\partial/\partial\mathbf{p}$ ($\mathbf{r} = i\,\partial/\partial\mathbf{p}$) and was used in Eq. (2.16). This is another demonstration of the equivalency of the quasimomentum $\mathbf{k}$ for a Bloch electron to the momentum $\mathbf{p}$ for a free electron. The fact that this rule works also with respect to the quasimomentum is a straightforward result of the *kq*-representation and turns out to be of great importance in the whole theory of effective Hamiltonians for a Bloch electron in a magnetic field. In order to get a feeling of the importance of rule (6.7), let us compare the free electron case with the Bloch electron. For a free electron, the energy spectrum is

$$p^2/2m \tag{6.8}$$

and the Hamiltonian in a magnetic field is

$$\left(\mathbf{p} + \frac{e}{2c}\mathbf{H} \times i\frac{\partial}{\partial \mathbf{p}}\right)^2 \Big/ 2m \tag{6.9}$$

For a Bloch electron the one-band spectrum is

$$\epsilon_n(\mathbf{k}) \tag{6.10}$$

and according to rule (6.7) one should expect that in some one-band approximation the Hamiltonian will be given by (6.10) with $\mathbf{k}$ replaced by (6.7):

$$\epsilon_n(\mathbf{k} + (\epsilon/2c)\,\mathbf{H} \times \mathbf{i}\,\partial/\partial\mathbf{k}). \tag{6.11}$$

This is a well-known result and represents the effective Hamiltonian in the lowest order of the magnetic field. It turns out that the same functional structure (the dependence on $\mathbf{k} + (e/2c)\mathbf{H} \times i\,\partial/\partial\mathbf{k}$) holds also to higher-order approximations and in what follows an effective Hamiltonian theory will be developed to any order in the magnetic field.

The main idea of an effective Hamiltonian is to represent Eq. (6.5) in a one-band form. Let us assume that one performs a transformation from the wave function $U(\mathbf{kq})$ in the kq-representation to a wave function $B_n(k)$ in a Bloch-type representation (nk-representation; see connection between representations in Appendix A). By denoting the transformation matrix by $(\mathbf{kq}|n\mathbf{k}')$, one has

$$U(\mathbf{kq}) = \sum_n \int dk'\,(\mathbf{kq}|n\mathbf{k}')\,B_n(\mathbf{k}'), \tag{6.12}$$

where the integration is over the first Brillouin zone. The transformation matrix $(\mathbf{kq}|n\mathbf{k}')$ (or the expansion functions) will play an important role in the whole effective Hamiltonian theory. The problem will be to choose $(\mathbf{kq}|n\mathbf{k}')$ in such a way as to make the equation for the functions $B_n(\mathbf{k})$ a one-band equation.

The transformation matrix $(\mathbf{kq}|n\mathbf{k}')$ will be assumed to be unitary, as it should be in order to connect wave functions $U(\mathbf{kq})$ and $B_n(\mathbf{k}')$ in two different quantum-mechanical representation. The unitarity of $(\mathbf{kq}|n\mathbf{k}')$ is given by the following relations

$$\int d\mathbf{k}''\,d\mathbf{q}\,(n\mathbf{k}|\mathbf{k}''\mathbf{q})(\mathbf{k}''\mathbf{q}|n'\mathbf{k}') = \delta_{nn'}\sum_m \delta(\mathbf{k} - \mathbf{k}' - \mathbf{K}_m), \tag{6.13}$$

$$\sum_{nk} (\mathbf{k}'\mathbf{q}|n\mathbf{k})(n\mathbf{k}|k''\mathbf{q}') = \sum_{mn} \delta(\mathbf{k}' - \mathbf{k}'' - \mathbf{K}_m)\,\delta(\mathbf{q} - \mathbf{q}' - \mathbf{R}_n). \tag{6.14}$$

As in relations (2.6a) and (2.6b), there are sums over the δ-functions on the right hand side of (6.13) and (6.14). Relation (6.12) can also be looked at as an expansion of $U(\mathbf{kq})$ in a set of functions $(\mathbf{kq}|n\mathbf{k}')$ with $B_n(\mathbf{k}')$ as expansion coefficients. Then (6.13) expresses the orthogonality of the functions $(\mathbf{kq}|n\mathbf{k}')$, while (6.14) gives their completeness. It should be pointed out that in the transformation matrix (or the expansion functions) $(\mathbf{kq}|n\mathbf{k}')$ in (6.12), the labels $n\mathbf{k}'$ denote a Bloch-type state and not a real Bloch eigenstate of the Hamiltonian (6.6). Bloch-type states were used in effective Hamiltonian approaches[3,6] and examples of them are the Kohn-Luttinger functions[3] and the Roth functions.[6] What is meant by a Bloch-type state is a state that is labeled by a band index n and a $\mathbf{k}$ vector.

By using expansion (6.12) in Schrödinger's equation (6.5), one arrives at the equation

$$\sum_{n'k'} H_{nn'}(\mathbf{kk}')\, B_{n'}(\mathbf{k}') = \epsilon B_n(\mathbf{k}), \tag{6.15}$$

where the following notation was used

$$H_{nn'}(\mathbf{kk}') = \int d\mathbf{k}''\, d\mathbf{q}\, (n\mathbf{k}|\mathbf{k}''\mathbf{q}) \left[\left(-i\frac{\partial}{\partial \mathbf{q}} + \mathbf{k} + \frac{e}{2c}\mathbf{H} \times i\frac{\partial}{\partial \mathbf{k}}\right)^2 \Big/ 2m + V(\mathbf{q})\right] (\mathbf{k}''\mathbf{q}|n'\mathbf{k}'). \tag{6.16}$$

Equation (6.15) is an exact multiband equation in the nk-representation. In order to arrive at a one-band effective Hamiltonian, a transformation matrix has to be found that makes expression (6.16) diagonal in the band index. For choosing a suitable transformation matrix, let us again compare Eq. (6.5) in the presence of a magnetic field with Eq. (6.6) in its absence. For the latter equation, the solution is well known and is just the periodic part of the Bloch function [See Eq. (3.2) with $\mathbf{k}'$ instead of $\mathbf{k}_B$]

$$U_{nk'}(\mathbf{kq}) = u_{nk}(\mathbf{q}) \sum_m \delta(\mathbf{k} - \mathbf{k}' - \mathbf{K}_m) = (\mathbf{kq}|n\mathbf{k}'). \tag{6.17}$$

The functions (6.17) can be given a two-fold meaning. First, they are solutions of Eq. (6.6) corresponding to a band index n and a wave vector $\mathbf{k}'$. In addition they have the meaning of a transformation matrix (the second equality in (6.17)) that diagonalizes Eq. (6.15) (or the Hamiltonian (6.16)) in the absence of the magnetic field. That $U_{nk'}(\mathbf{kq})$

in (6.17) diagonalizes the Bloch equation is very easily shown. When inserted in (6.12), the transformation matrix (6.17) leads to

$$U(\mathbf{kq}) = \sum_n u_{nk}(\mathbf{q})\, B_n(\mathbf{k}). \tag{6.18}$$

By substituting (6.18) into Schrödinger's equation (6.6) one finds the diagonalized equation for $B_n(\mathbf{k})$

$$\epsilon_n(\mathbf{k})\, B_n(\mathbf{k}) = \epsilon B_n(\mathbf{k}). \tag{6.19}$$

This is the equation for a Bloch electron in its own (Bloch) representation. Since Eq. (6.5) differs from (6.6) only by the substitution (6.7), one should expect to achieve a diagonalization procedure by replacing $\mathbf{k}$ in the function $u_{nk}(\mathbf{q})$ of the transformation matrix (6.17) according to (6.7). In what follows it will be shown that the transformation matrix obtained in such a way leads to the lowest order approximation of the effective Hamiltonian.

Before further developing the diagonalization procedure let us give a rule for replacing $\mathbf{k}$ by $\mathbf{k} + (e/2c)\mathbf{H} \times i\, \partial/\partial\mathbf{k}$ in a function of $\mathbf{k}$. It is clear that such a rule is needed because, in general, the replacement (6.7) is not well defined (different components of the operators

$$\mathbf{k} + (e/2c)\,\mathbf{H} \times i\, \partial/\partial\mathbf{k}$$

do not commute with one another).

Let $A(\mathbf{k})$ be an arbitrary function of $\mathbf{k}$ with the Fourier transform

$$A(\mathbf{k}) = \int A(\boldsymbol{\lambda}) \exp(i\boldsymbol{\lambda} \cdot \mathbf{k})\, d\boldsymbol{\lambda} \tag{6.20}$$

Then an operator $[A(\mathbf{k})]$ can be defined[6]

$$[A(\mathbf{k})] = \int A(\boldsymbol{\lambda}) \exp\{i\boldsymbol{\lambda} \cdot (\mathbf{k} + (e/2c)\,\mathbf{H} \times i\, \partial/\partial\mathbf{k})\}\, d\boldsymbol{\lambda} \tag{6.21}$$

The rectangular brackets (as in (6.21)) are used to indicate that one deals with an operator depending on $\mathbf{k} + (e/2c)\mathbf{H} \times i\, \partial/\partial\mathbf{k}$. It should be pointed out that relation (6.21) defines $[A(\mathbf{k})]$ in a symmetric way with respect to the three components of the wave vector $\mathbf{k}$. Such a symmetric form makes the operator (6.21) well defined an the noncommutativity of the components of $\mathbf{k} + (e/2c)\mathbf{H} \times i\, \partial/\partial\mathbf{k}$

$$\begin{aligned}(\mathbf{k} + (e/2c)\,\mathbf{H} \times i\, \partial/\partial\mathbf{k})_\alpha\, (\mathbf{k} + (e/2c)\,\mathbf{H} \times i\, \partial/\partial\mathbf{k})_\beta \\ -(\mathbf{k} + (e/2c)\,\mathbf{H} \times i\, \partial/\partial\mathbf{k})_\beta\, (\mathbf{k} + (e/2c)\,\mathbf{H} \times i\, \partial/\partial\mathbf{k})_\alpha = -i(e/c) \sum_{\gamma=1}^{3} \epsilon_{\alpha\beta\gamma} H^\gamma\end{aligned} \tag{6.22}$$

does not lead to any ambiguity. In (6.22) $\epsilon_{\alpha\beta\gamma}$ is a unit antisymmetric tensor in all three indices and H^γ is the γ component of $\mathbf{H}$. The use of the Fourier transform for defining the operator (6.21) is only one possible way to construct a well defined operator $[A(\mathbf{k})]$. It is clear that, in general, it is not necessary to use the Fourier transform in defining $[A(\mathbf{k})]$. The only important point is to symmetrize in $A(\mathbf{k})$ all products of different components of $\mathbf{k}$ before performing the substitution (6.7). Thus, for a function $A(\mathbf{k}) = k_x k_y$, the symmetrical function is $\frac{1}{2}(k_x k_y + k_y k_x)$ and one can write the operator $[A(\mathbf{k})]$ without using its Fourier transform. The result is

$$[A(\mathbf{k})] = \tfrac{1}{2}[k_x k_y + k_y k_x]. \tag{6.22a}$$

In applications one very often has to know how to multiply two or more operators of the form (6.21). The meaning of multiplying two operators $[A(\mathbf{k})]$ and $[B(\mathbf{k})]$ (in a given order) is to find a third operator $[C(\mathbf{k})]$ which equals their product

$$[C(\mathbf{k})] = [A(\mathbf{k})][B(\mathbf{k})]. \tag{6.23}$$

The function $C(\mathbf{k})$ in powers of the magnetic field was found by Roth[6] (see the derivation in Appendix B):

$$[A(\mathbf{k})][B(\mathbf{k})] = [A(\mathbf{k})\,B(\mathbf{k})] - i\frac{e}{2c}\epsilon_{\alpha\beta\gamma}H^\gamma\left[\frac{\partial A(\mathbf{k})}{\partial k_\alpha}\frac{\partial B(\mathbf{k})}{\partial k_\beta}\right]$$
$$-\frac{1}{2}\left(\frac{e}{2c}\right)^2\epsilon_{\alpha\beta\gamma}\epsilon_{\alpha'\beta'\gamma'}H^\gamma H^{\gamma'}\left[\frac{\partial^2 A(\mathbf{k})}{\partial k_\alpha\,\partial k_{\alpha'}}\frac{\partial^2 B(\mathbf{k})}{\partial k_\beta\,\partial k_{\beta'}}\right] + \cdots. \tag{6.24}$$

In (6.24), $\epsilon_{\alpha\beta\gamma}$ is again the antisymmetric unit tensor as in (6.22) and the square brackets have the meaning expressed by formula (6.21). The generalization of formula (6.24) for three and more functions is straightforward.

Let us now come back to the construction of an effective Hamiltonian. In notation (6.21) the existence of an effective Hamiltonian for Eq. (6.15) can be expressed by the requirement that $H_{nn'}(\mathbf{kk}')$ in (6.16) has the following one-band form

$$H_{nn'}(\mathbf{kk}') = \delta_{nn'}\,[H_n(\mathbf{k})]\sum_m \delta[\mathbf{k} - \mathbf{k}' - \mathbf{K}_m), \tag{6.25}$$

where $[H_n(\mathbf{k})]$ operates on the δ-function. That this is so can be seen by substituting (6.25) into equation (6.15). As a result one obtains a one-band effective Hamiltonian equation

$$[H_n(\mathbf{k})]\, B_n(\mathbf{k}) = \epsilon B_n(\mathbf{k}). \tag{6.26}$$

Equation (6.26) expresses the main idea of an effective Hamiltonian for a Bloch electron in a magnetic field: it is a one-band equation and the Hamiltonian depends on $\mathbf{k} + (e/2c)\mathbf{H} \times i\, \partial/\partial\mathbf{k}$. Having this idea in mind an effective Hamiltonian theory will be developed in powers of the magnetic field $\mathbf{H}$ (the meaning of developing in powers of $\mathbf{H}$ will become clear below).

Let us first show that the transformation matrix

$$(\mathbf{kq}|n\mathbf{k}') = [u_{nk}(\mathbf{q})] \sum_m \delta(\mathbf{k} - \mathbf{k}' - \mathbf{K}_m) \tag{6.27}$$

leads in the lowest order of the magnetic field, to the effective Hamiltonian equation

$$[\epsilon_n(\mathbf{k})]\, B_n(\mathbf{k}) = \epsilon B_n(\mathbf{k}), \tag{6.28}$$

where $\epsilon_n(\mathbf{k})$ is the energy spectrum of the nth band. That the transformation matrix (6.27) should lead to Eq. (6.28) was expected before because in the absence of a magnetic field the matrix (6.17) led to Eq. (6.19) and the difference between the equations (6.5) and (6.6) is only in the substitution (6.7). In order to find out to what approximation does the transformation matrix (6.27) lead to Eq. (6.28), the former is substituted into expression (6.16). In notation (6.21) one has

$$H_{nn'}(\mathbf{kk}') = \int d\mathbf{q}\, [u_{nk}^*(\mathbf{q})][H(\mathbf{kq})][u_{nk}(\mathbf{q})] \sum_m \delta(\mathbf{k} - \mathbf{k}' - \mathbf{K}_m), \tag{6.29}$$

where $H(\mathbf{kq})$ is the Hamiltonian of Eq. (6.6). For arriving at (6.29), the expression for the function $(n\mathbf{k}|\mathbf{k}''\mathbf{q})$ was used (formula (B.9), Appendix B)

$$(n\mathbf{k}|\mathbf{k}''\mathbf{q}) = (\mathbf{k}''\mathbf{q}|n\mathbf{k})^* = [u_{nk}^*(\mathbf{q})] \sum_m \delta(\mathbf{k}'' - \mathbf{k} - \mathbf{K}_m) \tag{6.30}$$

and an integration was performed over $\mathbf{k}''$. By using now the multiplication formula (6.24) for the triple product in (6.29), one finds that to the lowest order in the magnetic field (the lowest order means taking only the first term in the multiplication formula (6.24)) the Hamiltonian in (6.29) becomes

$$H_{nn'}(\mathbf{kk}') = \delta_{nn'}[\epsilon_n(\mathbf{k})] \sum_m \delta(\mathbf{k} - \mathbf{k}' - \mathbf{K}_m). \tag{6.31}$$

In obtaining result (6.31), integration over $\mathbf{q}$ was interchanged with the operation (6.7). Such an interchange is legitimate because the integration is over $\mathbf{q}$ while the operation (6.7) has to do with the quasimomentum k. It sould be pointed out that expression (6.31) has the right form (form (6.25)) for an effective Hamiltonian to exist. By substituting (6.31) into (6.15), one obtains Eq. (6.28) which is the effective Hamiltonian equation to the zero order in the magnetic field[4–6] (the order of expansion is called the "zero order" when all the terms but the first one are neglected in the multiplication formula (6.24)).

It can also be checked that the transformation matrix (6.27) is unitary to the zero order in magnetic field (Relations (6.13) and (6.14)). Relation (6.13) is a particular case of (6.16) with $H(\mathbf{kq}) = 1$ and result (6.31) verifies it. As to relation (6.14) for the transformation matrix (6.27) it can be written as follows

$$\sum_{nk} (\mathbf{k}'\mathbf{q}|n\mathbf{k})(n\mathbf{k}|\mathbf{k}''\mathbf{q}') = \sum_{n} [u_{nk}(\mathbf{q})][u^*_{nk}(\mathbf{q}')] \sum_{m} \delta(\mathbf{k}' - \mathbf{k}'' - \mathbf{K}_m) \tag{6.32}$$

By using the multiplication formula (6.24) in the right-hand side of (6.32) the latter becomes relation (6.14). It is therefore seen that the transformation matrix (6.27) leads to an effective Hamiltonian equation (6.27) in the zero order of the magnetic field.

It is again interesting to compare the transformation matrix (6.17) in the absence of the magnetic field with (6.27) in the presences of the field. As was previously remarked, in the former case, the transformation matrix (6.17) itself is a solution of the Schrödinger equation (6.6). This should be the case because it diagonalizes Schrödinger's equation completely (it leads to Eq. (6.19)). In the presence of a magnetic field, the transformation matrix (6.27) leads to a one-band equation (Eq. (6.28)) which is diagonal only in the band index but not in the wavevector. In the wave vector, the effective Hamiltonian equation is a difference equation. One should not therefore expect the transformation matrix (6.27) to be a solution of Eq. (6.5).

Before going on to higher-order terms in the effective Hamiltonian, let us find the transformation matrix (6.27) in the coordinate representation (the r-representation). For this one should use formula (2.18)

$$(\mathbf{r}|n\mathbf{k}') = \int d\mathbf{k}\, d\mathbf{q}\, \exp(i\mathbf{kq})(\mathbf{kq}|n\mathbf{k}')\, \psi_{kq}(\mathbf{r}). \tag{6.33}$$

The exponent in (6.33) appears because the connection (2.18) between the r-representation and the kq-representation is given for the function $C(\mathbf{kq})$ while the transformation matrix (6.27) is written for the U-function (see relation (6.2)). By substituting expression (6.27) for $(\mathbf{kq}|n\mathbf{k}')$ and (2.13) for $\psi_{kq}(\mathbf{r})$ in (6.33), one finds (see Appendix C)

$$(r|nk') = \sum_{R_p} \exp\{i(\mathbf{k}' + (e/2c)\,\mathbf{H} \times \mathbf{r})\,\mathbf{R}_p\}\, a_n(\mathbf{r} - \mathbf{R}_p), \tag{6.34}$$

where $a_n(\mathbf{r})$ is the Wannier function for band n. In the absence of a magnetic field (6.34) is a Bloch function $\psi_{nk'}(\mathbf{r})$. When the magnetic field is present, (6.34) are the Roth functions [Roth,[6] Eq. (10)] which were used as a basis in expanding solutions of Schrödinger's equation for a Bloch electron in a magnetic field. From what was proven about the transformation matrix (6.27), it is clear that Roth's functions (6.34) form an orthogonal and a complete set of functions to the zero order in the magnetic field. The advantage of having the transformation matrix written in the *kq*-representation is because of the appearance of the magnetic field in the combination $\mathbf{k} + (e/2c)\mathbf{H} \times i\,\partial/\partial\mathbf{k}$. This is the combination in which the magnetic field appears in the final form of the effective Hamiltonian.

For developing an effective Hamiltonian theory to some higher order in the magnetic field, a transformation matrix has to be found that is unitary to the same given order and that leads to a Hamiltonian of the form (6.25) (again, to the same order in field). As was already shown, the transformation matrix (6.27) satisfies the requirements mentioned here to the zero order in magnetic field. We shall start the diagonalization procedure to higher orders by making the transformation matrix (6.27) unitary to any order in the magnetic field. Once the transformation matrix is unitary one does not have to worry about the right-hand side of the Schrödinger equation (Eq. (6.5)) which will automatically remain diagonal. In order to make the transformation matrix (6.27) unitary to any order in the field, let us define the following matrix

$$[N_{nn'}(\mathbf{k})] = \int dq\, [u^*_{nk}(\mathbf{q})][u_{n'k}(\mathbf{q})]. \tag{6.35}$$

The matrix (6.35) looks like the "norm" of the transformation matrix (6.27) and one should therefore expect that $[N(\mathbf{k})]^{-1/2}$ (when properly defined) could be used as a "normalization factor" for normalizing the matrix (6.27). It turns out that this is actually the case if one is able to define an inverse matrix $[N(\mathbf{k})]^{-1}$ to the matrix (6.25). The matrix $[N(\mathbf{k})]$ can also be given another form

$$[N(\mathbf{k})] = [S(\mathbf{k})]^+\, [S(\mathbf{k})], \tag{6.36}$$

where

$$[S_{mn}(\mathbf{k})] = \int dq\, u^*_{m0}(\mathbf{q})[u_{nk}(\mathbf{q})] \tag{6.37}$$

and $[S(\mathbf{k})]^+$ is the Hermitian conjugate of $[S(\mathbf{k})]$. As can easily be checked, $[S(\mathbf{k})]^+ = [S^\dagger(\mathbf{k})]$. Matrix elements $[N(\mathbf{k})]_{nm}$ in (6.26) should be understood as $[N_{nn'}(\mathbf{k})]$ (it means, the subscripts inside the rectangular brackets). Instead of deriving (6.36) from (6.35) it is easier to check that (6.36) goes over into (6.35) by using definition (6.37) and the completeness of the functions $u_{n0}(\mathbf{q})$. By means of the multiplication rule (6.24) one finds that

$$[N(\mathbf{k})] = I + [N^{(1)}(\mathbf{k})] + [N^{(2)}(\mathbf{k})] + \cdots, \tag{6.38}$$

with

$$N^{(1)}(\mathbf{k}) = -ih_{\alpha\beta}\frac{\partial S(\mathbf{k})^+}{\partial k_\alpha}\frac{\partial S(\mathbf{k})}{\partial k_\beta} \tag{6.39}$$

$$N^{(2)}(\mathbf{k}) = -\frac{1}{2}h_{\alpha\beta}h_{\alpha'\beta'}\frac{\partial^2 S(\mathbf{k})^+}{\partial k_\alpha\,\partial k_{\alpha'}}\frac{\partial^2 S(\mathbf{k})}{\partial k_\beta\,\partial k_{\beta'}}\,. \tag{6.40}$$

In (6.39) and (6.40) the summation is assumed on repreated indices and the following notation was used[6]

$$h_{\alpha\beta} = (e/2c)\sum_{\gamma=1}^{3}\epsilon_{\alpha\beta\gamma}H^\gamma \tag{6.41}$$

where $\epsilon_{\alpha\beta\gamma}$ is as before ((6.22), (6.24)), the antisymmetric unit tensor. The superscripts on the left-hand side of (6.39), (6.40) denote the order of the magnetic field in the expansion. If expansion (6.38) holds, it is possible to define an inverse matrix

$$\begin{aligned}[N(k)]^{-1} &= \{I + [N^{(1)}(\mathbf{k})] + [N^{(2)}(\mathbf{k})] + \cdots\}^{-1}\\ &= I - [N^{(1)}(\mathbf{k})] - [N^{(2)}(\mathbf{k})] + [N^{(1)}(\mathbf{k})]^2 + \cdots,\end{aligned} \tag{6.42}$$

which should be correct for sufficiently small magnetic fields. By using the multiplication rule (6.24), it is easy to check that (6.42) and (6.38) are inverse matrices to each other to any order in the magnetic field.

The inverse matrix (6.42) can now be used for defining a transformation matrix $(\mathbf{kq}|n\mathbf{k}')$ which is unitary to any order in the magnetic field[28]

$$(\mathbf{kq}|n\mathbf{k}') = \sum_l [u_{lk}(\mathbf{q})][N_{ln}(\mathbf{k})]^{-1/2}\sum_m \delta(\mathbf{k}-\mathbf{k}'-\mathbf{K}_m) \tag{6.43}$$

[28] J. Zak, *Phys. Rev.* **177**, 1151 (1969).

where

$$[N(\mathbf{k})]^{-1/2} = \{I + [N^{(1)}(\mathbf{k})] + [N^{(2)}(\mathbf{k})] + \cdots\}^{-1/2}$$
$$= I - \tfrac{1}{2}[N^{(1)}(\mathbf{k})] - \tfrac{1}{2}[N^{(2)}(\mathbf{k})] + \tfrac{3}{8}[N^{(1)}(\mathbf{k})]^2 + \cdots . \qquad (6.44)$$

The matrix $[N(\mathbf{k})]^{-1/2}$ in (6.43) serves as an "orthonormalization factor" for making the transformation matrix (6.27) unitary. As can be checked the transformation matrix $(\mathbf{kq}|n\mathbf{k}')$ in (6.43) is unitary to any order in the magnetic field. For this one has to check relations (6.13) and (6.14). The first of them (relation (6.13)) follows at once

$$\int d\mathbf{k}''\, d\mathbf{q}''\, (n\mathbf{k}|\mathbf{k}''\mathbf{q}'')(\mathbf{k}''\mathbf{q}''|n'\mathbf{k}')$$
$$= \{[N(\mathbf{k})]^{-1/2}\, [S(\mathbf{k})]^{+}\, [S(\mathbf{k})][N(\mathbf{k})]^{-1/2}\}_{nn'} \sum_m \delta(\mathbf{k} - \mathbf{k}' - \mathbf{K}_m)$$
$$= \delta_{nn'} \sum_m \delta(\mathbf{k} - \mathbf{k}' - \mathbf{K}_m). \qquad (6.45)$$

It can also be shown that relation (6.14) holds to any order of the magnetic field.[28]

The unitary transformation matrix (6.43) will now be used to transform Schrödinger's equation (6.5) to the form (6.15). By substituting (6.43) in (6.16), the following expressions are obtained for relations (6.15) and (6.16):

$$\sum_{n'} \{[U(\mathbf{k})]^{+}\, [H'(\mathbf{k})][U(\mathbf{k})]\}_{nn'}\, B_{n'}(\mathbf{k}) = \epsilon B_n(\mathbf{k}), \qquad (6.46)$$

$$H(\mathbf{kk}') = [U(\mathbf{k})]^{+}\, [H'(\mathbf{k})][U(\mathbf{k})] \sum_m \delta(\mathbf{k} - \mathbf{k}' - K_m), \qquad (6.47)$$

where

$$[U(\mathbf{k})] = [S(\mathbf{k})][N(\mathbf{k})]^{-1/2} \qquad (6.48)$$

and

$$H'_{mn}(\mathbf{k}) = \int d\mathbf{q}\, u^{*}_{m0}(\mathbf{q})\, H(\mathbf{kq})\, u_{n0}(\mathbf{q}). \qquad (6.49)$$

In (6.49), $H(\mathbf{kq})$ is the Hamiltonian of Eq. (6.6). As before, any matrix element $[A(\mathbf{k})]_{nm}$ in (6.46)–(6.48) is understood as $[A_{nm}(\mathbf{k})]$ with the subscripts inside the brackets.

Equation (6.46) is a multiband effective Hamiltonian equation for a Bloch electron in a magnetic field. Its form is convenient for deriving a one-band Hamiltonian to any order in the magnetic field. Any unitary transformation which diagonalizes the left-hand side of Eq. (6.46) will automatically keep the right-hand side diagonal.

We turn now to the diagonalization of Eq. (6.46). To the zero order in magnetic field it is already diagonal and the result is given in (6.28). For arriving at higher-order effective Hamiltonians the following procedure can be used. Assume that one knows the expansion of the Hamiltonian in (6.46) to any order in the magnetic field (the Hamiltonian in the left-hand side of (6.47) is denoted by $[H(\mathbf{k})]$):

$$H(\mathbf{k}) = H^{(0)}(\mathbf{k}) + H^{(1)}(\mathbf{k}) + H^{(2)}(\mathbf{k}) + \cdots, \tag{6.50}$$

where the superscripts on the right-hand side of (6.50) denote the order of the magnetic field. The Hamiltonian in (6.50) is already diagonal to zero order, $H_{mn}^{(0)}(\mathbf{k}) = \delta_{nm}\epsilon_n(\mathbf{k})$, while the higher-order terms contain nondiagonal elements. The purpose of the diagonalization procedure is to arrive at an equation of the form (6.26) to some given order of the magnetic field. Let us start by removing nondiagonal terms in Eq. (6.46) to first order in the magnetic field. For this purpose we define a transformation

$$\exp(i[T^{(1)}(\mathbf{k})]) = I + i[T^{(1)}(\mathbf{k})] + \cdots \tag{6.51}$$

and apply it to Eq. (6.46). In (6.51), I is a unit matrix and $T^{(1)}(\mathbf{k})$ is an unknown matrix which will be chosen to make the Hamiltonian in (6.46) diagonal to first order. After applying transformation (6.51), the Hamiltonian of Eq. (6.46) to first order in the magnetic field will assume the form (the square brackets are ommitted for the expression as a whole)

$$H_1(\mathbf{k}) = H^{(0)}(\mathbf{k}) + i(H^{(0)}(\mathbf{k})\, T^{(1)}(\mathbf{k}) - T^{(1)}(\mathbf{k})\, H^{(0)}(\mathbf{k})) + H^{(1)}(\mathbf{k}). \tag{6.52}$$

The notation $H_1(\mathbf{k})$ in (6.52) is explained below. One can now choose $T^{(1)}(\mathbf{k})$ in such a way that the nondiagonal terms in (6.52) vanish to first order in the magnetic field

$$T_{nn'}^{(1)}(\mathbf{k}) = iH_{nn'}^{(1)}(\mathbf{k})/\{\epsilon_n^{(0)}(\mathbf{k}) - \epsilon_{n'}^{(0)}(\mathbf{k})\}, \qquad n' \neq n \tag{6.53}$$

$$T_{nn}^{(1)}(\mathbf{k}) = 0. \tag{6.54}$$

In (6.53) $\epsilon_n^{(0)}(\mathbf{k}) = \epsilon_n(\mathbf{k})$, and is the energy spectrum in the absence of $\mathbf{H}$. The matrix $T^{(1)}(\mathbf{k})$ defined by (6.53) and (6.54) can be easily checked to be Hermitian. It can also be shown that the bracketed matrix $[T^{(1)}(\mathbf{k})]$ is Hermitian. In general, if $A(\mathbf{k})$ is a Hermitian matrix then $[A(\mathbf{k})]$ is also Hermitian. This follows at once from the fact that $[A(\mathbf{k})]^\dagger = [A^\dagger(\mathbf{k})]$. Since $[T^{(1)}(\mathbf{k})]$ is Hermitian, the transformation matrix (6.51) is unitary.

With the choice (6.53), (6.54), it makes the Hamiltonian (6.52) diagonal to first order in the magnetic field:

$$H_1(\mathbf{k}) = \epsilon^{(0)}(\mathbf{k}) + \epsilon^{(1)}(\mathbf{k}) + \cdots, \tag{6.55}$$

where again $\epsilon_n^{(0)}(\mathbf{k}) = \epsilon_n(\mathbf{k})$ and $\epsilon^{(1)}(\mathbf{k})$ is the diagonal part of $H^{(1)}(\mathbf{k})$ in (6.52) (or (6.46)). The notation $H_1(\mathbf{k})$ with the subscript "1" in (6.52) and (6.55) means that the transformed Hamiltonian is diagonal to first order in the magnetic field. Since the transformation matrix (6.51) is unitary, the right-hand side of (6.46) will remain unchanged and the effective Hamiltonian equation to first order in the magnetic field will become

$$([\epsilon_n^{(0)}(\mathbf{k})] + [\epsilon_n^{(1)}(\mathbf{k})])\, B_n(\mathbf{k}) = \epsilon B_n(\mathbf{k}). \tag{6.56}$$

Before deriving the explicit form for $\epsilon_n^{(1)}(\mathbf{k})$ let us first point out how one continues with the diagonalization procedure to higher orders in the magnetic field. The diagonality of the Hamiltonian (Eq. (6.56)) to first order in the magnetic field was achieved by starting with a Hamiltonian (6.50) that was diagonal to zero and by transforming it by means of the unitary transformation (6.51). It is clear that the same process can now be used for diagonalizing the Hamiltonian (6.52) (which is diagonal to first order in the field) to second order in the magnetic field and so on. Assume, for example, that the Hamiltonian is already diagonal to the pth order in the magnetic field

$$H_p(\mathbf{k}) = \epsilon^{(0)}(\mathbf{k}) + \epsilon^{(1)}(\mathbf{k}) + \cdots + \epsilon^{(p)}(\mathbf{k}) + H^{(p+1)'}(\mathbf{k}) + \cdots, \tag{6.57}$$

where $\epsilon^{(0)}(\mathbf{k}), \epsilon^{(1)}(\mathbf{k}), \ldots, \epsilon^{(p)}(\mathbf{k})$ are the diagonal parts of the Hamiltonian up to the pth order and $H^{(p+1)'}(\mathbf{k})$ contains nondiagonal terms of the order $(p + 1)$ in the magnetic field. It can be shown that the unitary matrix

$$\exp(i[T^{(p+1)}(\mathbf{k})]) = I + i[T^{(p+1)}(\mathbf{k})] + \cdots \tag{6.58}$$

will lead to a Hamiltonian diagonal to the order $(p + 1)$ in the magnetic field. Under transformation (6.58), the Hamiltonian (6.57) will become (to the order $(p + 1)$)

$$H_{p+1}(\mathbf{k}) = \epsilon^{(0)}(\mathbf{k}) + \epsilon^{(1)}(\mathbf{k}) + \cdots + \epsilon^{(p)}(\mathbf{k}) \\ + i(\epsilon^{(0)}(\mathbf{k})\, T^{(p+1)}(\mathbf{k}) - T^{(p+1)}(\mathbf{k})\, \epsilon^{(0)}(\mathbf{k})) + H^{(p+1)'}(\mathbf{k}) + \cdots . \tag{6.59}$$

Again, $T^{(p+1)}(\mathbf{k})$ can be chosen in such a way that the nondiagonal elements of $H_{p+1}(\mathbf{k})$ vanish to the order $(p+1)$ in magnetic field

$$T_{nn'}^{(p+1)}(\mathbf{k}) = iH_{nn'}^{(p+1)'}(\mathbf{k})/\{\epsilon_n^{(0)}(\mathbf{k}) - \epsilon_{n'}^{(0)}(\mathbf{k})\}, \qquad n' \neq n \tag{6.60}$$

$$T_{nn}^{(p+1)}(\mathbf{k}) = 0. \tag{6.61}$$

The matrix $T^{(p+1)}(\mathbf{k})$ defined by (6.60) and (6.61) is Hermitian and the transformation (6.58) is therefore unitary. This completes the proof that a unitary matrix exists (Relation (6.58)) that transforms the Hamiltonian (6.57) (which is diagonal to the pth order in the magnetic field) into a Hamiltonian

$$H_{p+1}(\mathbf{k}) = \epsilon^{(0)}(\mathbf{k}) + \epsilon^{(1)}(\mathbf{k}) + \cdots + \epsilon^{(p)}(\mathbf{k}) + \epsilon^{(p+1)}(\mathbf{k}) + \cdots \tag{6.62}$$

that is diagonal to the order $(p+1)$ in the field (the subscript $p+1$ in $H_{p+1}(\mathbf{k})$ means that the Hamiltonian is diagonal to the order $(p+1)$). Since p is arbitrary, the above procedure can be used for diagonalizing the Hamiltonian in (6.46) step by step to any power in the magnetic field. It is to be pointed out that the unitarity of the transformation (6.58) makes the diagonalization procedure relatively simple because one does not have to worry at all about the right-hand side of Eq. (6.46).

Up to now the diagonalization procedure was outlined without explicitly giving the terms of the Hamiltonian before and after the transformation. In order to arrive at an explicit form of different order terms let us start with the expansion of the Hamiltonian on the left-hand side of Eq. (6.46). This Hamiltonian contains a product of five bracketed quantities and its expansion requires a lot of algebra. The expression for the Hamiltonian in (6.46) is

$$[H(\mathbf{k})] = [N(\mathbf{k})]^{-1/2}\,[S(\mathbf{k})]^{+}\,[H'(\mathbf{k})][S(\mathbf{k})][N(\mathbf{k})]^{-1/2} \tag{6.63}$$

By using the multiplication rule (6.24) (and its generalization to any order in the field given in Appendix B) once can derive an expression for (6.63) to any order of the magnetic field. In Appendix D a description of this expansion is given. In particular, a general formula to any order in the magnetic field is given for the triple product $[S^{\dagger}(\mathbf{k})][H'(\mathbf{k})][S(\mathbf{k})]$. In order to write down this formula, let us generalize the definitions (6.39) and (6.40). Namely, let us define a function

$$N_{\beta_1\beta_2\cdots\beta_m}^{(n)}(\mathbf{k}) = \frac{(-i)^n}{n!}\, h_{\alpha_1\beta_1} h_{\alpha_2\beta_2} \cdots h_{\alpha_n\beta_n} \frac{\partial^n S^{\dagger}(\mathbf{k})}{\partial k_{\alpha_1}\partial k_{\alpha_2}\cdots\partial k_{\alpha_n}} \frac{\partial^{n-m} S(\mathbf{k})}{\partial k_{\beta_{m+1}}\cdots\partial k_{\beta_n}}, \tag{6.64}$$

where $h_{\alpha\beta}$ is given in (6.41). It is clear that the function in (6.64) is a generalization of the definitions (6.39), (6.40) and the former goes over into the latter when $m = 0$. In Appendix D it is shown that the above mentioned triple product can be expanded as follows

$$[S^+(\mathbf{k})][H'(\mathbf{k})][S(\mathbf{k})]$$
$$= \sum_{m=0}^{n} \sum_{n=0}^{\infty} C_n{}^m [N_{\beta_1\beta_2\cdots\beta_m}^{(n)}(\mathbf{k})\, \partial^m E(\mathbf{k})/\partial k_{\beta_1}\, \partial k_{\beta_2} \cdots \partial k_{\beta_m}], \tag{6.65}$$

where a summation is understood on repeated indices and the following notations were used: $C_n{}^m$ is the binomial coefficient of the $(m+1)$ term, and

$$E(\mathbf{k}) = \epsilon^{(0)}(\mathbf{k}) + E^{(1)}(\mathbf{k}) + E^{(2)}(\mathbf{k}). \tag{6.66}$$

In the last formula $\epsilon^{(0)}(\mathbf{k})$ is the energy spectrum in the absence of the magnetic field, and

$$E^{(1)}(\mathbf{k}) = -h_{\alpha\beta} v_\alpha(\mathbf{k})\, X^\beta(\mathbf{k}) \tag{6.67}$$

$$E^{(2)}(\mathbf{k}) = (1/2m)\, h_{\alpha\beta} h_{\alpha\beta'} (i\, \partial X^\beta(\mathbf{k})/\partial k_{\beta'} + X^{\beta'}(\mathbf{k})\, X^\beta(\mathbf{k})) \tag{6.68}$$

with

$$v_\alpha(\mathbf{k})_{nn'} = (1/m) \int u_{nk}^*(\mathbf{q})(-i\,\partial/\partial q_\alpha + k_\alpha)\, u_{n'k}(\mathbf{q})\, d\mathbf{q} \tag{6.69}$$

and $X^\alpha(\mathbf{k})$ is given by Formula (5.17). With the aid of formula (6.65) the expansion of the Hamiltonian (6.63) to different orders in the magnetic fields becomes straightforward but the result even to second order in the field is rather long and does not have any simple physical meaning. To the second order in the magnetic field one has

$$[H(\mathbf{k})] = [H^{(0)}(\mathbf{k})] + [H^{(1)}(\mathbf{k})] + [H^{(2)}(\mathbf{k})], \tag{6.70}$$

where

$$H^{(0)}(\mathbf{k}) = \epsilon^{(0)}(\mathbf{k}) \tag{6.71}$$

$$H^{(1)}(\mathbf{k}) = \tfrac{1}{2} N^{(1)}(\mathbf{k})\, \epsilon^{(0)}(\mathbf{k}) - \tfrac{1}{2} \epsilon^{(0)}(\mathbf{k})\, N^{(1)}(\mathbf{k}) + N_\beta^{(1)}(\mathbf{k}) \frac{\partial \epsilon^{(0)}(\mathbf{k})}{\partial k_\beta} + E^{(1)}(\mathbf{k}) \tag{6.72}$$

$$\begin{aligned} H^{(2)}(\mathbf{k}) = {} & \tfrac{1}{2} N^{(2)}(\mathbf{k})\, \epsilon^{(0)}(\mathbf{k}) - \tfrac{1}{2} \epsilon^{(0)}(\mathbf{k})\, N^{(2)}(\mathbf{k}) \\ & \tfrac{1}{2} N^{(1)}(\mathbf{k})\, E^{(1)}(\mathbf{k}) - \tfrac{1}{2} E^{(1)}(\mathbf{k})\, N^{(1)}(\mathbf{k}) \\ & + \tfrac{3}{8} \epsilon^{(0)}(\mathbf{k}) (N^{(1)}(\mathbf{k}))^2 - \tfrac{1}{8} (N^{(1)}(\mathbf{k}))^2\, \epsilon^{(0)}(\mathbf{k}) \\ & - \tfrac{1}{2} N^{(1)}(\mathbf{k})\, N_\beta^{(1)}(\mathbf{k}) \frac{\partial \epsilon^{(0)}(\mathbf{k})}{\partial k_\beta} - \tfrac{1}{2} N_\beta^{(1)}(\mathbf{k}) \frac{\partial \epsilon^{(0)}(\mathbf{k})}{\partial k_\beta} N^{(1)}(\mathbf{k}) \\ & - \tfrac{1}{2} N^{(1)}(\mathbf{k})\, \epsilon^{(0)}(\mathbf{k})\, N^{(1)}(\mathbf{k}) + 2 N_\beta^{(2)}(\mathbf{k}) \frac{\partial \epsilon^{(0)}(\mathbf{k})}{\partial k_\beta} + E^{(2)}(\mathbf{k}) \\ & + N_{\beta\beta'}^{(2)}(\mathbf{k}) \frac{\partial^2 \epsilon^{(0)}(\mathbf{k})}{\partial k_\beta\, \partial k_{\beta'}} + N_\beta^{(1)}(\mathbf{k}) \frac{\partial E^{(1)}(\mathbf{k})}{\partial k_\beta} \\ & + \frac{i}{2} h_{\alpha\beta} \frac{\partial N^{(1)}(\mathbf{k})}{\partial k_\alpha} \frac{\partial \epsilon^{(0)}(\mathbf{k})}{\partial k_\beta} + \frac{i}{2} h_{\alpha\beta} \frac{\partial \epsilon^{(0)}(\mathbf{k})}{\partial k_\alpha} \frac{\partial N^{(1)}(\mathbf{k})}{\partial k_\beta}. \end{aligned} \tag{6.73}$$

In the above expressions $\epsilon^{(0)}(\mathbf{k})$ is the energy spectrum of the unperturbed solid; $N^{(1)}(\mathbf{k})$ and $N^{(2)}(\mathbf{k})$ are defined by (6.39) and (6.40); $N_\beta^{(1)}(\mathbf{k})$, $N_\beta^{(2)}(\mathbf{k})$ and $N_{\beta\beta'}^{(2)}(\mathbf{k})$ are given by (6.64); $E^{(1)}(\mathbf{k})$ and $E^{(2)}(\mathbf{k})$ are given by (6.67) and (6.68) and $h_{\alpha\beta}$ is defined in (6.41). The expressions (6.71)–(6.73) give the Hamiltonian (6.63) up to second order in the magnetic field. The "order" of the field is defined here according to the commutator expansion (6.24).

The Hamiltonian given by (6.71)–(6.73) is diagonal only to the zero order in the magnetic field. The procedure outlined above (Eqs. (6.51)–(6.62)) will now be used to diagonalize the Hamiltonian (6.70) up to the second order in the magnetic field. The first step is to make H diagonal to the first order in the field. For this, the transformation (6.51) (with $T(\mathbf{k})$ given by (6.53) and (6.54)) is applied to the Hamiltonian (6.70) and terms up to the second order are kept. The result is (the subscript "1" means that the new Hamiltonian is diagonal to first order in the magnetic field):

$$[H_1(\mathbf{k})] = [H_1^{(0)}(\mathbf{k})] + [H_1^{(1)}(\mathbf{k})] + [H_1^{(2)}(\mathbf{k})] + \cdots, \tag{6.74}$$

where

$$H_1^{(0)}(\mathbf{k}) = \epsilon^{(0)}(\mathbf{k}) \tag{6.75}$$

$$H_1^{(1)}(\mathbf{k}) = i\epsilon^{(0)}(\mathbf{k})\, T^{(1)}(\mathbf{k}) - iT^{(1)}(\mathbf{k})\, \epsilon^{(0)}(\mathbf{k}) + H^{(1)}(\mathbf{k}) \tag{6.76}$$

$$\begin{aligned} H_1^{(2)}(\mathbf{k}) = {} & -\tfrac{1}{2}(T^{(1)}(\mathbf{k}))^2\, \epsilon^{(0)}(\mathbf{k}) - \tfrac{1}{2}\epsilon^{(0)}(\mathbf{k})(T^{(1)}(\mathbf{k}))^2 \\ & + H^{(2)}(\mathbf{k}) - ih_{\alpha\beta}(\partial T^{(1)}(\mathbf{k})/\partial k_\alpha)(\partial H^{(1)}(\mathbf{k})/\partial k_\beta) \end{aligned} \tag{6.77}$$

The notations are as before: $T^{(1)}(\mathbf{k})$ is given by (6.53) and (6.54); $H^{(1)}(\mathbf{k})$ and $H^{(2)}(\mathbf{k})$ are defined in (6.72) and (6.73) respectively. From the diagonalization procedure it should follow that the Hamiltonian (6.74) is diagonal up to first order in the magnetic field. This also follows from expression (6.76) for $H_1^{(1)}(\mathbf{k})$ and from the formulas (6.53) and (6.54) for $T^{(1)}(\mathbf{k})$. Having the Hamiltonian (6.74), diagonalized to first order in the field, the diagonalization procedure can now be used for obtaining the effective Hamiltonian to second order in the magnetic field. For this we use the transformation (6.58) with $p = 1$

$$\exp(i[T^{(2)}(\mathbf{k})]) = I + i[T^{(2)}(\mathbf{k})] + \cdots, \tag{6.78}$$

where according to (6.60) and (6.61)

$$T_{nn'}^{(2)}(\mathbf{k}) = i\,\frac{(H_1^{(2)}(\mathbf{k}))_{nn'}}{\epsilon_n^{(0)}(\mathbf{k}) - \epsilon_{n'}^{(0)}(\mathbf{k})}, \qquad n' \neq n \tag{6.79}$$

$$T_{nn}^{(2)}(\mathbf{k}) = 0. \tag{6.80}$$

In (6.79), $H_1^{(2)}(\mathbf{k})$ is given by (6.77). By applying the transformation (6.78) to the Hamiltonian (6.74) we have

$$[H_2(\mathbf{k})] = [H_2^{(0)})\mathbf{k})] + [H_2^{(1)}(\mathbf{k})] + [H_2^{(2)}(\mathbf{k})], \tag{6.81}$$

where

$$(H_2^{(0)}(\mathbf{k}))_{nn'} = \delta_{nn'}\epsilon_n^{(0)}(\mathbf{k}), \tag{6.82}$$

$$(H_2^{(1)}(\mathbf{k}))_{nn'} = \delta_{nn'}H_{nn}^{(1)}(\mathbf{k}), \tag{6.83}$$

$$(H_2^{(2)}(\mathbf{k}))_{nn'} = \delta_{nn'}(H_1^{(2)}(\mathbf{k}))_{nn}\,, \tag{6.84}$$

where $H_{nn}^{(1)}(\mathbf{k})$ is found from (6.72) (the first two terms do not contribute to the diagonal elements) and $H_1^{(2)}(\mathbf{k})$ is given by equation (6.77). In the expressions (6.82)–(6.84) it is explicitly pointed out that the Hamiltonian (6.81) is diagonal to second order in the magnetic field (the appearance of the $\delta_{nn'}$-function). The diagonality of the Hamiltonian follows from the fact that the transformation (6.78) contains only second order terms (the diagonal character of the Hamiltonian (6.70) to the first order remains unchanged) and from the explicit form of $T^{(2)}(\mathbf{k})$. This completes the diagonalization procedure to the second order in the magnetic field. The effective Hamiltonian equation (Eq. (6.26)) to the second order will be

$$[H_2(\mathbf{k})]\, B_n(\mathbf{k}) = \epsilon B_n(\mathbf{k}) \tag{6.85}$$

where $[H_2(\mathbf{k})]$ is the Hamiltonian (6.81) given by (6.82)–(6.84) (the subscript "2" means that the Hamiltonian is diagonalized to the second order in the field).

From the way the diagonalization was performed up to second order in the magnetic field it is clear how to continue the process to higher orders. If one is interested in the diagonalization to order p, for example, one should keep terms of order p on each step of the procedure. For example, in the case of the diagonalization to second order in the field, terms up to second order were kept also on the first step of the diagonalization (e.g., Eqs. (6.74)–(6.77)). To pth order of the magnetic field, the effective Hamiltonian equation is

$$[H_p(\mathbf{k})]\, B_n(\mathbf{k}) = \epsilon B_n(\mathbf{k}) \tag{6.86}$$

where $H_p(\mathbf{k})$ is diagonal up to the order p in the magnetic field.

As is seen from the effective Hamiltonian (6.81)–(6.84), even to second order in the field, the expressions become rather complicated and their connection to the original band structure of the solid is not very straightforward.[29] Because of their complexity, the usefulness of effective Hamiltonians to orders higher than the second is therefore doubtful.

It is interesting to point out that the effective Hamiltonian $H_p(\mathbf{k})$ to any order p in the field (e.g., Eq. (6.86)) is periodic in $\mathbf{k}$ with the periodicity of the reciprocal lattice vectors. To see it we first refer to the expansion given by formula (6.65). As shown in Appendix D, each term in (6.65) is periodic in $\mathbf{k}$. Similarly one can show that the matrix $[N(\mathbf{k})]^{-1/2}$ entering the Hamiltonian (6.46) is periodic in $\mathbf{k}$. It follows therefore that the starting Hamiltonian (6.50) (to any order of the magnetic field) is periodic in $\mathbf{k}$. Since the diagonalization procedure uses the matrix $T(\mathbf{k})$ which by itself is periodic in $\mathbf{k}$, it is clear that the effective Hamiltonians to any order in the magnetic field are periodic functions of $\mathbf{k}$ with the periodicity of the reciprocal lattice vectors. As a result of this fact the effective Hamiltonian equation (Eq. (6.86)) to any order in the field is a difference equation.

In conclusion of this section it should be pointed out that the use of the *kq*-representation has the advantage of having the combination $\mathbf{k} + (e/2c)\mathbf{H} \times i\, \partial/\partial\mathbf{k}$ at each step of the diagonalization procedure including the starting Hamiltonian (6.5). Since the purpose of the effective Hamiltonian theory is to arrive at an equation of the form (6.86), the *kq*-representation turns out to be useful.

VII. Fundamental Dynamics in the Bloch Theory of Solids

In applications, in particular, in transport theory two theorems have been widely used. These are the acceleration theorems for a Bloch electron in external electric and magnetic fields. In case of the electric field, the acceleration theorem was already discussed in Section V (Eq. (5.1)). It was shown that this theorem follows straightforwardly from Schrödinger's equation (5.5). The second theorem is connected with the acceleration of a Bloch electron in a magnetic field and is often given in the following form[10]

$$\dot{\mathbf{k}} = -(e/c)\, \mathbf{v} \times \mathbf{H} \tag{7.1}$$

where $\dot{\mathbf{k}}$ is the time derivative of the quasimomentum $\mathbf{k}$, $\mathbf{v}$ is the velocity of the electron in a Bloch state specified by $\mathbf{k}$, and $\mathbf{H}$ is the magnetic field. Because of their importance, many proofs of the theorems (5.1) and (7.1) have been presented in the literature. Here we will discuss these theorems and also the case when both the electric and magnetic field are present in the light of the *kq*-representation.

[29] L. M. Roth, *Phys. Rev.* **145**, 434 (1966).

When there are no external fields the quasimomentum $\mathbf{k}$ which defines the eigenvalues $\exp(i\mathbf{k}\cdot\mathbf{a})$ of the translation $\exp(i\mathbf{p}\cdot\mathbf{a})$ is a constant of motion. When an electric field is present the eigenvalues $\exp(i\mathbf{k}\cdot\mathbf{a})$ are no longer constants of motion and are given by

$$\exp(i[\mathbf{k} - e\mathbf{E}t]\cdot\mathbf{a}). \tag{7.2}$$

Equation (7.2) is equivalent to the acceleration theorem (5.1) and follows from Schrödinger's equation for a Bloch electron in an electric field in the kq-representation. As was already mentioned in Section V, theorem (5.1) (or (7.2)) is exact and has nothing to do with a one-band approximation.

Before deriving the acceleration theorem in the magnetic field the following remark should be made. If we would try to calculate $\dot{\mathbf{k}}$ in the presence of a magnetic field, this would lead us to a gauge dependent result.[17] The reason for this is that the gauge independent combination is $\mathbf{k} + (e/2c)\mathbf{H} \times i\,\partial/\partial\mathbf{k}$ and it is therefore natural to look for the variation in time of this combination and not of $\mathbf{k}$ itself. In fact, the classical acceleration theorem in a magnetic field is

$$m\dot{\mathbf{v}} = -(e/c)\,\mathbf{v} \times \mathbf{H} \tag{7.3}$$

with the velocity $\mathbf{v}$ on the left-hand side and not the $\mathbf{k}$ vector as in (7.1).

In order to find an equation for a Bloch electron in a magnetic field that corresponds to (7.3) for a free electron, we use the result of the preceding section. It was shown there that in the lowest order of the magnetic field, the one-band Hamiltonian is given by $[\epsilon_n(\mathbf{k})]$, where the rectangular brackets mean here that $\epsilon_n(\mathbf{k})$ is first symmetrized with respect to different components of the vector $\mathbf{k}$ and then $\mathbf{k}$ is replaced by $\mathbf{k} + (e/2c)\mathbf{H} \times i\,\partial/\partial\mathbf{k}$. The derivative with respect to time of the vector $\mathbf{k} + (e/2c)\mathbf{H} \times i\,\partial/\partial\mathbf{k}$ (which corresponds to the velocity operator for an electron in a magnetic field) is as follows:

$$d[\mathbf{k}]/dt = i[[\epsilon_n(\mathbf{k}), [\mathbf{k}]], \tag{7.4}$$

where on the right-hand side the outside brackets mean a commutation relation. By using formula (6.24) one finds that to first order in the magnetic field, relation (7.4) will be

$$d[\mathbf{k}]/dt = -(e/c)[\partial\epsilon_n(\mathbf{k})/\partial\mathbf{k}] \times \mathbf{H}. \tag{7.5}$$

In deriving Eq. (7.5), which is correct to first order in the magnetic field, we took only the zeroth-order term in the effective one-band Hamiltonian, because higher-order terms in the effective Hamiltonian would lead to higher-order terms in (7.5).

It follows therefore that the correct equation for describing the motion of a Bloch electron in a magnetic field is (7.5) and not (7.1). The reason that Eq. (7.1) leads to correct results when used in a semiclassical theory of transport[10] is easily seen by comparing (7.1) with (7.5). The only difference between these two equations is that in the latter $\mathbf{k}$ is replaced by $\mathbf{k} + (e/2c)\mathbf{H} \times i\, \partial/\partial\mathbf{k}$ which is just a relabeling of the variable. It is clear that one has to have in mind the velocity components $\mathbf{k}+(e/2c)\mathbf{H} \times i\, \partial/\partial\mathbf{k}$ and not $\mathbf{k}$ when using the equation of motion for a Bloch electron in a magnetic field. The acceleration theorem to higher order in the magnetic field can be obtained by using higher order effective Hamiltonians[29] in relation (7.4).

The next question is what kind of an equation does one get for a Bloch electron in both a magnetic and electric field. We saw that for the electric case the vector $\mathbf{k}$ appears in the equation of motion, while in the magnetic case the vector $\mathbf{k} + (e/2c)\mathbf{H} \times i\, \partial/\partial\mathbf{k}$ appears. To find the equation of motion when both fields are present we need an effective one-band Hamiltonian. Let us show that the functions (6.27) in expansion (6.12) diagonalize the equation for a Bloch electron in both an electric and magnetic field to zero order of the magnetic field (as before) and to first order in the electric field. By applying this transformation to the part of the Hamiltonian that does not depend on the electric field, we get the same result as before to zero order in magnetic field, namely $[\epsilon_n(\mathbf{k})]$. The part of the Hamiltonian that depends on $\mathbf{E}$ will become ($S(\mathbf{k})$ is given in (6.37)

$$[S^{\dagger}(\mathbf{k})]\, e\mathbf{E} \cdot i(\partial/\partial\mathbf{k})[S(\mathbf{k})] = e\mathbf{E} \cdot i(\partial/\partial\mathbf{k}) + [S^{\dagger}(\mathbf{k})]\, e\mathbf{E} \cdot (i(\partial/\partial\mathbf{k})[S(\mathbf{k})]) \tag{7.6}$$

In the second term the derivative with respect to $\mathbf{k}$ can be replaced by the derivative with respect to $\mathbf{k} + (e/2c)\mathbf{H} \times i\, \partial/\partial\mathbf{k}$ and we can therefore apply formula (6.24) to this term. In zero order of the magnetic field and first order in the electric field the effective one-band Hamiltonian is

$$[\epsilon_n(\mathbf{k})] + e\mathbf{E} \cdot (i\, \partial/\partial\mathbf{k} + [\mathbf{X}_{nn}(\mathbf{k})]), \tag{7.7}$$

where $\mathbf{X}_{nn}(\mathbf{k})$ is given in (5.17). In the absence of the magnetic field when $\mathbf{H} = 0$, expression (7.7) goes over into the known one-band effective Hamiltonian for a Bloch electron in an electric field only.[26] We can now find the time derivative of $\mathbf{k} + (e/2c)\mathbf{H} \times i\, \partial/\partial\mathbf{k}$. By using (7.7) and (6.24) one gets in the lowest order of magnetic and electric fields

$$(d/dt)[\mathbf{k}] = -e(\mathbf{E} + (1/c)[\partial\epsilon_n(\mathbf{k})/\partial\mathbf{k}] \times \mathbf{H}). \tag{7.8}$$

It is again to be noted that formula (7.8) differs from the one commonly used in transport theory. In the latter no rectangular brackets appear. As was already mentioned this difference is not essential in the semi-classical transport theory because it mean just using a different notation for the vector **k**. Note, however, that quantum mechanically Eqs. (7.5) and (7.8) without the rectangular brackets are meaningless. In quantum mechanics the noncommutativity of the components of $\mathbf{k} + (e/2c)\mathbf{H} \times i\,\partial/\partial\mathbf{k}$ becomes essential. This is best demonstrated in the derivation of Onsager's relation ($\mathbf{E} = 0$). Let us assume that **H** is in the z direction and denote

$$\boldsymbol{\chi}^+ = \mathbf{k} + (e/2c)\mathbf{H} \times i\,\partial/\partial\mathbf{k} \tag{7.9}$$

$\chi_z^+ = k_z$, and according to (7.5) is a constant of motion. The other two components χ_x^+ and χ_y^+ satisfy relation (6.22), and the Bohr–Sommerfeld quantization rule for them is

$$\oint \chi_x^+ \, d\chi_y^+ = (2\pi eH/\hbar c)(n + \gamma) \tag{7.10}$$

On the left-hand side we have an area in the χ space $S(\epsilon, \chi_z^+)$ for a given energy ϵ and χ_z^+. Relation (7.10) can therefore be written

$$S(\epsilon, \chi_z^+) = (2\pi eH)\hbar c)(n + \gamma), \tag{7.11}$$

which is Onsager's relation.[30]

It should be noted that by using the coordinates χ^+ we did not take into account all the degrees of freedom of the three-dimensional problem. The couple of conjugate coordinates χ_x^+, χ_y^+ describes one degree of freedom, while $\chi_z^+ = k_z$ describes another one. The third degree of freedom can be described by the x and y components of the vector[31]

$$\boldsymbol{\chi}^- = \mathbf{k} - (e/2c)\mathbf{H} \times i\,\partial/\partial\mathbf{k}. \tag{7.12}$$

Their commutation relation is

$$[\chi_x^-, \chi_y^-] = ieH/c. \tag{7.13}$$

Both χ_x^- and χ_y^- commute with the Hamiltonian $[\epsilon_n(\mathbf{k})]$ and the commuting set of operators [see expressions (2.2) and (2.4)]

$$\exp(i\chi_x^- A), \qquad \exp(i\chi_y^- B) \tag{7.14}$$

with

$$AB = 2\pi c/eH \tag{7.15}$$

[30] L. Onsager, *Phil. Mag.* **43**, 1006 (1952).
[31] R. Kubo, H. Hasegawa, and N. Hashitsume, *J. Phys. Soc. Jap.* **14**, 56 (1959).

can be used for specifying eigenstates of $[\epsilon_n(\mathbf{k})]$. The complete specification of the eigenstates of $[\epsilon_n(\mathbf{k})]$ will therefore be given by the number n in (7.11), by k_z and by the eigenvalues of the operators (7.14). Since the Hamiltonian $[\epsilon_n(\mathbf{k})]$ does not depend on the latter, this leads to the known degeneracy of Landau levels which in our description will be given by the number of states contained in the area of variation of χ_x^-, χ_y^-. This area equals $2\pi/AB$ and the number of states is therefore[30]

$$eH/2\pi\hbar c. \tag{7.16}$$

In concluding this section we should like to make a number of remarks. First, it should again be mentioned that the acceleration theorem in the electric field (Eq. (5.1)) is exact, while in the magnetic field and in the mixed case (Eqs. (7.5) and (7.8) respectively) the theorem is given to the lowest order in the magnetic and electric fields. Another remark is with respect to the acceleration in the quasicoordinate $\mathbf{q}$. This coordinate does not appear at all in the effective Hamiltonians and therefore its derivative vanishes. Such a behavior with respect to the quasi coordinate is to be expected in the effective Hamiltonian approximation where variation inside a unit cell are assumed to be negligible. It should also be pointed out that the effective Hamiltonian in the presence of both a magnetic and an electric field (Eq. (7.7)) cannot hold for the whole band because as was shown in Section V interband terms in the presence of an electric field can only be neglected for a part of the Brillouin zone.

VIII. Symmetry of a Bloch Electron in the Presence of an External Magnetic Field

In this section the kq-representation is used to discuss the symmetry for a Bloch electron in a magnetic field. It will be shown that symmetry arguments can be used for choosing the most suitable functions in the central expansion (6.12) of the effective Hamiltonian theory. As one should expect the best functions in expansion (6.12) should be the functions with the symmetry of the problem. In section VI the idea was to choose the functions $(\mathbf{kq}|n\mathbf{k}')$ in such a way as to make the Hamiltonian (6.16) assume the form of an effective Hamiltonian (6.25). The question of how to choose the most suitable functions is not trivial at all. One can however always use the symmetry of Schrödinger's equation for a Bloch electron in a magnetic field in order to simplify the problem of choosing the right functions. In the absence of the magnetic field the functions that diagonalize the Hamiltonian in (6.6) (or transform (6.6) into (6.19)) are the Bloch functions (6.17). The latter are specified according to the translation symmetry of the unperturbed crystal. When the magnetic field is turned on the symmetry of the Hamiltonian is given by the magnetic translation operators[32–34a] which in the r-representation are

$$\tau(\mathbf{R}_n) = \exp\{(i/\hbar)(\mathbf{p} - (e/2c)\mathbf{H} \times \mathbf{r}) \cdot \mathbf{R}_n\}. \tag{8.1}$$

In (8.1) $\mathbf{R}_n$ is a Bravais lattice vector. In the kq-representation these translation operators will become (after the phase transformation (6.2) has been performed)

$$\tau(\mathbf{R}_n) = \exp\{i(-i\,\partial/\partial\mathbf{q} + \mathbf{k} - (e/2c\hbar)\mathbf{H} \times i\,\partial/\partial\mathbf{k}) \cdot \mathbf{R}_n\}. \tag{8.2}$$

As can easily be shown these operators commute with the Hamiltonian of Eq. (6.5) for a Bloch electron in a magnetic field. Before discussing the symmetry of the transformation matrix in (6.12) with respect to the magnetic translations (8.2) we shall first have a closer look at the structure of the latter in the view of the kq-representation. As can easily be checked, two magnetic translations, $\tau(\mathbf{R}_n)$ and $\tau(\mathbf{R}_n')$, commute if the magnetic field $\mathbf{H}$ satisfies the rationality relation[32–34]

$$\mathbf{H} \cdot \mathbf{R}_n \times \mathbf{R}_n'/(hc/e) = m,$$

where m is an integer. The commutativity of the operators $\tau(\mathbf{R}_n)$ and $\tau(\mathbf{R}_n')$ is of particular interest because these operators depend on both the radius vector $\mathbf{r}$ and the momentum operator $\mathbf{p}$. It is clear that only two very special functions of $\mathbf{r}$ and $\mathbf{p}$ can commute with each other since $\mathbf{r}$ does not commute with $\mathbf{p}$. A very simple connection can be established between the structure of the magnetic translation operators (8.1) (or (8.2)) and the operators (2.2) and (2.4) that define the kq-representation. For simplicity let us assume that the magnetic field is in the direction of the unit cell vector $\mathbf{a}_3$ in the Bravais lattice and that the two other unit cell vectors $\mathbf{a}_1$ and $\mathbf{a}_2$ are in a plane perpendicular to $\mathbf{a}_3$ and are perpendicular to each other. Then the magnetic translations (8.1) in the direction $\mathbf{a}_3$ do not contain the magnetic field and they reduce to ordinary translations. One is left with magnetic translations of the form

$$e^{iP_1a_1}, \tag{8.3}$$

$$e^{iP_2a_2}, \tag{8.4}$$

[32] E. Brown, *Phys. Rev.* **133**, A1038 (1964).
[33] J. Zak, *Phys. Rev.* **134**, A1602 (1964).
[34] J. Zak, *Phys. Rev.* **134**, A1607 (1964).
[34a] W. Opechowski and W. G. Tam, *Physica* **42**, 529 (1969).

where

$$\mathbf{P} = (1/\hbar)(\mathbf{p} - (e/2c)\,\mathbf{H} \times \mathbf{r}) \tag{8.5}$$

and P_1 and P_2 are its components in the a_1 and a_2 directions. All the other magnetic translations are functions of (8.3) and (8.4). P_1 and P_2 satisfy the following commutation relation

$$[P_1\,, P_2] = ieH/\hbar c \tag{8.6}$$

and in order for the operators (8.3) and (8.4) to commute the magnetic field has to satisfy the rationality condition

$$Ha_1a_2/(\hbar c/e) = 2\pi. \tag{8.7}$$

It is now easy to compare the magnetic translation operators (8.3) and (8.4) with the operators (2.2) and (2.4) that define the *kq*-representation. One could write the latter in a more general form

$$e^{ipa}, \tag{8.8}$$

$$e^{ixb}. \tag{8.9}$$

In order for them to commute one has to require

$$ab = 2\pi. \tag{8.10}$$

The analogy between the magnetic translation operators (8.3) and (8.4) and *kq*-operators (8.8) and (8.9) is therefore complete. The "rationality" condition (8.10) for the operators (8.8) and (8.9) is very simple because the basic operators p and x satisfy a very simple commutation relation $[p, x] = -i\hbar$. Since the operators (2.2) and (2.4) form a complete set of commuting operators, it is clear that the magnetic translations (8.3) and (8.4) can also be used as a complete set of operators to define a degree of freedom. This argument was already used in the discussion of Onsager's theorem in the previous section.

As a digression let us point out that the whole Bloch theory is based on the fact that functions of $\mathbf{p}$ (the translation operators (2.11)) commute with the periodic potential of the Bloch Hamiltonian (Hamiltonian of Eq. (2.16) with $\mathbf{H} = v = 0$). As can be seen from relation (2.19), the periodic potential is a function of the operators (2.12) that together with (2.11) define the *kq*-representation. The operators (2.11) and (2.12) that define the *kq*-representation form therefore the basis for the commuting operators in both the Bloch problem (in the absence of external fields) and the problem for a Bloch electron in a magnetic field.

Let us now check the behavior of the functions $(\mathbf{kq}|n\mathbf{k}')$ in expansion (6.12) with respect to the magnetic translations (8.2). On each step of the diagonalization procedure in Section VI the transformation matrix $(\mathbf{kq}|n\mathbf{k}')$ has the general structure (6.27) (when one goes to a higher-order diagonalization, the only thing that will change is the function in the rectangular brackets but not the structure of the function (6.27)). One therefore has to check the behavior under magnetic translation of the functions

$$(\mathbf{kq}|n\mathbf{k}') = [W_{nk}(\mathbf{q})] \sum_{K} \delta(\mathbf{k} - \mathbf{k}' - \mathbf{K}), \tag{8.11}$$

where $W_{nk}(\mathbf{q})$ is a function with the same periodic properties as the Bloch function $u_{nk}(\mathbf{q})$ is. In applying the translation operators (8.2) to the functions (8.11) one should keep in mind that the former depend on $\mathbf{k} - (e/2c)\mathbf{H} \times i\,\partial/\partial\mathbf{k}$ while the $[W_{nk}(\mathbf{q})]$ depends on $\mathbf{k} + (e/2c)\mathbf{H} \times i\,\partial/\partial\mathbf{k}$. Since $\mathbf{k} - (e/2c\hbar)\mathbf{H} \times i\,\partial/\partial\mathbf{k}$ commutes with $\mathbf{k} + (e/2ch)\mathbf{H} \times i\,\partial/\partial\mathbf{k}$ and since also $W_{nk}(q)$ is periodic in $\mathbf{q}$ with the period $\mathbf{R}_n$, it follows

$$\begin{aligned}\tau(\mathbf{R}_n)(\mathbf{kq}|n\mathbf{k}') &= \tau(\mathbf{R}_n)[W_{nk}(\mathbf{q})] \sum_{K} \delta(\mathbf{k} - \mathbf{k}' - \mathbf{K}) \\ &= [W_{nk}(\mathbf{q})] \exp\{i(\mathbf{k} - (e/2c\hbar)\,\mathbf{H} \times i\,\partial/\partial\mathbf{k}) \cdot \mathbf{R}_n\} \sum \delta(\mathbf{k} - \mathbf{k}' - \mathbf{K}) \\ &= \exp(i\mathbf{k}' \cdot \mathbf{R}_n)(\mathbf{kq}|n\mathbf{k}' + (e/2c\hbar)\mathbf{H} \times \mathbf{R}_n). \end{aligned} \tag{8.12}$$

We see, therefore, that the functions $(\mathbf{kq}|n\mathbf{k}')$ with the structure (8.11) form a basis for a representation of the magnetic translation group.[34] In general, this representation will be of infinite dimension; it will be finite-dimensional for magnetic fields that satisfy the rationality condition. A more general form of this condition is

$$\mathbf{H} = (hc/e)(m/NV)\,\mathbf{a}_3\,, \tag{8.13}$$

where V is the volume of a unit cell in the Bravais lattice and m, N are integers with no common factor. Condition (8.13) requires the magnetic field to be in a direction of a lattice vector and this particular direction is for definiteness given in (8.13) by $\mathbf{a}_3$, one of the unit cell vectors. When the magnetic field satisfies condition (8.13), the behavior of the functions $(\mathbf{kq}|n\mathbf{k}')$ under the magnetic translations (relation (8.12)) can be expressed as follows

$$\tau(\mathbf{R}_n)(\mathbf{kq}|n\mathbf{k}') = \exp(i\mathbf{k}' \cdot \mathbf{R}_n)(\mathbf{kq}|n\mathbf{k}' + (m/2N)(n_1\mathbf{K}_1 - n_2\mathbf{K}_2)), \tag{8.14}$$

where $\mathbf{K}_1$, $\mathbf{K}_2$ are unit cell vectors of the reciprocal lattice and n_1, n_2 come from $\mathbf{R}_n = n_1\mathbf{a}_1 + n_2\mathbf{a}_2 + n_3\mathbf{a}_3$. Two cases can be distinguished.

One for m even and the other for odd m. When m is even, m and $2N$ have a common factor and then when either n_1 or n_2 equals N a vector of a reciprocal lattice is added to $\mathbf{k}'$ in (8.14) (for odd m this will happen when n_1 or n_2 equal $2N$). The structure of the function $(\mathbf{kq}|n\mathbf{k}')$ in (8.14) is such that when a reciprocal vector is added to $\mathbf{k}'$, the function goes over to itself (see relation (8.11)). From (8.14) it is clear that for "rational" magnetic fields the functions $(\mathbf{kq}|n\mathbf{k}')$ will induce a representation of finite dimension. This dimension for even m is N^2 and for odd m is $(2N)^2$. For the general symmetry problem here it does not matter what the dimension is. All that matters is that functions with the structure (8.11) form a basis for a representation of the magnetic translation group. It can therefore be said that from the point of view of symmetry the most proper functions in expansion (6.12) are functions with the structure (8.11). The Kohn–Luttinger functions have the symmetry of the crystal in the absence of the magnetic field, and when they are used in expansion (6.12) for the problem of a Bloch electron in a magnetic field one arrives at a rather complicated algebra.[4]

In the diagonalization procedure of Section VI the functions $(\mathbf{kq}|n\mathbf{k}')$ in expansion (6.12) had the structure (8.11) to each order of the magnetic field. It was also shown that the functions $(\mathbf{kq}|n\mathbf{k}')$ form an orthonormal system to any order of the field. Together with their property (8.12) it follows that the functions $(\mathbf{kq}|n\mathbf{k}')$ used in the diagonalization procedure of Section VI form bases for *irreducible* representations of the magnetic translation group. It is a general feature of any problem in quantum mechanics that the most suitable classification of its states can be achieved by using the irreducible representations of the symmetry of the problem.[35] It follows therefore that the simplicity of the diagonalization procedure in Section VI was achieved by using properly symmetrized functions in expansion (6.12)

Up to now the symmetry of the exact Hamiltonian (Hamiltonian of Eq. (6.5)) for the Bloch electron in a magnetic field was discussed. It is interesting also to consider the symmetry of the effective Hamiltonians. As was shown in Section VI the effective Hamiltonian to any order in the magnetic field is a function of the operator $\mathbf{k} + (e/2c)\mathbf{H} \times i\,\partial/\partial\mathbf{k}$:

$$H_{\text{eff}} = H_n(\mathbf{k} + (e/2c)\mathbf{H} \times i\,\partial/\partial\mathbf{k}). \tag{8.15}$$

Since $\mathbf{k} - (e/2c)\mathbf{H} \times i\,\partial/\partial\mathbf{k}$ commutes with $\mathbf{k} + (e/2c)\mathbf{H} \times i\,\partial/\partial\mathbf{k}$, this means that any operator

$$\exp[i(\mathbf{k} - (e/2c)\mathbf{H} \times i\,\partial/\partial\mathbf{k}) \cdot \mathbf{a}] \tag{8.16}$$

[35] Eugene P. Wigner, "Group Theory." Academic Press, New York, 1959.

with an arbitrary **a** will commute with the Hamiltonian (8.15). In comparing the operators (8.16) with the magnetic translation operators (8.2) it is seen that while the latter allow translations in **k** space only by discrete vectors (corresponding to the Bravais lattice vectors) the former contain any infinitesimal translations. This means that the symmetry of the effective Hamiltonian is higher than that of the exact Hamiltonian (6.5). The symmetry of the effective Hamiltonian (8.15) is actually the same as the symmetry of an electron in a magnetic field in the absence of the crystal. As is well known,[7] the motion of an electron in a magnetic field in the plane perpendicular to the field is quantized and the energy spectrum consists of discrete Landau levels. These levels are infinitely degenerate, the degeneracy being a consequence of the symmetry (8.16).[36] When a crystalline field is turned on the symmetry is lowered and as a consequence of this it can be shown that the Landau levels broaden.[36–41] Since the symmetry of the effective Hamiltonian (8.15) is the same as the symmetry in the absence of the crystalline field, the effective Hamiltonian approach will, in general, lead to discrete Landau levels without any broadening. A different way to see that the effective Hamiltonian approach will lead to no broadening is by noticing that the **q** coordinate is missing from the effective Hamiltonian (8.15). Qualitatively, Landau level broadening can be explained by the energy difference between Landau orbits centered at different points in a unit cell of the crystal. The averaging on the **q** coordinate (the coordinate describing different points in a unit cell) that led to the effective Hamiltonian therefore washed out the Landau level broadening.[42]

IX. Conclusions

The *kq*-representation is defined in this article on a natural basis, a crystalline lattice, and provides us with a system of coordinates, **k** and **q**, that reflect most closely the translational symmetry of solids. The sense in which the *kq*-coordinates are related to the translational symmetry is the same as that in which spherical coordinates are related to problems with spherical symmetry. In both cases, the use of the most suitable coordinate makes the description of the problem as simple as possible.

[36] J. Zak, *Phys. Rev.* **136**, A776 (1964).
[37] D. G. Harper, *Proc. Phys. Soc. (London)* **A68**, 879 (1955).
[38] A. D. Brailsford, *Proc. Phys. Soc. (London)* **A70**, 275 (1957).
[39] H. J. Fishbeck, *Phys. Status Solidi* **22**, 235 (1967).
[40] W. G. Chambers, *Phys. Rev.* **140**, A135 (1965).
[41] D. Langbein, *Phys. Rev.* **180**, 633 (1969).
[42] An exception of this rule can be obtained in the case of open orbits (see Roth[29]).

In this article the *kq*-representation was applied to a number of problems in the dynamics of electrons in solids. The problems were of quite general character and the usefulness of the *kq*-representation was demonstrated on them. Because of its universality, the *kq*-representation will undoubtedly turn out to be useful in applications to many other excitations in solids (in addition to electrons), like excitons, polarons, magnons, etc. and also in different theoretical approaches to problems in solid state physics.[43–45]

Being a general quantum mechanical representation, the significance of the *kq*-representation is not limited exclusively to solids. It was already used in a number of fundamental problems in quantum mechanics.[13,13a] Although from most points of view the *kq*-representation is a regular quantum mechanical representation, there is nevertheless something very particular about its eigenstates (2.6). One can easily verify that the averages of both the coordinate x and the momentum p, $\bar{x}$, and $\bar{p}$, are completely undefined. This means that in the eigenstates of the operators (2.2) and (2.4) the uncertainties of x and p, Δx, and Δp, can be as large as one likes. It is interesting to compare the states (2.6) with what are known as the most "classical states" in quantum mechanics.[46] (The latter are states with the minimum product of the uncertainties Δx, Δp, $\Delta x\, \Delta p = \hbar/2$). Since in the eigenstates (2.6) of the *kq*-operators (2.2) and (2.4) the product $\Delta x\, \Delta p$ is infinite, these states are the least classical states or the most quantum-mechanical states. In the states (2.6) one has no information about the coordinate and the momentum. In analogy with the coherent representation[46] which is the most classical representation, the *kq*-representation can be called the "most quantum-mechanical" representation.

Following this discussion the whole dynamics of electrons in solids could be looked upon as an extreme quantum-mechanical theory. As was mentioned before, the commutativity of the translation operator (2.11) with the periodic potential (2.19) is actually a consequence of the commutativity of the basic operators (2.11) and (2.12) that define the *kq*-representation. In the Bloch theory of solids the operator (2.11) is of fundamental importance because it defines the quasimomentum concept. As can easily be seen there is no classical analog of this operator. When written without assuming $\hbar = 1$, the operator (2.11) is

$$\exp((i/\hbar)\mathbf{p}\cdot\mathbf{a}_n) \tag{9.1}$$

[43] H. C. Praddaude, *Phys. Rev.* **140**, A1292 (1965).
[44] S. Teitler, *Solid State Commun.* **6**, 485 (1968).
[45] L. S. Schulman, *Phys. Rev.* **188**, 1139 (1969).
[46] R. J. Glauber, *Phys. Rev.* **131**, 2766 (1963).

The usual rule for obtaining a classical limit ($\hbar \to 0$) for the operator (9.1) does not lead to any meaningful result. As a physical quantity the expression (9.1) is therefore purely quantum without any classical limit. This is not a surprising observation because the whole band structure of solids is a consequence of Schrödinger's equation and has no counterpart in classical mechanics.

Appendix A: Wavefunctions in Different Representations

The representations used in this article are the r-representation, the Bloch representation (nk-representation), the kq-representation, the Rq-representation and the nR-representation. In this Appendix relations between these representations are given.

Formula (2.18) of Section II gives the relation between the wave function $\psi(r)$ in the r-representation and the wave function $C(\mathbf{kq})$ in the kq-representation

$$\psi(\mathbf{r}) = \int d\mathbf{k}\, d\mathbf{q}\, C(\mathbf{kq})\, \psi_{kq}(\mathbf{r}), \tag{A.1}$$

where $\psi_{kq}(\mathbf{r})$ are the eigenfunctions (2.13) of the operators (2.11) and (2.12) that define the kq-representation. Formula (2.24) of Section II gives the relation between $C(\mathbf{kq})$ and the Bloch functions $\psi_{nk}(\mathbf{q})$

$$C(\mathbf{kq}) = \sum_n B_n(\mathbf{k})\, \psi_{nk}(\mathbf{q}). \tag{A.2}$$

In (A.2), $B_n(\mathbf{k})$ is the wave function in the nk-representation (the Bloch representation). Expansion (A.2) contains no integration on $\mathbf{k}$ and it should be compared with the relation between $\psi(\mathbf{r})$ and $\psi_{nk}(\mathbf{r})$

$$\psi(r) = \sum_n \int d\mathbf{k}\, B_n(\mathbf{k})\, \psi_{nk}(\mathbf{r}). \tag{A.3}$$

In the latter expansion there is an integration on $\mathbf{k}$, which often makes it less convenient than expansion (A.2). In Section III, the functions $\psi(\mathbf{q} + \mathbf{R})$ and $F_n(\mathbf{R})$ are used which are the wave functions in the Rq-representation and nR-representation respectively. Let us present here a derivation for the relation between $\psi(\mathbf{q} + \mathbf{R})$ and $F_n(\mathbf{R})$ (Relation (3.5) in the text). The Bloch function in (A.3) can be expanded in Wannier functions $a_n(\mathbf{r})$

$$\psi_{nk}(\mathbf{r}) = \sum_R \exp(i\mathbf{k} \cdot \mathbf{R})\, a_n(\mathbf{r} - \mathbf{R}) \tag{A.4}$$

where the summation is on Bravais lattice vectors $\mathbf{R}$. By using this expansion (A.3) becomes

$$\psi(\mathbf{r}) = \sum_{n\mathbf{R}} \int d\mathbf{k}\, B_n(\mathbf{k}) \exp(i\mathbf{k} \cdot \mathbf{R})\, a_n(\mathbf{r} - \mathbf{R})$$

$$= \sum_{nR} F_n(\mathbf{R})\, a_n(\mathbf{r} - \mathbf{R}), \qquad \text{(A.5)}$$

where

$$F_n(\mathbf{R}) = \int d\mathbf{k} \exp(i\mathbf{k} \cdot \mathbf{R})\, B_n(\mathbf{k}). \qquad \text{(A.6)}$$

(A.5) can also be written for $\psi(\mathbf{r})$ as a function of $\mathbf{q} + \mathbf{R}'$

$$\psi(\mathbf{q} + \mathbf{R}') = \sum_{nR} F_n(\mathbf{R})\, a_n(\mathbf{q} + \mathbf{R}' - \mathbf{R})$$

$$= \sum_{nR''} F_n(\mathbf{R}' + \mathbf{R}'')\, a_n(\mathbf{q} - \mathbf{R}'') \qquad \text{(A.7)}$$

$$= \sum_{nR''} \exp\left(\mathbf{R}'' \cdot \frac{\partial}{\partial \mathbf{R}'}\right) a_n(\mathbf{q} - \mathbf{R}'')\, F_n(\mathbf{R}')$$

which is relation (3.5).

Appendix B: Commutator Expansions

In developing the effective Hamiltonian theory one had to know how to multiply functions in rectangular brackets, e.g., $[A(\mathbf{k})][B(\mathbf{k})]$ (the rectangular brackets express an operation given by formula (6.21)). As was shown by Roth[6] the following formula holds for any two functions $A(\mathbf{k})$ and $B(\mathbf{k})$

$$[A(\mathbf{k})][B(\mathbf{k})] = [C(\mathbf{k})], \qquad \text{(B.1)}$$

with

$$C(\mathbf{k}) = \exp\{-i(e/2c)\mathbf{H} \cdot \nabla_k \times \nabla_{k'}\}\, A(\mathbf{k})\, B(\mathbf{k}')|_{k'=k}$$

$$= A(\mathbf{k})\, B(\mathbf{k}) - ih_{\alpha\beta}(\partial A(\mathbf{k})/\partial k_\alpha)(B(\mathbf{k})/\partial k_\beta)$$

$$-\tfrac{1}{2} h_{\alpha\beta} h_{\alpha'\beta'}(\partial^2 A(\mathbf{k})/\partial k_\alpha\, \partial k_{\alpha'})(\partial^2 B(\mathbf{k})/\partial k_\beta\, \partial k_{\beta'}) + \cdots, \qquad \text{(B.2)}$$

where in the first line one has to differentiate with respect to $\mathbf{k}$ and $\mathbf{k}'$ respectively and then put $\mathbf{k}' = \mathbf{k}$; also

$$h_{\alpha\beta} = \epsilon_{\alpha\beta\gamma} e H^\gamma / 2c, \qquad \text{(B.3)}$$

with $\epsilon_{\alpha\beta\gamma}$ being the antisymmetric unit tensor and summation is understood on repeated indices. Similarly, the following formula can be proven

$$[A(\mathbf{k})]^- \, [B(\mathbf{k})]^- = [C(\mathbf{k})]^- \tag{B.4}$$

with

$$\begin{aligned} C(\mathbf{k}) &= \exp\{i(e/2c)\mathbf{H} \cdot \nabla_k \times \nabla_{k'}\} \, A(\mathbf{k}) \, B(\mathbf{k}')|_{k'=k} \\ &= A(\mathbf{k}) \, B(\mathbf{k}) + ih_{\alpha\beta}(\partial A(\mathbf{k})/\partial k_\alpha)(\partial B(\mathbf{k})/\partial \mathbf{k}_\beta) \\ &\quad -\tfrac{1}{2}h_{\alpha\beta}h_{\alpha'\beta'}(\partial^2 A(\mathbf{k})/\partial \mathbf{k}_\alpha \, \partial k_{\alpha'})(\partial^2 B(\mathbf{k})/\partial k_\beta \, \partial k_{\beta'}) + \cdots \end{aligned} \tag{B.5}$$

In (B.4) the brackets $[\ ]^-$ for a function $A(\mathbf{k})$ are defined in the same way as in formula (6.21) but with $\mathbf{k}$ replaced by $\mathbf{k} - (e/2c)\mathbf{H} \times i \, \partial/\partial\mathbf{k}$ (instead of the replacement (6.5) in (6.21)). Let us give here the derivation of (B.4). For this we write explicitly the left hand side and the right hand side of (B.4):

$$\begin{aligned} &[A(\mathbf{k})]^- \, [B(\mathbf{k})]^- \\ &\quad = \int d\boldsymbol{\lambda} \, d\boldsymbol{\lambda}' \, A(\boldsymbol{\lambda}) \, B(\boldsymbol{\lambda}') \exp\left\{i\left(\mathbf{k} - \frac{e}{2c}\mathbf{H} \times i\frac{\partial}{\partial\mathbf{k}}\right) \cdot \boldsymbol{\lambda}\right\} \\ &\quad\quad \times \exp\left\{i\left(\mathbf{k} - \frac{e}{2c}\mathbf{H} \times i\frac{\partial}{\partial\mathbf{k}}\right) \cdot \boldsymbol{\lambda}'\right\} \\ &\quad = \int d\boldsymbol{\lambda} \, d\boldsymbol{\lambda}' \, A(\boldsymbol{\lambda}) \, B(\boldsymbol{\lambda}') \exp\left\{i\left(\mathbf{k} - \frac{e}{2c}\mathbf{H} \times i\frac{\partial}{\partial\mathbf{k}}\right) \cdot (\boldsymbol{\lambda} + \boldsymbol{\lambda}')\right\} \\ &\quad\quad \times \exp\left\{i\frac{e}{2c}\mathbf{H} \times \boldsymbol{\lambda} \cdot \boldsymbol{\lambda}'\right\}, \end{aligned} \tag{B.6}$$

where the exponentials were multiplied by using the formula

$$e^A e^B = e^{A+B+\frac{1}{2}[A,B]} \tag{B.7}$$

that holds for any operators A and B for which the commutator $[A, B]$ commutes with A and B (Messiah[47]). The right hand side of (B.4) is easily checked to be equal the final expression of (B.6).

Another formula that was used in text is that if

$$(\mathbf{k}''\mathbf{q}|n\mathbf{k}) = [u_{nk}(\mathbf{q})] \sum_{\mathbf{K}} \delta(\mathbf{k}'' - \mathbf{k} - \mathbf{K}) \tag{B.8}$$

then

$$(\mathbf{n}\mathbf{k}|k''\mathbf{q}) = (\mathbf{k}''\mathbf{q}|n\mathbf{k})^* = [u_{nk}^*(\mathbf{q})] \sum_{\mathbf{k}} \delta(\mathbf{k}'' - \mathbf{k} - \mathbf{K}). \tag{B.9}$$

[47] A. Messiah, "Quantum Mechanics." North-Holland Publ., Amsterdam, 1969.

The proof of (B.9) is as follows

$$(\mathbf{k}''\mathbf{q}|n\mathbf{k})^* = [u_{nk}(\mathbf{q})]^* \sum_k \delta(\mathbf{k}'' - \mathbf{k} - \mathbf{K})$$

$$= \int u_{n\lambda}^*(\mathbf{q}) \exp\{-i(\mathbf{k} + (e/2c)\mathbf{H} \times i\,\partial/\partial\mathbf{k}) \cdot \boldsymbol{\lambda}\} \sum_{\mathbf{K}} \delta(\mathbf{k}'' - \mathbf{k} - \mathbf{K}),$$

which is the right hand side of (B.9). In the last expression, $u_{n\lambda}(\mathbf{q})$ is the Fourier transform of $u_{nk}(\mathbf{q})$.

Let us also show that

$$\exp(i\mathbf{k} \cdot \mathbf{q})[u_{nk}(\mathbf{q})] \sum_k \delta(\mathbf{k} - \mathbf{k}' - \mathbf{K}) = [u_{nk'}(\mathbf{q})]^- \exp(i\mathbf{k}' \cdot \mathbf{q}) \sum_k \delta(\mathbf{k} - \mathbf{k}' - \mathbf{K}). \tag{B.10}$$

The proof of (B.10) follows at once if one writes an explicit expression for the right-hand side:

$$[u_{nk'}(\mathbf{q})]^- \exp(i\mathbf{k}' \cdot \mathbf{q}) \sum_k \delta(\mathbf{k} - \mathbf{k}' - \mathbf{K})$$

$$= \exp(i\mathbf{k} \cdot \mathbf{q}) \int d\boldsymbol{\lambda}\, u_{n\lambda}(\mathbf{q}) \exp\{i(\mathbf{k}' - (e/2c)\,\mathbf{H} \times i\,\partial/\partial\mathbf{k}') \cdot \boldsymbol{\lambda}\}$$

$$\times \sum_{\mathbf{K}} \exp(-i\mathbf{q} \cdot \mathbf{K})\, \delta(\mathbf{k} - \mathbf{k}' - \mathbf{K}). \tag{B.11}$$

In the last expression the derivative with respect to $\mathbf{k}'$ can be replaced by the derivative with respect to $\mathbf{k}$ with a change in sign. $\mathbf{k}'$ itself can be replaced by $\mathbf{k} - \mathbf{K}$ (because of the δ-function). Expression (B.11) becomes therefore

$$\exp(i\mathbf{k} \cdot \mathbf{q}) \sum_{\mathbf{K}} [u_{n\mathbf{k}-\mathbf{K}}(\mathbf{q})] \exp(-i\mathbf{q} \cdot \mathbf{K})\, \delta(\mathbf{k} - \mathbf{k}' - \mathbf{K}).$$

By using the property of $u_{nk}(\mathbf{q})$, that $u_{n\mathbf{k}-\mathbf{K}}(\mathbf{q}) = u_{nk}(\mathbf{q}) \exp(i\mathbf{q} \cdot \mathbf{K})$, one arrives at the left-hand side of (B.10).

Appendix C: Transformation Matrix $(\mathbf{kq}|n\mathbf{k}')$ in the r-Representation

In this Appendix the transformation matrix $(\mathbf{kq}|n\mathbf{k}')$ (Formula (6.27) in the text) is transformed into the r-representation. For this, Formula (2.18) is used:

$$(\mathbf{r}|n\mathbf{k}') = \int d\mathbf{k}\, d\mathbf{q} \exp(i\mathbf{k} \cdot \mathbf{q})(\mathbf{kq}|n\mathbf{k}')\, \psi_{kq}(\mathbf{r}). \tag{C.1}$$

As was explained in Section III the exponent $\exp(i\mathbf{k} \cdot \mathbf{q})$ in (C.1) appears because formula (2.18) is written for the function $C(\mathbf{kq})$ while the transformation matrix $(\mathbf{kq}|n\mathbf{k}')$ is written for the $U(\mathbf{kq})$-function. By using expression (2.13) for $\psi_{kq}(\mathbf{r})$ and by noticing that $(\mathbf{kq}|n\mathbf{k}')$ is periodic in $\mathbf{q}$, one has

$$(\mathbf{r}|n\mathbf{k}') = [V_0/(2\pi)^3]^{1/2} \int d\mathbf{k} \exp(i\mathbf{k}\cdot\mathbf{r})[u_{nk}(\mathbf{r})] \sum_{\mathbf{K}} \delta(\mathbf{k}-\mathbf{k}'-\mathbf{K}) \quad (C.2)$$

Now using formula (B.10), (C.2) becomes

$$(\mathbf{r}|n\mathbf{k}') = [V_0/(2\pi)^3]^{1/2} [u_{nk'}(\mathbf{r})]^- \exp(i\mathbf{k}'\cdot\mathbf{r}). \quad (C.3)$$

(C.3) is already the transformation matrix in the r-representation. A different form for this matrix can be obtained by using an expansion of $u_{nk}(\mathbf{r})$ in Wannier functions (expansion (A.4) in Appendix A for $u_{nk}(\mathbf{r})$)

$$\begin{aligned}(\mathbf{r}|n\mathbf{k}') &= \sum_R \exp\{-i(\mathbf{k}' - (e/2c)\mathbf{H}\times i\,\partial/\partial\mathbf{k}')\cdot(\mathbf{r}-\mathbf{R})\}\exp(i\mathbf{k}'\cdot\mathbf{r})\,a_n(\mathbf{r}-\mathbf{R})\\ &= \sum_R \exp\{i(\mathbf{k}' + (e/2c)\mathbf{H}\times\mathbf{r})\cdot\mathbf{R}\}\,a_n(\mathbf{r}-\mathbf{R}). \end{aligned} \quad (C.4)$$

Expression (C.4) for the transformation matrix is a Bloch-type function with $\mathbf{k}'$ replaced by $\mathbf{k}' + (e/2c)\mathbf{H}\times\mathbf{r}$. These functions were used as starting functions in the effective Hamiltonian theory of Ref. 6.

Appendix D: General Formula for the Expansion of the Hamiltonian

Let us give here the expansion of the triple product (6.65):

$$[S^+(\mathbf{k})][H'(\mathbf{k})][S(\mathbf{k})] \quad (D.1)$$

In (D.1) (see formulas (6.37) and (6.49) respectively)

$$S(\mathbf{k})_{mn} = \int d\mathbf{q}\, u^*_{m0}(\mathbf{q})\, u_{nk}(\mathbf{q}), \quad (D.2)$$

$$H'(k)_{mn} = \int d\mathbf{q}\, u^*_{m0}(\mathbf{q})\, H(\mathbf{kq})\, u_{n0}(\mathbf{q}), \quad (D.3)$$

where

$$H(\mathbf{kq}) = (\mathbf{p}+\mathbf{k})^2/2m + V(\mathbf{q}). \quad (D.4)$$

Since $H'(\mathbf{k})$ in (D.1) is a quadratic function of $\mathbf{k}$ the following expression is obtained (according to (6.24))

$$\begin{aligned}[H'(\mathbf{k})][S(\mathbf{k})] &= [H'(\mathbf{k})\,S(\mathbf{k})] - ih_{\alpha\beta}\left[\frac{\partial H'(\mathbf{k})}{\partial k_\alpha}\frac{\partial S(\mathbf{k})}{\partial k_\beta}\right]\\ &\quad -\tfrac{1}{2}h_{\alpha\beta}h_{\alpha'\beta'}\left[\frac{\partial^2 H'(\mathbf{k})}{\partial k_\alpha\,\partial k_{\alpha'}}\frac{\partial^2 S(\mathbf{k})}{\partial k_\beta\,\partial k_{\beta'}}\right], \end{aligned} \quad (D.5)$$

where $h_{\alpha\beta}$ is given by (B.3). There are no higher-order terms in (D.5). The matrix $S(\mathbf{k})$ is unitary which follows at once from definition (D.2)

$$S^{+}(\mathbf{k})\, S(\mathbf{k}) = I. \tag{D.6}$$

Having this in mind any term in (D.5) can be multiplied inside the brackets by the product (D.6). The first term will become $S(\mathbf{k})\,\epsilon^{(0)}(\mathbf{k})$, because as can easily be checked (see (6.18) and (6.19)),

$$S^{+}(\mathbf{k})\, H(\mathbf{kq})\, S(\mathbf{k}) = \epsilon^{(0)}(\mathbf{k}), \tag{D.7}$$

where $\epsilon^{(0)}(\mathbf{k})$ is the energy spectrum of the solid in the absence of the magnetic field. For the second and third terms in (D.5) let us define the following expressions (see formulas (5.17) and (6.69) respectively)

$$\begin{aligned} X^{\alpha}(\mathbf{k})_{nn'} &= i(S^{+}(\mathbf{k})\, \partial S(\mathbf{k})/\partial k_{\alpha})_{nn'} \\ &= i \int u^{*}_{nk}(\mathbf{q})(\partial u_{n'k}(\mathbf{q})/\partial k_{\alpha})\, d\mathbf{q}, \end{aligned} \tag{D.8}$$

$$\begin{aligned} v_{\alpha}(\mathbf{k})_{nn'} &= (S^{+}(\mathbf{k})(\partial H'(\mathbf{k})/\partial k_{\alpha})\, S(\mathbf{k}))_{nn'} \\ &= (1/m) \int u^{*}_{nk}(\mathbf{q})(-i(\partial/\partial q_{\alpha}) + k_{\alpha})\, u_{n'k}(\mathbf{q})\, d\mathbf{q}. \end{aligned} \tag{D.9}$$

Expression (D.5) can now be written in the following form

$$\begin{aligned} [H'(\mathbf{k})][S(\mathbf{k})] &= [S(\mathbf{k})\, \epsilon^{(0)}(\mathbf{k})] + [S(\mathbf{k})\, E^{(1)}(\mathbf{k})] \\ &\quad + [S(\mathbf{k})\, E^{(2)}(\mathbf{k})] \end{aligned} \tag{D.10}$$

where

$$E^{(1)} = -h_{\alpha\beta} v_{\alpha}(\mathbf{k})\, X^{(\beta)}(\mathbf{k}) \tag{D.11}$$

$$E^{(2)} = (1/2m)\, h_{\alpha\beta} h_{\alpha\beta'} (i\, \partial X^{\beta}(\mathbf{k})/\partial k_{\beta'} + X^{\beta'}(\mathbf{k})\, X^{(\beta)}(\mathbf{k})), \tag{D.12}$$

and where in (D.11) and (D.12) summation is understood over repeated indices. The terms in (D.10) are of zero, first and second order in magnetic field, correspondingly. For obtaining expression (D.12), the following relation was used

$$S^{+}(\mathbf{k})\, \partial^2 S(\mathbf{k})/\partial k_{\alpha}\, \partial k_{\beta} = -i\, \partial X^{\alpha}(\mathbf{k})/\partial k_{\beta} - X^{\beta}(\mathbf{k})\, X^{\alpha}(\mathbf{k}). \tag{D.13}$$

The last equality is obtained from definition (D.8). By using expression (D.10) one has

$$[S^+(\mathbf{k})][H'(\mathbf{k})][S(\mathbf{k})] = \Big[\epsilon^{(0)}(\mathbf{k}) - ih_{\alpha\beta}\frac{\partial S^+(k)}{\partial k_\alpha}\frac{\partial}{\partial k_\beta}(S(\mathbf{k})\,\epsilon^{(0)}(\mathbf{k})) - \tfrac{1}{2}h_{\alpha_1\beta_1}h_{\alpha_2\beta_2}$$

$$\times \frac{\partial^2 S^+(k)}{\partial k_{\alpha_1}\,\partial k_{\alpha_2}}\frac{\partial^2}{\partial k_{\beta_1}\,\partial k_{\beta_2}}(S(\mathbf{k})\,\epsilon^{(0)}(\mathbf{k})) + \cdots + E^{(1)}(\mathbf{k})$$

$$- ih_{\alpha\beta}\frac{\partial S^+(k)}{\partial k_\alpha}\frac{\partial}{\partial k_\beta}(S(\mathbf{k})\,E^{(1)}(\mathbf{k})) - \tfrac{1}{2}h_{\alpha_1\beta_1}h_{\alpha_2\beta_2}$$

$$\times \frac{\partial^2 S^+(\mathbf{k})}{\partial k_{\alpha_1}\,\partial k_{\alpha_2}}\frac{\partial^2}{\partial k_{\beta_1}\,\partial k_{\beta_2}}(S(\mathbf{k})\,E^{(1)}(\mathbf{k})) + \cdots + E^{(2)}(\mathbf{k})$$

$$- ih_{\alpha\beta}\frac{\partial S^+(k)}{\partial k_\beta}(S(\mathbf{k})\,E^{(2)}(\mathbf{k})) - \tfrac{1}{2}h_{\alpha_1\beta_1}h_{\alpha_2\beta_2}$$

$$\times \frac{\partial^2 S^+(k)}{\partial k_{\alpha_1}\,\partial k_{\alpha_2}}\frac{\partial^2}{\partial k_{\beta_1}\,\partial k_{\beta_2}}(S(\mathbf{k})\,E^{(2)}(\mathbf{k})) + \cdots\Big]. \qquad \text{(D.14)}$$

In order to write expression (D.14) in a compact form let us define a matrix (see (6.64))

$$N^{(n)}_{\beta_1\beta_2\cdots\beta_m}(\mathbf{k}) = \frac{(-i)^n}{n!}h_{\alpha_1\beta_1}h_{\alpha_2\beta_2}\cdots h_{\alpha_n\beta_n} \times \frac{\partial^n S^+(\mathbf{k})}{\partial k_{\alpha_1}\cdots\partial k_{\alpha_n}}\frac{\partial^{n-m}S(\mathbf{k})}{\partial k_{\beta_{m+1}}\cdots\partial k_{\beta_n}}, \qquad \text{(D.15)}$$

where summation is understood on repeated indices. The matrix (D.15) is of the nth order in magnetic field and is a generalization of the definition (6.39) and (6.40) in the text. By using definition (D.15), the expression (D.14) can be given the following simple form:

$$[S^+(\mathbf{k})][H'(\mathbf{k})][S(\mathbf{k})] = \sum_{m=0}^{n}\sum_{n=0} C_n{}^m[N^{(n)}_{\beta_1\beta_2\cdots\beta_m}(\mathbf{k})\,\partial^m E(\mathbf{k})/\partial k_{\beta_1}\,\partial k_{\beta_2}\cdots\partial k_{\beta_m}], \qquad \text{(D.16)}$$

where $C_n{}^m$ is the binomial coefficient of the $(m+1)$st term and

$$E(\mathbf{k}) = \epsilon^{(0)}(\mathbf{k}) + E^{(1)}(\mathbf{k}) + E^{(2)}(\mathbf{k}). \qquad \text{(D.17)}$$

The form (D.16) is very useful and enables one to write a general expression for the nth-order term in $[S^{+}(\mathbf{k})][H'(\mathbf{k})][S(k)]$:

$$\sum_{n=0} C_{nm} N_{\beta_1\beta_1\cdots\beta_m}(\mathbf{k})\, \partial^m \epsilon^{(0)}(\mathbf{k})/\partial k_{\beta_1}\, \partial k_{\beta_2} \cdots \partial k_{\beta_m}$$

$$+ \sum_{m=0}^{n-1} C_{n-1}^{m} N_{\beta_1\beta_2\cdots\beta_m}^{(n-1)}(\mathbf{k})\, \partial^m E^{(1)}(\mathbf{k})/\partial k_{\beta_1}\, \partial k_{\beta_2} \cdots \partial k_{\beta_m}$$

$$+ \sum_{m=0}^{n-2} C_{n-2}^{m} N_{\beta_1\beta_2\cdots\beta_m}^{(n-2)}(\mathbf{k})\, \partial^m E^{(2)}(\mathbf{k})/\partial k_{\beta_1}\, \partial k_{\beta_2} \cdots \partial k_{\beta_m}. \tag{D.18}$$

For example, to second order in magnetic field, expression (D.16) becomes

$$(S^{+}HS)^{(0)} = \epsilon^{(0)}(\mathbf{k}), \tag{D.19}$$

$$(S^{+}HS)^{(1)} = N^{(1)}(\mathbf{k})\, \epsilon^{(0)}(\mathbf{k}) + N_{\beta_1}^{(1)}(\mathbf{k})(\partial\epsilon^{(0)}(\mathbf{k})/\partial k_{\beta_1}) + \mathbf{E}^{(1)}(\mathbf{k}), \tag{D.20}$$

$$\begin{aligned}(S^{+}HS)^{(2)} &= N^{(2)}(\mathbf{k})\, \epsilon^{(0)}(\mathbf{k}) + 2N_{\beta_1}^{(2)}(\mathbf{k})(\partial\epsilon^{(0)}(\mathbf{k})/\partial k_{\beta_1}) \\ &\quad + N_{\beta_1\beta_2}^{(2)}(\mathbf{k})(\partial^2\epsilon^{(0)}(\mathbf{k})/\partial k_{\beta_1}\, \partial k_{\beta_2}) + N^{(1)}(\mathbf{k})\, E^{(1)}(\mathbf{k}) \\ &\quad + N_{\beta}^{(1)}(\mathbf{k})(\partial E^{(1)}(\mathbf{k})/\partial k_{\beta}) + E^{(2)}(\mathbf{k}).\end{aligned} \tag{D.21}$$

Having the general formula (D.16) and by using the expansion (6.45) for $[N(\mathbf{k})]^{-1/2}$ one can get an expression for the Hamiltonian (6.63) to any order in magnetic field. To second order in magnetic field this is given in text [formulas (6.65)–(6.73)].

In conclusion of this Appendix let us show that $N_{\beta_1\beta_2\cdots\beta_m}(\mathbf{k})$ in (D.15) and $E(\mathbf{k})$ in (D.17) are periodic in $\mathbf{k}$ with the periodicity of the reciprocal lattice vectors. By using the definition of $S(\mathbf{k})$ [formula (D.2)] one has

$$\left(\frac{\partial^n S^{+}(\mathbf{k})}{\partial k_{\alpha_1}\, \partial k_{\alpha_2} \cdots \partial k_{\alpha_n}} \frac{\partial^{n-m} S(\mathbf{k})}{\partial k_{\beta_{m+1}} \cdots \partial k_{\beta_n}}\right)_{ll'} = \int \frac{\partial^n u_{lk}(\mathbf{q})}{\partial k_{\alpha_1}\, \partial k_{\alpha_2} \cdots \partial k_{\alpha_n}} \frac{\partial^{n-m} u_{l'k}(\mathbf{q})}{\partial k_{\beta_{m+1}} \cdots \partial k_{\beta_n}}\, d\mathbf{q}. \tag{D.22}$$

From the behavior of the periodic part of the Bloch function, $u_{lk}(\mathbf{q})$, as a function of $\mathbf{k}$, it follows that the expression (D.22) is periodic in $\mathbf{k}$. In a similar way one shows that $E(\mathbf{k})$ [(D.17) and (D.12)] is periodic in $\mathbf{k}$ [see definition (D.8) and (D.9)].

Acknowledgments

The author would like to thank Professors M. Revzen and L. Shulman for reading the manuscript and for useful remarks.

It is the author's pleasure to thank Mrs. Gila Ezion, Secretary at the Physics Department of the Technion, for careful typing and correcting the manuscript.

Thermoelectricity in Metals and Alloys*

R. P. HUEBENER

Argonne National Laboratory, Argonne, Illinois

* Based on work performed under the auspices of the U.S. Atomic Energy Commission.

I. Introduction

The three thermoelectric phenomena which are associated with the names Seebeck, Peltier, and Thomson were all discovered in the last century. However, only in the last 30 years or so have the thermoelectric properties of solids been investigated in detail. While the initial experiments had established the main physical behavior of thermoelectricity in many materials, very recently thermoelectricity has been used as a more and more sophisticated tool for studying electron and phonon scattering in solids and even for probing the Fermi surface in metals and alloys. For many metals a qualitative understanding of the thermoelectric properties has now emerged, taking into account the details of the Fermi surface and of the electron–phonon interaction. In this article we shall summarize our present knowledge about thermoelectricity in metals and alloys, both experimentally and theoretically. Rather short accounts of thermoelectricity have been given earlier in a number of texts.[1–8] An excellent survey of thermoelectricity has been presented about ten years ago by MacDonald.[9] It is this book by MacDonald which comes perhaps closest to the aim of the present article. Needles to say, during the past decade the field of thermoelectricity has seen progress on numerous sides.

In the following we shall concentrate mainly on the results, both of experiment and theory. Thus we shall say nothing about experimental techniques nor shall we present a complete theoretical treatment of thermoelectricity from general transport theory. Detailed accounts on the general transport theory and its application to thermoelectricity

[1] N. F. Mott and H. Jones, "The Theory of the Properties of Metals and Alloys." Oxford Univ. Press (Clarendon), London and New York, 1936.

[2] A. H. Wilson, "The Theory of Metals." Cambridge Univ. Press, London and New York, 1953.

[3] F. J. Blatt, *Solid State Phys.* **4**, 199 (1957).

[4] F. J. Blatt, "Physics of Electronic Conduction in Solids." McGraw-Hill, New York, 1968.

[5] J. M. Ziman, "Electrons and Phonons." Oxford Univ. Press (Clarendon), London and New York, 1960.

[6] J. M. Ziman, *in* "The Physics of Metals. I. Electrons" (J. M. Ziman, ed.). Cambridge Univ. Press, London and New York, 1969.

[7] J. L. Olsen, "Electronic Transport in Metals." Wiley, New York, 1962.

[8] H. M. Rosenberg, "Low Temperature Solid State Physics." Oxford Univ. Press (Clarendon), London and New York, 1963.

[9] D. K. C. MacDonald, "Thermoelectricity: An Introduction to the Principles." Wiley, New York, 1962.

can be found elsewhere.[3,5,10] As the title indicates we concern ourselves here only with metals and alloys, and we completely omit semiconductors, semimetals and metallic compounds. We do not include in this review the giant thermoelectric phenomena associated with magnetic impurities (Kondo effect). Magnetic impurities and their thermoelectric properties clearly should be discussed in close context with their many other interesting properties. Recently, this subject has been treated in a number of summaries.[11-16] Further, we do not discuss the influence of pressure, structural lattice defects, and magnetic fields, nor shall we treat liquid metals. Finally, we restrict ourselves exclusively to the basic physical phenomena and completely ignore the practical applications.

II. Seebeck, Peltier, and Thomson

Three thermoelectric effects occur when an electric field and a temperature gradient exist simultaneously in an electrical conductor. If two different metals A and B are connected as shown in Fig. 1(a)

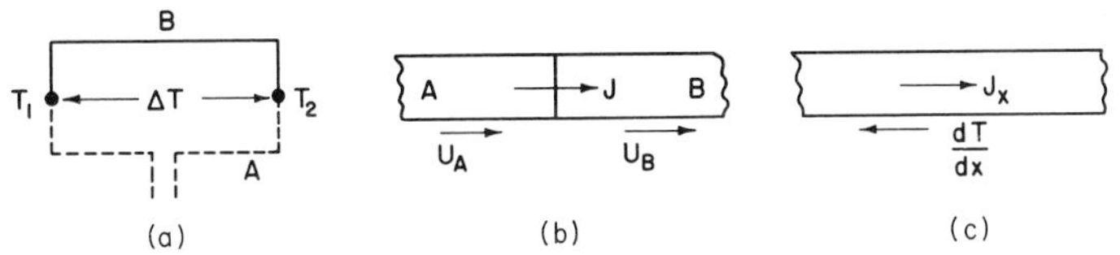

FIG. 1. The thermoelectric effects: (a) Seebeck effect, (b) Peltier effect, (c) Thomson effect.

and if the two junctions are kept at different temperatures T_1 and T_2, a thermoelectric emf is developed in the circuit which can be measured with a voltmeter attached to the extension of, say, metal A. This effect is known as the *Seebeck effect*, and was discovered in 1822.[17] The thermoelectric voltage V is generally found to be proportional to the

[10] A. Sommerfeld and H. Bethe, *in* "Handbuch der Physik" (H. Geiger and K. Scheel, eds.), Vol. XXIV, part 2. Springer, Berlin, 1933.
[11] M. Bailyn, *Advan. Phys.* **15**, 179 (1966).
[12] M. D. Daybell and W. A. Steyert, *Rev. Mod. Phys.* **40**, 380 (1968).
[13] P. W. Anderson, *in* "Many Body Physics" (C. DeWitt and R. Balian, eds.), p. 229. Gordon and Breach, New York, 1968.
[14] A. J. Heeger, *Solid State Phys.* **23**, 284 (1969).
[15] J. Kondo, *Solid State Phys.* **23**, 184 (1969).
[16] K. Fischer, *in* "Springer Tracts in Modern Physics" (G. Höhler, ed.), Vol. 54, p. 1. Springer Verlag, Berlin, 1970.
[17] A. Seebeck, *Pogg. Ann.* **6**, 133 (1826).

temperature difference ΔT between the two junctions. The derivative

$$S_{AB} \equiv S_A - S_B = dV/dT \tag{II.1}$$

is defined as the thermoelectric power S_{AB} of metal A relative to metal B; S_A and S_B in Eq. (II.1) are the *absolute thermoelectric power* of metal A and B, respectively. In the following we write the absolute thermoelectric power as S. S_{AB} is taken to be positive ($S_A > S_B$) if the higher potential appears at the lead of metal A connected to the junction at the higher temperature. From Fig. 1(a) we see that the Seebeck effect disappears because of symmetry reasons if metals A and B are identical.

Next we consider two metals A and B connected in series and kept at uniform temperature [Fig. 1(b)]. If an electrical current with the current density J passes through the two conductors, reversible heat will be generated or absorbed at the junction thus heating or cooling the junction area. This is the *Peltier effect*, discovered in 1834.[18] The Peltier effect is linear in the electric current and changes its sign if the current direction is reversed. This must be contrasted with the irreversible Joule heating which depends on the square of the current. The Peltier effect is caused by the fact that with an electrical current there is always associated the transport of thermal energy. If the heat current density, coupled with the electrical current, changes suddenly at the junction between two different metals, heat will be generated or absorbed at the junction. The ratio of the heat current density U_x to the electrical current density J_x is defined as the *absolute Peltier coefficient* π of the material:

$$\pi = U_x/J_x \, . \tag{II.2}$$

It is clear that the Peltier effect disappears if the two metals in contact are identical. The net Peltier heat at the junction between two conductors A and B is given by $\pi_{AB} = \pi_A - \pi_B$; π_{AB} is positive if heat is generated at the junction when the current flows from A to B.

Although the Seebeck and Peltier effect do not occur if the two metals in contact are identical, both effects can be observed if the metals A and B are the same but in a different state. If metal A is strained, plastically deformed, exposed to elevated pressure, or if A contains structural lattice defects due to irradiation etc., whereas B consists of the same material as A in a well annealed state at room pressure, both Seebeck and Peltier effects do exist.

The third thermoelectric effect occurs if a current flows in an electrical conductor in which a temperature gradient along the current direction

[18] J. C. Peltier, *Ann. Chim. Phys.* **56**, 371 (1834).

is maintained [Fig. 1(c)]. The rate of heat generation per volume in the conductor is

$$\dot{Q} = \frac{J_x^2}{\sigma} + \frac{d}{dx}\left(\kappa \frac{dT}{dx}\right) - \mu J_x \frac{dT}{dx}, \tag{II.3}$$

where σ and κ are the electrical conductivity and the heat conductivity, respectively. The first term represents the irreversible Joule heat, whereas the second term contains the divergence of the heat current density. The third term, which is linear in J_x and dT/dx, describes the *Thomson effect* and represents a reversible generation or absorption of heat. The Thomson effect was predicted theoretically in 1854 and observed experimentally in 1856.[19] The coefficient μ in the third term of Eq. (II.3) is defined as the *Thomson coefficient* of the material. The Thomson coefficient is taken positive if heat is generated reversibly when the current flows from the high- to the low-temperature end of the sample. The heat production due to the Thomson effect will overcome the Joule heat if the electrical current density is sufficiently small since the first is linear in J whereas the second is proportional to J^2. Whereas the Seebeck and Peltier effect can only be observed between two different metals (or between two sections of the same material if they are in a different state), the Thomson effect can be measured directly for one homogeneous material.

Thomson already recognized that the absolute thermoelectric power S, the Peltier coefficient π, and the Thomson coefficient μ are related to each other, and from thermodynamic arguments he derived the following equations:

$$\mu = T\, dS/dT \tag{II.4}$$

and

$$\pi = T \cdot S. \tag{II.5}$$

Although it was questionable whether the thermodynamic arguments leading to relations (II.4) and (II.5) were strictly valid, recently these relations have been derived rigorously from the theory of irreversible thermodynamics.[20–22] Equations (II.4) and (II.5) are then a direct consequence of the Onsager reciprocity relations.[23] From Eq. (II.4)

[19] Lord Kelvin (W. Thomson), *Proc. Roy. Soc. Edinburgh* **3**, 255 (1854); Lord Kelvin, *Collect. Papers Cambridge* **1**, 316 (1882).

[20] H. B. Callen, *Phys. Rev.* **73**, 1349 (1948).

[21] S. R. De Groot, "Thermodynamics of Irreversible Processes." North–Holland Publ., Amsterdam, 1951.

[22] C. A. Domenicali, *Rev. Mod. Phys.* **26**, 237 (1954).

[23] L. Onsager, *Phys. Rev.* **37**, 405 (1931); **38**, 2265 (1931).

we find

$$S(T) - S(0) = \int_0^T \mu/T\, dT. \tag{II.6}$$

Because of the third law of thermodynamics we have $S(0) = 0$ and therefore

$$S(T) = \int_0^T \mu/T\, dT. \tag{II.7}$$

Although experimentally only *differences* between the thermoelectric powers and the Peltier coefficients for two substances can be measured, S and π are bulk properties in the same sense as μ or the electrical conductivity and the heat conductivity of the material. Absolute values of the thermoelectric power and the Peltier coefficient can always be obtained from measurements of the Thomson heat together with Eqs. (II.7) and (II.5). In this way an absolute thermoelectric scale has been constructed by Borelius *et al.*[24] and more recently by Pearson and Templeton[25] and Christian *et al.*[26] These authors have measured the Thomson coefficient of Pb and obtained an absolute scale for Pb up to temperatures as high as 300°K. The choice of Pb as a thermoelectric reference is convenient because of three reasons. First, below 7.2°K, its superconducting transition temperature, the absolute thermoelectric power of Pb is zero. Second, it can be purified and annealed relatively easily. Third, its thermoelectric power appears to be not very sensitive to small amounts of chemical impurities.[27] Since thermoelectricity disappears in a superconductor at zero magnetic field, using as a reference a superconductor with a very high transition temperature, e.g., Nb_3Sn, provides a convenient method for determining the absolute thermoelectric power up to about 18°K. Lander[28] measured the Thomson coefficients of Cu, Ag, Au, Pt, Pd, Mo, and W from 400°K up to their melting temperatures. From these data Cusack and Kendall[29] constructed tables for the absolute thermoelectric power for these metals up to their melting points, extending the absolute scale of thermoelectric power to a temperature as high as 2400°K.

[24] G. Borelius, W. H. Keesom, C. H. Johansson, and J. O. Linde, *Proc. Acad. Sci. Amsterdam* **35**, 3 (1932); G. Borelius, *in* "Handbuch der Metallphysik" (G. Masing, ed.), Vol. I, p. 385. Akademische Verlagsgesellschaft, Leipzig, 1935.

[25] W. B. Pearson and I. M. Templeton, *Proc. Roy. Soc. (London)* **A231**, 534 (1955).

[26] J. W. Christian, J. P. Jan, W. B. Pearson, and I. M. Templeton, *Proc. Roy. Soc. (London)* **A245**, 213 (1958).

[27] J. P. Jan, W. B. Pearson, and I. M. Templeton, *Can. J. Phys.* **36**, 627 (1958).

[28] J. J. Lander, *Phys. Rev.* **74**, 479 (1948).

[29] N. Cusack and P. Kendall, *Proc. Phys. Soc.* **72**, 898 (1958).

After establishing an absolute thermoelectric scale, measurements of S, π, and μ are completely equivalent because of Eqs. (II.5) and (II.7). However, by far most of the experimental work has been carried out through measurements of the absolute thermoelectric power, simply because voltages can be measured much more easily than heat generation. However, at very low temperatures the Peltier effect sometimes can be easier to determine than the Seebeck effect,[30] because here the Seebeck voltages become extremely small, whereas calorimetric measurements are not so difficult thanks to the very high sensitivity of, say, carbon resistor thermometers.

III. Electron Diffusion

We consider now the contribution to thermoelectricity arising from the deviation of the *electron system* from its equilibrium distribution in the presence of an electric field and a temperature gradient. However, throughout this section we assume the phonon system to be always in thermal equilibrium (Bloch assumption).

1. Thomson Coefficient and the Specific Heat of the Conduction Electrons

A crude estimate of the thermoelectric effects can be obtained as follows. From the definition of the Thomson coefficient in Eq. (II.3), we have per volume element $A\,dx$, where A is the cross sectional area of the conductor,

$$\frac{dQ}{dt}\frac{1}{A\,dx} = -\mu\frac{j_x\,dT}{A\,dx}. \tag{1.1}$$

Here Q is the thermal energy of the electron system and j_x the electric current in x direction. By rewriting Eq. (1.1) we obtain

$$\mu = -\frac{dQ}{dT}\Big/ j_x\,dt. \tag{1.2}$$

From this equation we can interpret the Thomson coefficient as the specific heat of the electron system per unit electric charge:

$$\mu = C_e/n\mathbf{e}, \tag{1.3}$$

where C_e is the electronic specific heat, n the number of electrons per

[30] H. J. Trodahl, *Rev. Sci. Instrum.* **40**, 648 (1969).

unit volume and $\mathbf{e}$ the charge per electron (in magnitude and sign). With $C_e = \pi^2 k_B{}^2 nT/2E_F$ for a free electron metal we find

$$\mu = \pi^2 k_B{}^2 T/2\mathbf{e}E_F \tag{1.4}$$

and

$$S = \int_0^T \mu/T\, dT = \pi^2 k_B{}^2 T/2\mathbf{e}E_F\,. \tag{1.5}$$

Here k_B is Boltzmann's constant and E_F the Fermi energy. From Eqs. (1.4) and (1.5) we expect μ and S to be negative (since $\mathbf{e}$ is negative) and to be proportional to the absolute temperature. From (1.5) we find approximately

$$S = -(3.66 \cdot 10^{-8}/E_F)\, T \quad [\mathrm{V}/°\mathrm{K}^2], \tag{1.6}$$

where E_F is measured in electron volts. Experimentally, the thermoelectric power is quite often found to be positive and to have a different order of magnitude than that given by Eq. (1.6). Further it is generally not linear in the absolute temperature.

2. Nonequilibrium in the Electron System

The treatment of thermoelectricity within the general transport theory consists of setting up the relevant Boltzmann equation assuming that a relaxation time exists.[3,5,10] The steady state electron distribution function $f(\mathbf{k})$ in the presence of a temperature gradient is[31]

$$\begin{aligned} f_1(\mathbf{k}) &= f(\mathbf{k}) - f_0(\mathbf{k}) \\ &= l(\mathbf{k})\left(-\frac{\partial f_0(\mathbf{k})}{\partial E(\mathbf{k})}\right)\left\{\frac{E(\mathbf{k}) - E_F}{T}(-\nabla T) + \mathbf{eE}\right\}. \end{aligned} \tag{2.1}$$

Here, $\mathbf{k}$ is the electron wave vector, $f_0(\mathbf{k})$ the Fermi function, $l(\mathbf{k})$ and $E(\mathbf{k})$ the electron mean free path and the electron energy and $\mathbf{E}$ the effective electric field. Further, we write

$$j_1(\mathbf{k}) = \mathbf{e}v(\mathbf{k})\, f_1(\mathbf{k}) \tag{2.2}$$

and

$$l(\mathbf{k}) = \tau(\mathbf{k})\, v(\mathbf{k}), \tag{2.3}$$

where $v(\mathbf{k})$ and $\tau(\mathbf{k})$ are the group velocity and the relaxation time, respectively. Since $f_0(\mathbf{k})$ is symmetric around $\mathbf{k} = 0$, the contribution

[31] J. M. Ziman, "Electrons and Phonons," Sects. 9.9 and 9.10. Oxford Univ. Press (Clarendon), London and New York, 1960.

of $f_0(\mathbf{k})$ to the current vanishes. The microscopic contribution to the current arising from carriers near $\mathbf{k}$ is then

$$2(2\pi)^{-3}\,d^3k \cdot j_1(\mathbf{k}) = 2(2\pi)^{-3}\,dE \cdot d\mathfrak{S}(E, \mathbf{k})\,j_1(\mathbf{k})/\hbar v(\mathbf{k})$$
$$= -(\mathbf{e}/4\pi^3\hbar)\,dE\,d\mathfrak{S}(E, \mathbf{k})\,f_1(\mathbf{k}). \tag{2.4}$$

Here $d\mathfrak{S}(E, \mathbf{k})$ is the area element of the Fermi surface within the solid angle d^2k around the direction $\mathbf{k}$. It is instructive to describe $f(\mathbf{k})$ in terms of electrons and holes, which are required to excite the system from equilibrium to the steady state distribution.[32] In Fig. 2 the distribu-

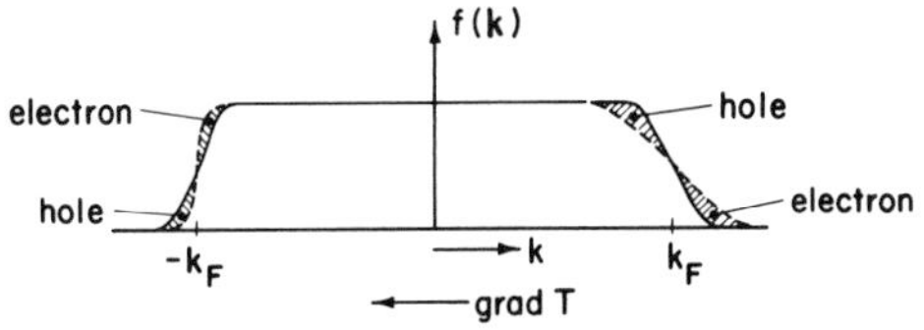

FIG. 2. The electron distribution functions. Solid line: equilibrium distribution $f_0(\mathbf{k})$; dashed line: steady state distribution $f(\mathbf{k})$ in a temperature gradient directed as shown.

tion functions $f(\mathbf{k})$ and $f_0(\mathbf{k})$ and the regions containing electrons and holes are shown. Since $f_1(\mathbf{k})$ is antisymmetric about $\mathbf{k} = 0$ because of the time reversal symmetry of $E(\mathbf{k})$ and $\tau(\mathbf{k})$, $j_1(\mathbf{k})$ is exactly symmetric about $\mathbf{k} = 0$. Because of this and of the $(-\partial f_0/\partial E)$ factor, we can confine ourselves to the vicinity of the Fermi level and to electrons and holes moving *down* the temperature gradient. We need discuss only the electrical current which would be produced by $(-\nabla T)$ acting alone, choosing the field such that the total current vanishes. Within this discussion there is, of course, no direct relation to the electronlike or holelike character of an underlying band structure. The influence of the band structure is only contained in the phase space element $d\mathfrak{S}(E, \mathbf{k})$.

From Eq. (2.1), we see that the behavior of $f_1(\mathbf{k})$ near the Fermi level is determined by the energy dependence of $l(E, \mathbf{k})$. As indicated in Fig. 3, we can distinguish three cases. If $(\partial l/\partial E)_{E_F} = 0$, we have exact symmetry between electrons and holes about the Fermi level for every $\mathbf{k}$. If, in addition, the phase space element were symmetric about E_F, the current due to $(-\nabla T)$ would vanish and the termoelectric power would be zero. If $(\partial l/\partial E)_{E_F} > 0$, then $f_1(\mathbf{k})$ is asymmetric in favor of electrons. If we neglect any asymmetry in phase space, the thermoelectric power will be negative. For $(\partial l/\partial E)_{E_F} < 0$, holes will be dominant and the

[32] J. E. Robinson, *Phys. Rev.* **161**, 533 (1967).

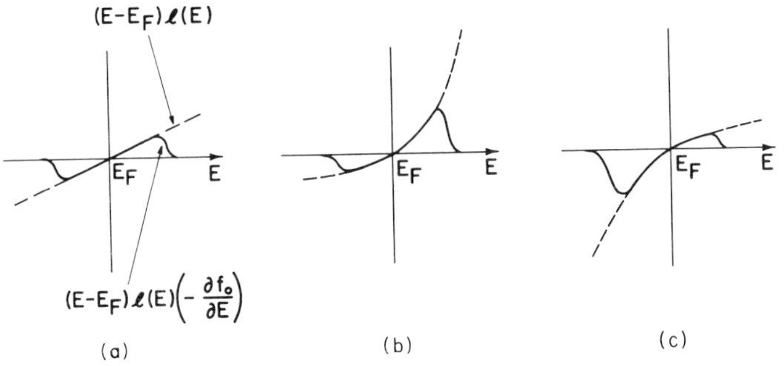

FIG. 3. The energy dependence of the electron scattering and its influence on the distribution $f_1(E) = f(E) - f_0(E)$. Solid lines: $(E - E_F)\, l(E)(-\partial f_0/\partial E)$; dashed lines: $(E - E_F)\, l(E)$. (a) $(\partial l/\partial E)_{E_F} = 0$, (b) $(\partial l/\partial E)_{E_F} > 0$, (c) $(\partial l/\partial E)_{E_F} < 0$.

thermoelectric power will be positive if we disregard again any asymmetry in phase space.

From the qualitative arguments given we see that the magnitude and sign of the thermoelectric quantities depend sensitively on the energy dependence of the electron scattering. The complete theory yields the following expression for the absolute thermoelectric power[1,2,5]:

$$S^{\mathrm{e}} = \frac{\pi^2 k_{\mathrm{B}}^2 T}{3e}\left(\frac{\partial \ln \rho(E)}{\partial E}\right)_{E_{\mathrm{F}}}$$

$$= -\frac{\pi^2 k_{\mathrm{B}}^2 T}{3e}\left(\frac{\partial \ln l(E)}{\partial E} + \frac{\partial \ln \mathfrak{S}(E)}{\partial E}\right)_{E_{\mathrm{F}}}. \tag{2.5}$$

Here e is the absolute magnitude of the elementary charge, ρ the electrical resistivity, and $\mathfrak{S}$ the Fermi surface area. As already mentioned, the result of Eq. (2.5) is obtained by neglecting any deviation of the phonon system from its equilibrium distribution. The quantity in Eq. (2.5) is called the *electron diffusion thermoelectric power* and is indicated here and in the following by the symbol S^{e}.

As seen from Eq. (2.5), the electron diffusion thermoelectric power is determined by the dependence of the electrical resistivity on the electron energy. To obtain S^{e}, we have to calculate the electrical resistivity for different electron energies and then find the derivative $\partial \ln \rho/\partial E$ at the Fermi energy. We note that S^{e} is proportional to the absolute temperature. In the derivation of Eq. (2.5) it is assumed that the electrical conductivity and the electronic component of the heat conductivity are determined by the same relaxation time, i.e., that the electron scattering is elastic (Wiedemann–Franz law). At low temperatures, inelasticity in the electron phonon interaction becomes appreciable. Its effect is to decrease the slope $\partial S^e/\partial T$ somewhat as we approach $T = 0$.[33] However, at low temperatures, this effect of inelasticity is usually masked by phonon drag, which we shall discuss in Part IV. For a single band of standard form, we find from Eq. (2.5)

$$S^e = -\frac{\pi^2 k_B{}^2 T}{3eE_F}\left(1 + \frac{\partial \ln l(E)}{\partial \ln E}\right)_{E_F}. \tag{2.6}$$

Measurements indicate that the absolute thermoelectric power in pure metals is quite often linear in T at high temperatures (Figs. 4 and 5).

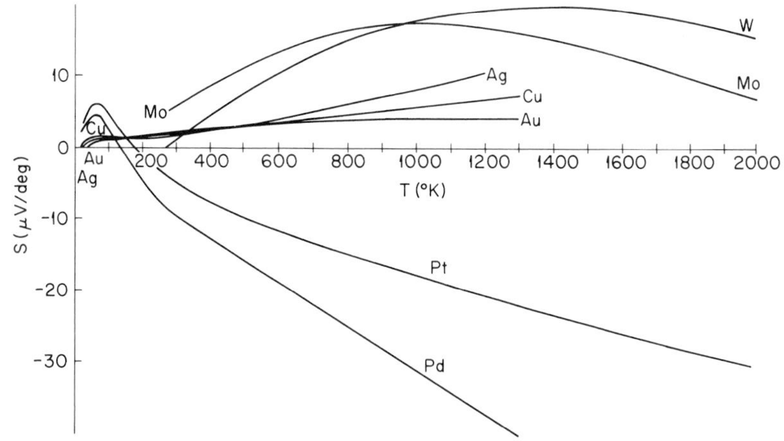

FIG. 4. Thermoelectric power of the refractory and noble metals over a wide temperature range [after N. Cusack and P. Kendall, *Proc. Phys. Soc.* **72**, 898 (1958)].

However, at lower temperatures a pronounced hump in the $S(T)$ curves is usually observed. This strong deviation of the $S(T)$ curves from linearity at low temperatures is predominantly due to the departure of the phonon system from the equilibrium distribution, which has been neglected in the theory so far through the use of the Bloch assumption. It is this invalidity of the Bloch assumption which will be the content of Part IV.

[33] J. M. Ziman, "Electrons and Phonons," Sec. 9.12. Oxford Univ. Press (Clarendon), London and New York, 1960.

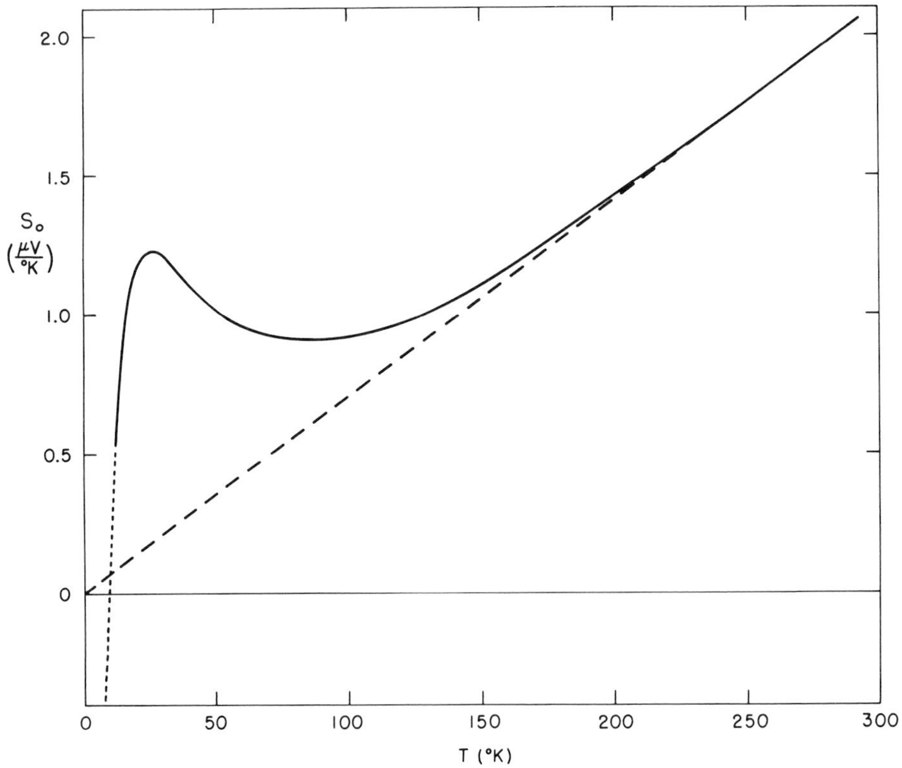

FIG. 5. Thermoelectric power of gold versus temperature [after R. P. Huebener, *Phys. Rev.* **135**, A1281 (1964)].

3. TWO CONDUCTION BANDS

The simple result of Eq. (2.5) is strictly valid only for a single isotropic band. Very often the conduction electron system consists of more than one band, or the Fermi surface is highly anisotropic. For describing the thermoelectric power in such a system, it is helpful to consider two parallel conduction bands and to find the thermoelectric power of the combined system in terms of the properties of the individual components (Fig. 6). In the Peltier picture we can argue as follows. If the electrical conductivities and the Peltier heat current densities in the two conductors are σ_1, σ_2 and U_1, U_2, respectively, the Peltier coefficients for both conductors are given by

$$\pi_1 = U_1/\sigma_1 E; \qquad \pi_2 = U_2/\sigma_2 E, \tag{3.1}$$

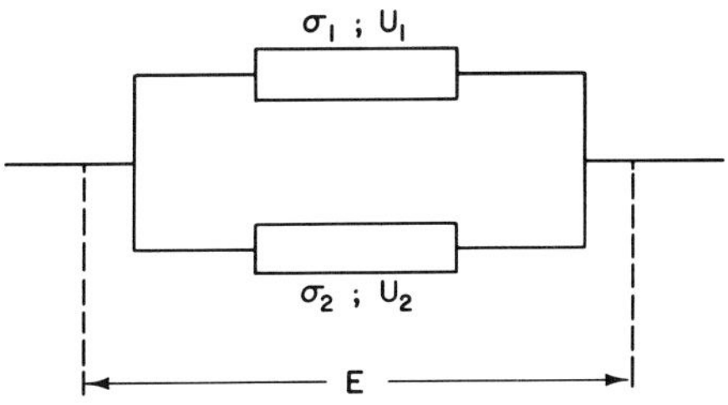

FIG. 6. Two conduction bands.

where E is the electric field across the circuit. The Peltier coefficient of the circuit is, with $\sigma = \sigma_1 + \sigma_2$,

$$\pi = (U_1 + U_2)/\sigma E = (\pi_1\sigma_1 + \pi_2\sigma_2)/\sigma. \tag{3.2}$$

With the Thomson relation (II.5) we have

$$S = (\sigma_1 S_1 + \sigma_2 S_2)/\sigma. \tag{3.3}$$

Equation (3.3) can easily be generalized for n parallel conductors

$$S = \sum_{j=1}^{n} (\sigma_j/\sigma)\, S_j\,, \tag{3.4}$$

where σ_j and S_j is the electrical conductivity and the thermoelectric power of the individual conductor j, respectively.

4. Two Scattering Mechanisms: Influence of Impurities

So far we have considered the case of only one electron scattering mechanism—the electron phonon interaction. What happens if more than one scattering mechanism is present, say phonons and impurities? To find the rule for obtaining the thermoelectric power if more than one scattering mechanism exist we make the following assumptions:

(1) one isotropic conduction band;
(2) the heat transported by the conduction electrons is independent of other heat transporting mechanisms (phonons); and
(3) the scattering of electrons by the impurity and by all other electron scattering events in the crystal are independent of each other (Matthiessen's rule).

Under the assumptions (1)–(3) we can separate a region in which a mixture of the two electron scattering mechanisms is active into two sections in which only one of the two mechanisms acts (Fig. 7). The

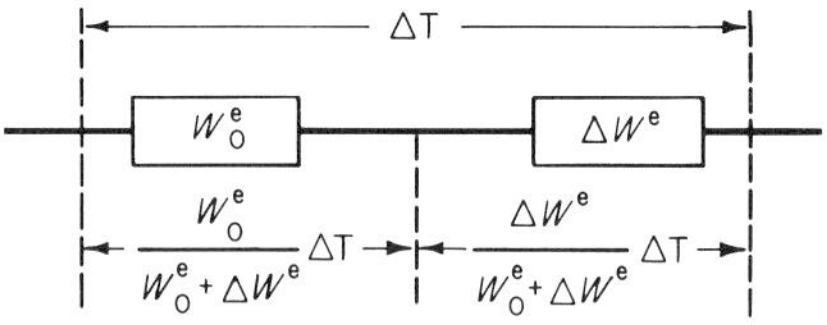

FIG. 7. Two scattering mechanisms.

electronic part W^e of the thermal resistance of the total system is the sum of the resistance W_0^e of the pure material and the resistance ΔW^e of the impurities

$$W^e = W_0^e + \Delta W^e. \tag{4.1}$$

A small temperature difference ΔT across the specimen can be divided in two parts according to the thermal resistances W_0^e and ΔW^e:

$$\Delta T = \frac{W_0^e}{W_0^e + \Delta W^e}\Delta T + \frac{\Delta W^e}{W_0^e + \Delta W^e}\Delta T. \tag{4.2}$$

The thermoelectric voltage ΔV across the sample is then given by the sum

$$\Delta V = -S^e\,\Delta T = -S_0^e \frac{W_0^e}{W_0^e + \Delta W^e}\Delta T - S_i^e \frac{\Delta W^e}{W_0^e + \Delta W^e}\Delta T. \tag{4.3}$$

Here (and in the following) S_0^e and S_i^e are the electron diffusion thermoelectric power of the pure material and of the impurities, respectively. By analogy to Eq. (2.5) S_i^e is given by

$$S_i^e = \frac{\pi^2 k_B{}^2 T}{3e}\left(\frac{\partial \ln \Delta\rho(E)}{\partial E}\right)_{E_F}, \tag{4.4}$$

where $\Delta\rho$ is the impurity resistivity, writing the electrical resistivity in the impure system as the sum $\rho = \rho_0 + \Delta\rho$. From Eq. (4.4) we find

$$S^e = \frac{W_0^e}{W_0^e + \Delta W^e} S_0^e + \frac{\Delta W^e}{W_0^e + \Delta W^e} S_i^e. \tag{4.5}$$

Equation (4.5) was derived originally by Kohler[34] from a variational solution of the Boltzmann equation. It was obtained from arguments

[34] M. Kohler, *Z. Phys.* **126**, 481 (1949).

similar to those given above by De Vroomen[35] and by Gold *et al.*[36] From Eq. (4.5) we find

$$\Delta S^{\mathrm{e}} \equiv S^{\mathrm{e}} - S_0^{\mathrm{e}}$$

$$= \left[S_0^{\mathrm{e}} \Big/ \left(\frac{W_0^{\mathrm{e}}}{\Delta W^{\mathrm{e}}} + 1\right)\right]\left(\frac{S_{\mathrm{i}}^{\mathrm{e}}}{S_0^{\mathrm{e}}} - 1\right). \tag{4.6}$$

Since the bracket in Eq. (4.6) is temperature independent in good approximation, the temperature dependence of ΔS^{e} is given by the function

$$F = S_0^{\mathrm{e}} \Big/ \left(\frac{W_0^{\mathrm{e}}}{\Delta W^{\mathrm{e}}} + 1\right). \tag{4.7}$$

If we require in addition to the assumptions (1)–(3) that the same relaxation time is valid in the electrical conductivity and in the electronic component of the heat conductivity (Wiedemann–Franz law), Eq. (4.6) can be replaced by

$$\Delta S^{\mathrm{e}} = \left[S_0^{\mathrm{e}} \Big/ \left(\frac{\rho_0}{\Delta \rho} + 1\right)\right]\left(\frac{S_{\mathrm{i}}^{\mathrm{e}}}{S_0^{\mathrm{e}}} - 1\right). \tag{4.8}$$

The temperature dependence of ΔS^{e} is then in good approximation given by the function

$$G = S_0^{\mathrm{e}} \Big/ \left(\frac{\rho_0}{\Delta \rho} + 1\right). \tag{4.9}$$

Equation (4.8) can also be derived directly, of course, from Eq. (2.5) if we replace ρ by $\rho_0 + \Delta\rho$. From Eq. (4.9) we see that at low temperatures, where $\Delta\rho \gg \rho_0$, G is proportional to S_0^{e} and therefore $G \sim T$. At intermediate temperatures, where, say, $\rho_0 \gg \Delta\rho$ and $\rho_0 \sim T^5$, we have $G \sim T^{-4}$. At high temperatures, where $\rho_0 \gg \Delta\rho$ and $\rho \sim T$, we find $G = \text{const}$. This temperature dependence of the function G, and thereby of ΔS^{e}, is qualitatively indicated in Fig. 8 for lattice vacancies in gold as the second electron scattering mechanism.

Generally, the electron scattering by impurities is elastic. In Eq. (4.6) the inelasticity in the electron scattering by phonons can be taken into account by replacing only the impurity contribution ΔW^{e} with the electrical resistance increment using the Wiedemann–Franz law, but

[35] A. R. DeVroomen, Thesis, Univ. of Leiden (1959) (unpublished).

[36] A. V. Gold, D. K. C. MacDonald, W. B. Pearson, and I. M. Templeton, *Phil. Mag.* **5**, 765 (1960).

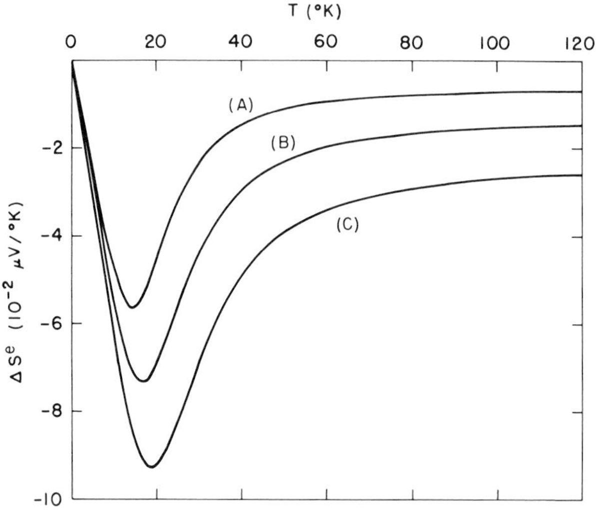

FIG. 8. The change ΔS^e due to lattice vacancies in gold versus temperature at the following vacancy concentrations: (A): $0.40 \cdot 10^{-2}$ at. %; (B): $0.87 \cdot 10^{-2}$ at. %; (C): $1.42 \cdot 10^{-2}$ at. % [after R. P. Huebener, *Phys. Rev.* **135**, A1281 (1964)].

keeping the phonon contribution W_0^e.[37] The function G in Eq. (4.9) is then replaced by

$$G' = S_0^e \Big/ \Big(\frac{L_0 T}{\Delta\rho\kappa_0^e} + 1 \Big), \tag{4.10}$$

where κ_0^e is the electronic heat conductivity in the pure substance and L_0 the Lorentz number: $L_0 = 2.45 \times 10^{-8}\ \mathrm{V^2/deg^2}$.

Equation (4.8) is sometimes written in the following form, with $S^e \equiv S_0^e + \Delta S^e$,

$$\begin{aligned} S^e &= S_0^e + (S_i^e - S_0^e) \frac{\Delta\rho}{\rho_0 + \Delta\rho} \\ &= S_i^e + (S_0^e - S_i^e) \frac{\rho_0}{\rho_0 + \Delta\rho} . \end{aligned} \tag{4.11}$$

Hence, if we plot the total thermoelectric power S^e versus the inverse of the total electrical resistivity at constant temperature for different impurity concentrations, we should obtain a straight line. From the

[37] R. P. Huebener, *Phys. Rev.* **135**, A1281 (1964).

slope and intercept of this line we should be able to determine directly S_0^e and S_i^e, i.e., the thermoelectric power of each individual scattering mechanism acting alone. This scheme is known as the *Nordheim–Gorter rule*,[38] and is illustrated in Fig. 9.

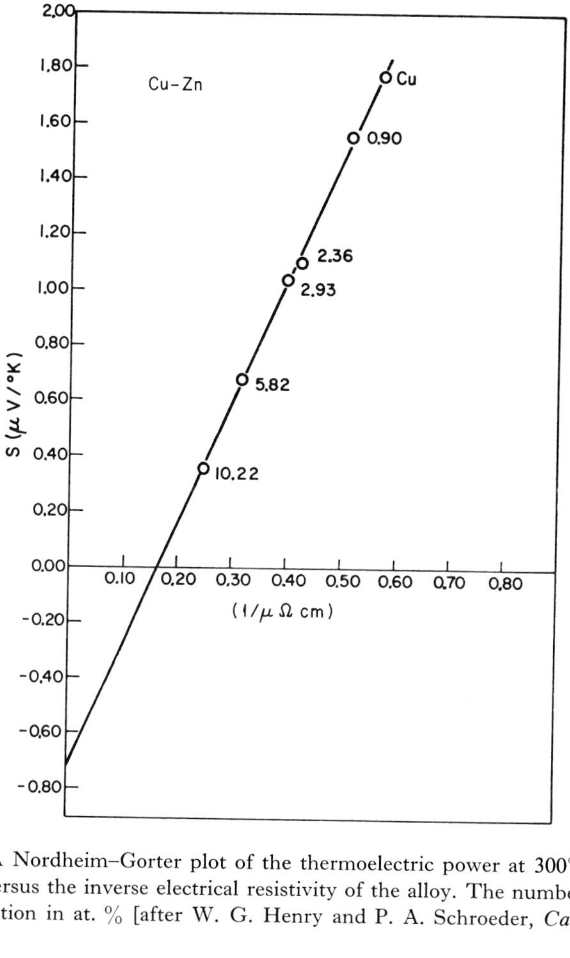

FIG. 9. A Nordheim–Gorter plot of the thermoelectric power at 300°K for copper–zinc alloys versus the inverse electrical resistivity of the alloy. The numbers indicate the Zn concentration in at. % [after W. G. Henry and P. A. Schroeder, *Can. J. Phys.* **41**, 1076 (1963)].

From Eqs. (4.6) or (4.8) we see that the measurement of ΔS^e yields information on the derivative $(\partial \Delta\rho(E)/\partial E)_{E_F}$, provided the quantities S_0^e, ρ_0, and $\Delta\rho$ are known. Again, for calculating ΔS^e or S_i^e we have

[38] L. Nordheim and C. J. Gorter, *Physica* **2**, 383 (1935).

to determine $\Delta\rho$ for different electron energies and then find the derivative $\partial\,\Delta\rho/\partial E$ at the Fermi energy. With

$$\Delta\rho = 3N_i\Sigma(E)/2e^2n(E)\,v(E), \tag{4.12}$$

where N_i is the number of impurities per volume, $\Sigma(E)$ the momentum transfer cross section of the impurity for electron scattering, $n(E)$ the electron density, and $v(E)$ the electron group velocity, we have

$$\frac{\partial \ln \Delta\rho(E)}{\partial E} = \frac{\partial \ln \Sigma(E)}{\partial E} - \frac{\partial \ln[n(E)\,v(E)]}{\partial E}\,. \tag{4.13}$$

Hence, if the functions $n(E)$ and $v(E)$ are known, measurements of ΔS^e or S_i^e yield information on the derivative $\partial\Sigma(E)/\partial E$ at the Fermi energy.

IV. Phonon Drag

5. Nonequilibrium in the Phonon System

We have seen in Part III that usually the thermoelectric behavior cannot be described correctly by a theory which is based entirely on the Bloch assumption. From the fact that electrical insulators do conduct heat we know that a temperature gradient induces heat transport via phonons. The phonon current present in a temperature gradient interacts with the electron system and "drags" the electrons along. Thus the term *phonon drag* is used to describe the effects in the electron system which are caused by the deviation of the phonon system from its equilibrium distribution.

The deviation of the phonon system from thermal equilibrium and its effect on the electronic transport properties was considered relatively early. Peierls[39] first pointed out that the electrical conductivity and the electronic heat conductivity go to infinity if there are only N-processes in the electron–phonon and phonon–phonon interaction. Finite conductivity is obtained only through the additional electron–phonon and phonon–phonon U-processes and, of course, through the interaction of electrons and phonons with impurities. These ideas have been further discussed by Sommerfeld and Bethe.[10] Bailyn[40] observed that, in principle, infinite conductivity occurs not merely when U-processes are neglected, but also whenever there is only one type of process possible for a phonon, whether that is of N-type or U-type. The fact that the

[39] R. Peierls, *Ann. Phys.* **4**, 121 (1930); **5**, 244 (1930).
[40] M. Bailyn, *Phys. Rev.* **112**, 1587 (1958).

nonequilibrium in the phonon system may cause a significant contribution to the thermoelectric power was first suggested by Gurevich.[41] This contribution is generally called *phonon drag thermoelectric power* and is indicated in the following by the symbol S^g. Qualitative discussions of the phonon drag thermoelectric power were given by Klemens,[42] MacDonald,[43] and Ter Haar and Neaves.[44] As pointed out by Sondheimer,[45] the correct treatment of the electronic transport properties requires the simultaneous solution of the two coupled Boltzmann equations for the electrons and the phonons. He showed that the Kelvin relations, Eqs. (II.4) and (II.5), are satisfied if the transport equations for the electrons and phonons are treated consistently. A simultaneous solution of the coupled Boltzmann equations for electrons and phonons has been given by Hanna and Sondheimer[46] and by Tsuji,[47] using a variational method and neglecting U-processes. The coupled transport equations for the combined electron phonon system have been treated consistently with the inclusion of U-processes by Bailyn[40,48,49] and Ziman[50–52] in a variational scheme. The influence of phonon drag on the transport properties of alloys has been discussed recently by Masharov.[53]

A crude estimate of the phonon drag thermoelectric power S^g, can be obtained using arguments very similar to those leading to Eqs. (1.3)–(1.5). In Eq. (1.3) we simply replace C_e by the lattice heat capacity C_g. In addition we must apply a factor, say α, which measures the relative probability for a phonon to interact with a conduction electron. Hence, we have

$$\mu^g \approx (C_g/n\mathbf{e})\alpha \qquad \text{and} \qquad S^g \approx (C_g/n\mathbf{e})\alpha. \tag{5.1}$$

For $T \ll$ Debye temperature θ, we have $C_g \sim T^3$ and $\alpha \approx 1$, since the

[41] L. E. Gurevich, *J. Phys. USSR* **9**, 477 (1945); **10**, 67 (1946).
[42] P. G. Klemens, *Aust. J. Phys.* **7**, 520 (1954).
[43] D. K. C. MacDonald, *Physica* **20**, 996 (1954).
[44] D. Ter Haar and A. Neaves, *Proc. Roy. Soc.* **A228**, 568 (1955).
[45] E. H. Sondheimer, *Proc. Roy. Soc.* **A234**, 391 (1956).
[46] I. I. Hanna and E. H. Sondheimer, *Proc. Roy. Soc.* **A239**, 247 (1957).
[47] M. Tsuji, *Mem. Fac. Sci. Kyushu Univ. Ser. B* **2**, 119 (1958); *J. Phys. Soc. Japan* **14**, 618 (1959).
[48] M. Bailyn, *Phys. Rev.* **120**, 381 (1960).
[49] M. Bailyn, *Phil. Mag.* **5**, 1059 (1960).
[50] J. M. Ziman, "Electrons and Phonons," Sect. 9.13. Oxford Univ. Press (Clarendon), London and New York, 1960.
[51] J. M. Ziman, *Phil. Mag.* **4**, 371 (1959).
[52] J. G. Collins and J. M. Ziman, *Proc. Roy. Soc. (London)* **A264**, 60 (1961).
[53] S. I. Masharov, *Phys. Status Solidi* **27**, 455 (1968).

interaction of phonons with electrons dominates. Therefore, we find in this temperature range $S^{g} \sim T^{3}$. For $T \gg \theta$, we have $C_{g} = \text{const.}$ and $\alpha \sim T^{-1}$ because of the increasing influence of phonon–phonon scattering. Hence, here we expect $S^{g} \sim T^{-1}$.

For an accurate treatment of the phonon drag thermoelectric power we must, of course, integrate over the phonon spectrum. In a general form one can write

$$S^{g} = D \sum_{s} \int C_{s}(\omega)\, \alpha(\omega)\, d\omega, \tag{5.2}$$

where D is a constant, and where the sum is taken over the phonon polarizations s and the integral over all phonon frequencies ω. Here, $C_{s}(\omega)$ is the contribution to the lattice specific heat of phonons with polarization s and frequency ω. The function $\alpha(\omega)$ is the relative probability for the phonon with frequency ω to interact with an electron. Neglecting phonon–impurity and phonon–boundary collisions, we have

$$\alpha(\omega) = \frac{1}{\tau_{\text{pe}}(\omega)} \Big/ \left(\frac{1}{\tau_{\text{pe}}(\omega)} + \frac{1}{\tau_{\text{pp}}(\omega)} \right) = 1 \Big/ \left(1 + \frac{\tau_{\text{pe}}(\omega)}{\tau_{\text{pp}}(\omega)} \right), \tag{5.3}$$

where τ_{pe} and τ_{pp} are the relaxation times for the phonon–electron and phonon–phonon scattering, respectively. In the Debye approximation, neglecting dispersion and assuming free electrons we have for a monovalent metal at sufficiently low temperatures (where only N-processes occur)[9,46–49]

$$S^{g} = \frac{3k_{\mathrm{B}}}{\mathbf{e}} \left(\frac{T}{\theta} \right)^{3} \int_{0}^{\theta/T} \frac{z^{4} e^{z}}{(e^{z} - 1)^{2}}\, \alpha(z)\, dz, \tag{5.4}$$

where $z = \hbar\omega / k_{\mathrm{B}} T$. A more general expression for S^{g}, which includes U-processes, has been given by Bailyn.[40,48,49] We note from Eq. (5.4), that the form of S^{g} is very similar to the expression for the lattice heat conductivity, κ_{g}, in an isotropic crystal in the Debye approximation,[54,55]

$$\kappa_{g} = \frac{k_{\mathrm{B}}}{2\pi^{2} v_{\mathrm{s}}} \left(\frac{k_{\mathrm{B}} T}{\hbar} \right)^{3} \int_{0}^{\theta/T} \frac{z^{4} e^{z}}{(e^{z} - 1)^{2}} \frac{\tau_{\text{pe}}(z) \cdot \tau_{\text{pp}}(z)}{\tau_{\text{pe}}(z) + \tau_{\text{pp}}(z)}\, dz, \tag{5.5}$$

where v_{s} is the sound velocity.

With the approximations used in Eq. (5.4) we see, that S^{g} is negative. For a more general discussion of the sign of S^{g}, we continue to restrict ourselves to sufficiently low temperatures, where the phonon wave vectors, $\mathbf{q}$, are small, and electrons can be scattered through phonons

[54] P. G. Klemens, *Solid State Phys.* **7**, 1 (1968).
[55] J. Callaway, *Phys. Rev.* **113**, 1046 (1959).

only by small angles. The phonon drag thermoelectric power can then be discussed in terms of the local differential geometry of the Fermi surface.[56] We have two cases as illustrated in Fig. 10. If the phonon wave

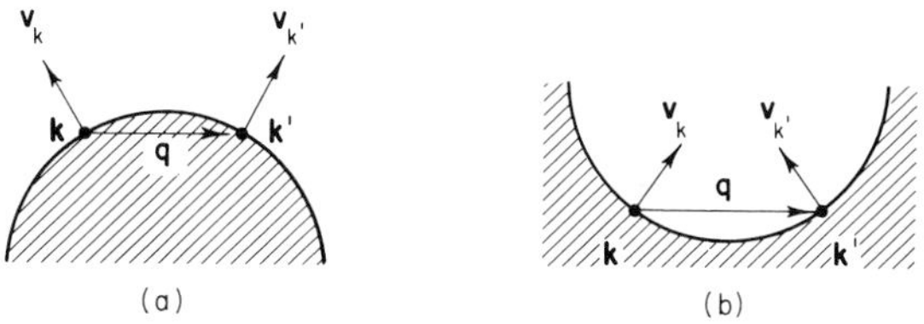

FIG. 10. Electron-like (a) and hole-like (b) transition at the Fermi surface.

vector, **q**, joining the two electron states **k** and **k**′, passes through a region in k-space, which is occupied with electrons, **k** and **k**′ are on an electronlike orbit (Fig. 10a). The group velocity of the electron **k** is

$$\mathbf{v}(\mathbf{k}) = (1/\hbar)\,\mathrm{grad}_k\,E. \tag{5.6}$$

The main component of the difference $\mathbf{v}_{k'} - \mathbf{v}_k$ is oriented in the direction of **q**. Because of the negative electronic charge, we then have $S^g < 0$. We note further that $|S^g|$ increases as the curvature of the Fermi surface becomes larger. On the other hand, if **q** passes a region in k-space which is unoccupied by electrons, **k** and **k**′ are on a holelike orbit (Fig. 10b). The main component of the difference $\mathbf{v}_{k'} - \mathbf{v}_k$ is then oriented opposite to **q**, and we have $S^g > 0$.

At higher temperatures, where larger values of **q** are allowed, we cannot discuss S^g in terms of the local differential geometry of the Fermi surface. For a normal band of standard form the contribution of N-processes to S^g is negative, whereas the contribution of U-processes in general is positive.[48] In the case of an inverted band of standard form, N-processes give a positive contribution to S^g, whereas U-processes yield in general a negative contribution.

The phonon drag effect was first experimentally observed in measurements of the thermoelectric power of germanium at low temperatures.[57,58] Already in 1956, Sondheimer[59] suggested that the phonon drag effect may contribute significantly to the thermoelectric power of Na. Con-

[56] J. M. Ziman, *Advan. Phys.* **10**, 1 (1961).
[57] H. P. R. Frederikse, *Phys. Rev.* **91**, 491 (1953); **92**, 248 (1953).
[58] T. H. Geballe, *Phys. Rev.* **92**, 857 (1953); T. H. Geballe and G. W. Hull, *Phys. Rev.* **94**, 1134 (1954); **98**, 940 (1955).
[59] E. H. Sondheimer, *Can. J. Phys.* **34**, 1246 (1956).

vincing experimental evidence for the effect in the thermoelectric power of metals was then obtained by MacDonald and co-workers,[36,60,61] De Vroomen *et al.*,[62] and Blatt and Kropschot.[63] Guénault and MacDonald[61] studied the suppression of the phonon drag thermoelectric power in K due to alloying with Na, Rb, and Cs. A similar experiment was carried out in Cu and dilute Cu alloys by Blatt and Kropschot.[63] De Vroomen *et al.*[62] studied the thermoelectric power S of Al at very low temperatures. By plotting S/T versus T^2 they showed that S consists of an electron diffusion component linear in T and a phonon drag component proportional to T^3 (Fig. 11).

Since in most metals the Fermi surface is far from a simple sphere, the contributions to S^g from the various parts of the Fermi surface can

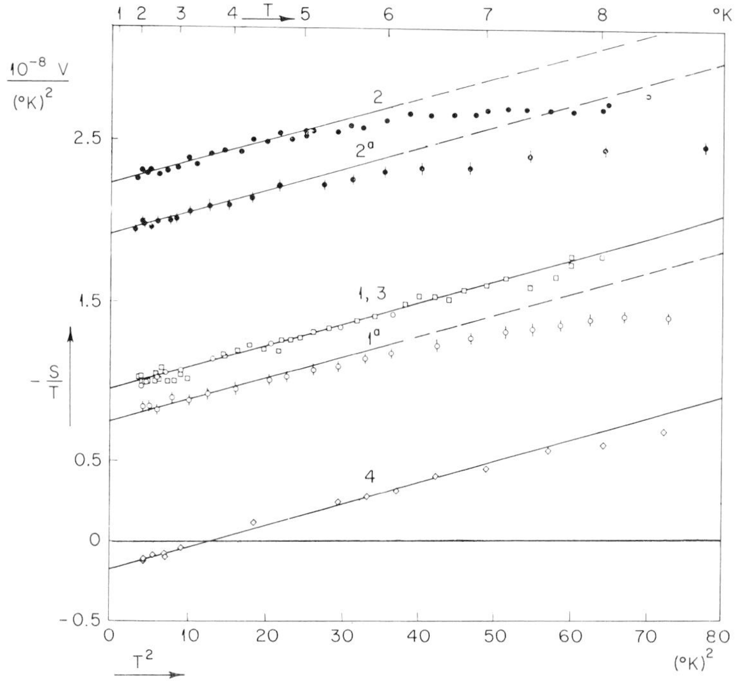

FIG. 11. The phonon drag contribution to the thermopower of aluminum. $-S/T$ is plotted versus T^2 for six aluminum samples [after A. R. De Vroomen, C. Van Baarle, and A. J. Cuelenaere, *Physica* **26**, 19 (1960)].

[60] D. K. C. MacDonald, W. B. Pearson, and I. M. Templeton, *Proc. Roy. Soc.* **A248**, 107 (1958).

[61] A. M. Guénault and D. K. C. MacDonald, *Proc. Roy. Soc.* **A264**, 41 (1961).

[62] A. R. DeVroomen, C. Van Baarle, and A. J. Cuelenaere, *Physica* **26**, 19 (1960).

[63] F. J. Blatt and R. H. Kropschot, *Phys. Rev.* **118**, 480 (1960).

differ strongly from each other in magnitude and sign. Bailyn[64] considered the phonon drag thermoelectric power for a highly anisotropic Fermi surface in detail. He showed that, after dividing the Fermi surface into different regions j, the total phonon drag thermoelectric power can be written in the form

$$S^{g} = \sum_{j} (\sigma_j/\sigma)\, S_j^{g}. \tag{5.7}$$

Here, σ_j and S_j^{g} are the electrical conductivity and the phonon drag thermopower of region j, respectively; σ is the total conductivity. Equation (5.7) is particularly useful if the Fermi surface can be divided such that σ_j is approximately constant for each region. At low temperatures, where only small phonon wave vectors are important, the phonon induced electron transitions between different regions j can be neglected. Here, S_j^{g} is the phonon drag thermopower one would find if region j would act alone. At higher temperatures the meaning of S_j^{g} is more general. The same applies to the partial conductivity σ_j, if it is limited by electron phonon scattering. If σ_j involves scattering by impurities, it includes, of course, at all temperatures a large number of transitions between regions. Equation (5.7) is identical with Eq. (3.4), which we derived earlier from the two-band model. However, in Eq. (5.7) the quantities σ_j and S_j^{g} are somewhat more generally defined than for a model consisting of different independent bands. At low temperatures and in the absence of impurities Eq. (5.7) can be obtained directly from the two-band model.

If N- and U-processes contribute simultaneously to S^{g} within the same region of the Fermi surface, the phonon drag thermopower can be treated using the same arguments which have led to Eq. (4.5). The overall phonon drag thermoelectric power can then be written as

$$S^{g}(T) = \frac{W_{N}^{e}(T)}{W_{N}^{e}(T) + W_{U}^{e}(T)}\, S_{N}^{g}(T) + \frac{W_{U}^{e}(T)}{W_{N}^{e}(T) + W_{U}^{e}(T)}\, S_{U}^{g}(T), \tag{5.8}$$

where the indices N and U indicate the contributions from N- and U-processes, respectively.

MacDonald and Pearson[65] noted that in all cases where it had been measured S^{g} was already negligible near room temperature, whereas rough estimates of the contribution from N-processes indicate that S^{g} and S^{e} should be comparable. They suggested that this must be due to

[64] M. Bailyn, *Phys. Rev.* **157**, 480 (1967).
[65] D. K. C. MacDonald and W. B. Pearson, *Proc. Phys. Soc.* **78**, 306 (1961).

a rather precise cancellation of both contributions to S^g from N- and U-processes. However, it remained unexplained, how such a cancellation can be a general feature in metals. Blatt[66] pointed out that the disappearance of S^g around or below room temperature can be well understood from the dispersion in the phonon system and is not the result of some cancellation between $S_N{}^g$ and $S_U{}^g$.

6. Effect of Phonon Drag on the Electrical Resistivity

Although we do not concern ourselves here primarily with the electrical resistivity, it is instructive to consider briefly the influence of phonon drag on the electrical resistivity. We write the electrical resistivity in the form

$$\rho = \rho_{\text{Bloch}} - \rho_g \,. \tag{6.1}$$

Here ρ_{Bloch} is the resistivity obtained with the Bloch assumption and ρ_g the reduction due to the nonequilibrium in the phonon system. Using simple physical arguments,[67] one can show that the following relation exists between the effects of phonon drag on the electrical resistivity and on the thermoelectric power:

$$\rho_g = (S^g)^2\, T/\kappa_g \,. \tag{6.2}$$

Here κ_g is the lattice heat conductivity. Equation (6.2) can also be obtained from Ziman's expressions for the various transport properties.[50–52] We note that ρ_g is always positive, independent of the sign of S^g. On the other hand, S^g can be positive or negative, depending on the shape of the Fermi surface and on whether N- or U-processes are dominant in the electron–phonon interaction. From the appearance of $(S^g)^2$ in Eq. (6.2), we see that the influence of phonon drag on the electrical resistivity is a *second-order effect*. The electron current, built up by an electric field, causes a current in the phonon system, which then acts back on the electrons. Since the basic interaction mechanism enters the effect twice, the electrical resistivity is always reduced by phonon drag.

7. Influence of Impurities

The admixture of impurities or structural lattice defects which act as scattering centers for phonons generally reduces the phonon drag thermoelectric power, simply because the factor α in Eqs. (5.2) and (5.4) becomes smaller. A phonon scattering process induced by an impurity

[66] F. J. Blatt, *Proc. Phys. Soc.* **83**, 1065 (1964); *Helv. Phys. Acta* **37**, 196 (1964).
[67] R. P. Huebener, *Phys. Rev.* **146**, 502 (1966).

is "lost" for phonon drag thermoelectricity which requires the interaction of phonons with electrons. The change in S^g due to the phonon scattering by an imperfection can directly be obtained from the same arguments leading to Eq. (4.6). In applying Eq. (4.6) to the phonon drag component, we just have to set the analogous quantity $S_i^g = 0$ and must write the expression as an integral over the phonon spectrum. If we define the function $s_g(q)$ as follows

$$S_0^g \equiv \int d^3q\, s_g(q), \tag{7.1}$$

we have

$$\begin{aligned} \Delta S^g &\equiv S^g - S_0^g \\ &= -\int d^3q\, s_g(q) \Big/ \left(1 + \frac{\tau_i(q)}{\tau_0(q)}\right). \end{aligned} \tag{7.2}$$

Here S_g and S_0^g are the phonon drag thermoelectric power in the impure and the pure metal, respectively; τ_0 is the total relaxation time for phonon scattering in the pure metal; τ_i is the relaxation time for phonon scattering at the impurity. We see that $\Delta S^g \to 0$ for $\tau_i/\tau_0 \to \infty$. For $\tau_i/\tau_0 \to 0$ we have $\Delta S^g \to -S_0^g$ and $S^g \to 0$.

We note that measurements of the change in the phonon drag thermopower caused by impurities can yield information on the phonon scattering property of the imperfection. If the functions $s_g(q)$ and $\tau_0(q)$ can be determined semiempirically, say, from measurements of $S_0^g(T)$ and $\kappa_g(T)$, then we can determine the strength of the phonon–impurity interaction if we assume a suitable scattering law for τ_i. In this way, for example, we can find the Rayleigh parameter a, assuming the Rayleigh scattering law $\tau_i^{-1} = a\omega^4$. Since measurements of ΔS^g allow the deduction of the phonon scattering properties of small concentrations of lattice imperfections, the phonon drag thermoelectric power plays in metals a role similar to that of the lattice heat conductivity in electrical insulators for investigating phonon scattering by impurities. Heat conductivity experiments are generally unsuitable for determining the phonon scattering by small impurity concentrations in metals, because of the dominant contribution of the electrons to the heat current.

8. Separation of the Components S^e and S^g

The total thermoelectric power in a metal is generally the sum of the electron diffusion component and the phonon drag component

$$S = S^e + S^g. \tag{8.1}$$

Hence, the change ΔS caused by lattice imperfections is the sum

$$\Delta S = \Delta S^{e} + \Delta S^{g}. \tag{8.2}$$

The electron diffusion component and the phonon drag component of the thermoelectric power can be separated from each other in the following way. At very low temperatures we expect $S^{g} \sim T^{3}$, hence we have

$$S = AT + BT^{3}, \tag{8.3}$$

where A and B are constants. A plot of S/T versus T^{2} should yield a straight line, the intercept and slope of which permit the determination of S^{e} and S^{g}. It is this procedure which was employed by De Vroomen *et al.*[62] to demonstrate the existence of the phonon drag effect in the thermopower of Al (Fig. 11). However, this scheme is sometimes handicapped by the giant thermoelectric effects associated with very small concentrations of magnetic impurities at very low temperatures.

At elevated temperatures one can use the fact that S_0^{g} vanishes and S is identical with S_0^{e} if T is sufficiently high. In the temperature range where S_0^{g} vanishes the thermoelectric power is then expected to be proportional to the absolute temperature (Equation (2.5)). Such a dependence on T is found in many metals above about 300°K. The function $S_0^{e}(T)$ can be obtained to a good approximation by extrapolating the high temperature branch of $S(T)$ linearly to $T = 0$. The function $S_0^{g}(T)$ is then found as the difference $S_0^{g}(T) = S(T) - S_0^{e}(T)$. In Fig. 5 the electron diffusion component obtained in this way is indicated by the dashed straight line.

In the high temperature range, where S^{g} vanishes, the change ΔS caused by impurities is identical with ΔS^{e}. Hence, the high temperature value of ΔS can be used to calculate the function $\Delta S^{e}(T)$ using Eq. (4.6) or (4.8), perhaps with the modified function G' of Eq. (4.10). The function $\Delta S^{g}(T)$ is then obtained from the difference $\Delta S^{g}(T) = \Delta S(T) - \Delta S^{e}(T)$. It appears that this procedure was employed first by Blatt and Kropschot.[63] In Fig. 12 we show some of their results for dilute Cu alloys. It is evident that the admixture of elements of the Ag group to Cu strongly reduces the phonon drag contribution, indicating strong phonon scattering by the added impurities. On the other hand, the impurities of the Cu group do not scatter phonons in Cu very effectively and therefore cause only little reduction in the phonon drag thermoelectric power.

At low temperatures, the extrapolation of the functions $S_0^{e}(T)$ and $\Delta S^{e}(T)$ from the high temperature results may be somewhat uncertain since, for example, the slope $\partial S_0^{e}/\partial T$ decreases slightly as we approach

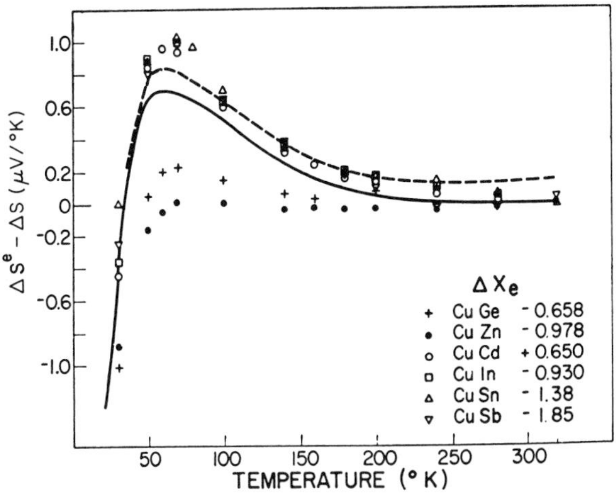

FIG. 12. The difference $\Delta S^e(T) - \Delta S(T)$ for dilute copper alloys. The dashed curve shows the assumed phonon drag contribution in pure copper [after F. J. Blatt and R. H. Kropschot, *Phys. Rev.* **118**, 480 (1960)].

$T = 0$,[33] and since Matthiessen's rule may not be satisfied completely. However, here S_0^g and ΔS^g are often much larger than the electron diffusion component, so that the phonon drag part is not very sensitive to a small error in the contribution from electron diffusion. If giant thermoelectric effects due to magnetic impurities are present at very low temperatures, the separation of the phonon drag component is, of course, more complicated at these temperatures.

9. Low Temperature Anomaly and Anisotropy in the Electron Scattering

As we have seen from Eq. (7.2), the phonon drag thermoelectric power is generally reduced in absolute magnitude by the admixture of impurities which act as scattering centers for phonons. At very low temperatures this reduction should be rather small since the phonon scattering cross section of point defects decreases strongly with decreasing phonon frequency. However, recently a number of experiments [68–72] have shown,

[68] A. M. Guénault, *Phil. Mag.* **15**, 17 (1967).
[69] C. Van Baarle, *Physica* **33**, 424 (1967).
[70] R. P. Huebener and C. Van Baarle, *Phys. Letters* **23**, 189 (1966); *Proc. Int. Conf. Low Temp. Phys.*, 10*th* (M. P. Malkov, ed.), Vol. III, p. 241. Viniti, Moscow, 1967.
[71] R. P. Huebener and C. Van Baarle, *Phys. Rev.* **159**, 564 (1967).
[72] T. Farrel and D. Greig, *Proc. Int. Conf. Low-Temp. Phys.*, 10*th* (M. P. Malkov, ed.), Vol. IV, p. 96. Viniti, Moscow, 1967.

that at low temperatures the magnitude of S^g can *increase* strongly (Fig. 13) and even, in some cases, S^g can change its sign due to the

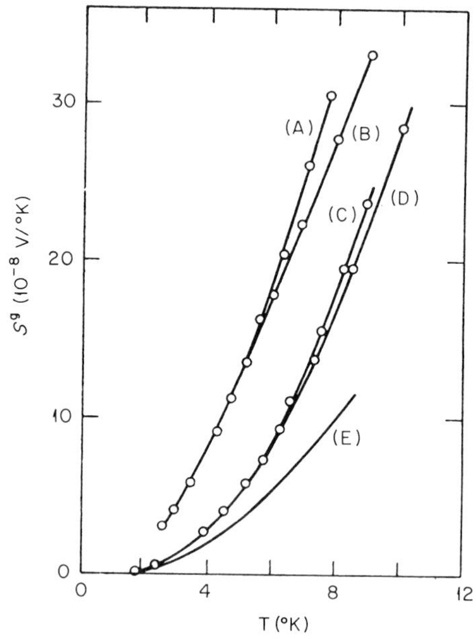

FIG. 13. The phonon drag thermopower S^g of some dilute silver alloys: (A) 0.3 at. % Au; (B) 1.0 at. % Au; (C) 0.03 at. % Sb; (D) 0.1 at. % Sb; (E) estimated phonon drag thermopower of pure silver [after C. Van Baarle, *Physica* **33**, 424 (1967)].

admixture of impurities. An explanation for these anomalous results has been suggested by Van Baarle[69] and by Dugdale *et al.*[73,74] in terms of the anisotropy in the electron scattering and can be summarized as follows. We have seen in Eq. (5.7) that for an anisotropic Fermi surface the total phonon drag thermopower can be written as a sum of the phonon drag contributions from the different sections of the Fermi surface, where each contribution is weighted by the factor σ_j/σ. The contributions S_j^g from the different parts of the Fermi surface can differ strongly from each other, both in magnitude and sign. In a sufficiently pure metal the factors σ_j/σ are determined by the electron-phonon interaction. However, in a dilute alloy, at low temperatures, the weighting factors σ_j/σ are given by the electron–impurity interaction. If the electron

[73] R. Fletcher and J. S. Dugdale, *Proc. Int. Conf. Low-Temp. Phys.*, 10*th* (M. P. Malkov, ed.), Vol. III, p. 246. Viniti, Moscow, 1967.

[74] J. S. Dugdale and M. Bailyn, *Phys. Rev.* **157**, 485 (1967).

scattering for the two scattering mechanisms varies *differently* over the Fermi surface, the factors σ_j/σ in the pure metal can be quite different from those in the metal containing a small impurity concentration. In this way the low temperature phonon drag thermopower can depend very sensitively on the electron scattering mechanism. A rather complete discussion of these effects in the noble metals has been given by Dugdale and Bailyn.[74]

V. Pure Metals

10. Compilation of Thermopower Measurements

Early surveys of the data on thermoelectricity were given by Borelius[75] and Meissner.[76] Later Nyström[77] summarized the thermoelectric measurements, up to about 1958, for pure metals and binary alloys. Recently Vedernikov[78] reviewed the thermoelectric power of the transition metals. In Table I we give a compilation of thermopower measurements in the solid pure metals since 1958.

Table I. Thermoelectric Measurements in Solid Pure Metals Since 1958

Group	Element	Reference and temperature range[a] (°K)
Ia	Li	(1): 0.05–1; (2): 0.2–3; (3): 2–20; (4): 50–250; (5): 273
	Na	(2): 0.2–2; (6): 15–43; (3): 2–55; (5): 273
	K	(2), (7): 0.2–3; (1): 0.05–5; (3): 2–65; (5): 273
	Rb	(2): 0.1–2; (8): 0.1–3; (9): 0.3–1.5; (3): 2–20; (5): 273
	Cs	(8): 0.1–3; (2): 0.2–3; (3): 2–20; (5): 273
	Fr	
Ib	Cu	(10): 0.1–350; (11): 1–4.2; (4): 1–250; (12): 40–300; (13): 78–400; (14): 100–1300; (15): 1100–1470
	Ag	(16): 0.1–1.2; (17): 2–10; (4): 1–250; (18): 10–280; (14): 100–1200; (15): 1100–1350
	Au	(16): 0.1–2.2; (19): 0.1–18; (20): 1.4–9; (21): 2.5–9; (4): 1–250; (18): 10–280; (22): 15–300; (14): 100–1300; (15): 1000–1470

[a] Following the reference, the approximate temperature range or temperature of the measurements is given in °K.

(continued)

[75] G. Borelius, *in* "Handbuch der Metallphysik" (G. Masing, ed.), Vol. I, part 1. Akademische Verlagsgesellschaft, Leipzig, 1935.

[76] W. Meissner, *in* "Handbuch der Experimental Physik" (W. Wien and F. Harms, eds.), Vol. XI, part 2. Akademische Verlagsgesellschaft, Leipzig, 1932.

[77] J. Nyström, *in* "Landolt-Börnstein, Zahlenwerte und Funktionen" (K. H. Hellwege and A. M. Hellwege, eds.), Vol. II, part 6, p. 929. Springer-Verlag, Berlin, 1959.

[78] M. V. Vedernikov, *Advan. Phys.* 18, 337 (1969).

TABLE I (*continued*)

Group	Element	Reference and temperature range[a] (°K)
IIa	Be	(23): 4–120
	Mg	(24): 5–300; (25): 1.2–300
	Ca	
	Sr	
	Ba	
	Ra	
IIb	Zn	(25): 1.2–300
	Cd	(25): 1.2–300
	Hg	
IIIa	Al	(26): 2–9; (27): 4.2–6; (28): 4.2–700; (29): 10–300; (15): 930–1100
	Ga	(30): 273; (31): 10–200
	In	(32): 80–340; (32a): 4.2–300
	Tl	
IIIb	Sc	(33): 10–300; (34): 80–1700; (35): 300
	Y	(36): 3–300; (37): 7–300; (34): 80–1700; (38): 273–1273
	Lu	(39): 8–300; (37): 7–300
IVa	Sn	(40): 3.7–9; (41): 4.2–280; (42): 120–273
	Pb	(43): 6–295; (44): 6–16; (45): 293–313
IVb	Ti	(34): 30–1800
	Zr	(34): 30–1800; (46): 600–1500
	Hf	(34): 80–1650
Vb	V	(47): 4.2–340; (34): 300–1600; (48): 373–1273; (49): 600–1600
	Nb	(34): 100–1700; (50): 300–1200; (46): 600–1500; (51): 9–330
	Ta	(34): 50–1800; (46): 600–1500
VIb	Cr	(47): 4.2–340; (52): 80–400; (34): 80–1600; (53): 303–313
	Mo	(23): 4–120; (14): 273–2400
	W	(23): 4–120; (34): 50–2500; (50): 500–1400; (14): 273–2400
VIIb	Mn	(54): 20–300; (34): 80–1500
	Tc	
	Re	(34): 80–1750; (55): 373–2700

TABLE I (*continued*)

Group	Element	Reference and temperature range[a] (°K)
VIII	Fe	(56): 4.2–1100; (57): 273–1273; (58): 273–1273; (59): 300–1250; (34): 300–1700; (60): 800–1175
	Co	(61): 1.2–6; (59): 300–1450; (34): 300–1650; (60): 1060–1475
	Ni	(9): 0.05–1.5; (62): 2–120; (63): 4.2–300; (56): 10–273; (45): 283–313; (60): 375–825; (59): 300–1500; (64): 620–635
	Ru	(34): 80–1750
	Rh	(34): 300–1700
	Pd	(19): 0.1–3; (65): 0.1–4.2; (66): 1–110; (14): 100–2000; (67): 10–360
	Os	(34): 80–1700
	Ir	(34): 300–1700; (55): 373–2700
	Pt	(19): 0.1–2; (14): 100–2000; (68): 10–500; (66): 1–110
Lanthanides	La	(37): 7–300; (34): 80–1100
	Ce	
	Pr	(37): 7–300
	Nd	(37): 7–300
	Pm	
	Sm	(37): 7–300
	Eu	(69): 10–300
	Gd	(37), (70): 7–300; (71): 15–92
	Tb	(37), (70): 7–300
	Dy	(37), (70): 7–300; (72): 90–190
	Ho	(37), (70): 7–300
	Er	(37), (70): 7–300
	Tm	(37), (73): 7–300
	Yb	(37): 7–300
Actinides	Ac	
	Th	(74): 4.2–300; (75): 5–115; (34): 300–1500
	Pa	
	U	(74): 4.2–300; (76): 300–1100; (77): 300–1200
	Np	(74): 4.2–300; (78): 300–900
	Pu	(74): 4.2–300; (79): 10–270; (77): 300–1200

Footnotes to Table I:

[1] D. K. C. MacDonald, W. B. Pearson, and I. M. Templeton, *Phil. Mag.* **3**, 917 (1958).

[2] D. K. C. MacDonald, W. B. Pearson, and I. M. Templeton, *Proc. Roy. Soc.* **A256**, 334 (1960).

[3] D. K. C. MacDonald, W. B. Pearson, and I. M. Templeton, *Proc. Roy. Soc.* **A248**, 107 (1958).

Footnotes to Table I (continued):

[4] W. B. Pearson, *Fiz. Tverd. Tela* **3**, 1411 (1961) [*English Transl.: Sov. Phys.-Solid State* **3**, 1024 (1961)].

[5] J. S. Dugdale and J. N. Mundy, *Phil. Mag.* **6**, 1463 (1961).

[6] J. Adler and S. B. Woods, *Can. J. Phys.* **40**, 550 (1962).

[7] A. M. Guénault and D. K. C. MacDonald, *Proc. Roy. Soc.* **A264**, 41 (1961).

[8] A. M. Guénault and D. K. C. MacDonald, *Proc. Roy. Soc.* **A274**, 154 (1963).

[9] D. K. C. MacDonald, W. B. Pearson, and I. M. Templeton, *Phil. Mag.* **4**, 380 (1959).

[10] A. V. Gold, D. K. C. MacDonald, W. B. Pearson, and I. M. Templeton, *Phil. Mag.* **5**, 765 (1960).

[11] E. R. Rumbo, *Phil. Mag.* **19**, 689 (1969).

[12] W. G. Henry, *Can. J. Phys.* **41**, 1094 (1963).

[13] J. P. Moore, D. L. McElroy, and R. S. Graves, *Can. J. Phys.* **45**, 3849 (1967).

[14] N. Cusack and P. Kendall, *Proc. Phys. Soc.* **72**, 898 (1958).

[15] T. Ricker and G. Schaumann, *Phys. Kondens. Mater.* **5**, 31 (1966).

[16] D. K. C. MacDonald, W. B. Pearson, and I. M. Templeton, *Phil. Mag.* **3**, 657 (1958).

[17] C. Van Baarle, G. J. Roest, M. K. Roest-Young, and F. W. Gorter, *Physica* **32**, 1700 (1966).

[18] W. B. Pearson, *Can. J. Phys.* **38**, 1048 (1960).

[19] D. K. C. MacDonald, W. B. Pearson, and I. M. Templeton, *Proc. Roy. Soc.* **A266**, 161 (1962).

[20] W. Worobey, P. Lindenfeld, and B. Serin, *Phys. Lett.* **16**, 15 (1965); *in* "Basic Problems in Thin Films" (R. Niedermayer and H. Mayer, eds.), p. 601. Vandenhoeck and Ruprecht, Göttingen, 1966.

[21] H. H. Andersen and M. Nielsen, *Phys. Lett.* **6**, 17 (1963); Risö Rep. No. 77, February 1964.

[22] R. P. Huebener, *Phys. Rev.* **135**, A1281 (1964).

[23] R. L. Powell, J. L. Harden, and E. F. Gibson, *J. Appl. Phys.* **31**, 1221 (1960).

[24] F. J. Blatt, *Helv. Phys. Acta* **41**, 693 (1968).

[25] V. A. Rowe and P. A. Schroeder, *J. Phys. Chem. Solids* **31**, 1 (1970).

[26] A. R. DeVroomen, C. Van Baarle, and A. J. Cuelenaere, *Physica* **26**, 19 (1960).

[27] L. Holwech and V. Sollien, *Phys. Status Solidi* **34**, 403 (1969).

[28] R. J. Gripshover, J. B. Van Zytveld, and J. Bass, *Phys. Rev.* **163**, 598 (1967).

[29] R. P. Huebener, *Phys. Rev.* **171**, 634 (1968).

[30] P. Horner, *Nature* **191**, 58 (1961); **193**, 58 (1962).

[31] R. F. Powell, *Brit. J. Appl. Phys.* **2**, 1467 (1969).

[32] M. Barisoni, R. K. Williams, and D. L. McElroy, *Proc. Conf. Thermal Conduct.*, 7th *Gaithersburg*, 1967, p. 279. NBS Publ. no. 302, 1968.

[32a] B. Bosacchi and R. P. Huebener, *J. Phys. F: Metal Physics* **1**, L27 (1971).

[33] G. T. Meaden and N. H. Sze, *J. Less-Common Metals* **19**, 444 (1969).

[34] M. V. Vedernikov, *Advan. Phys.* **18**, 337 (1969).

[35] O. P. Naumkin, V. F. Terekhova, and Ye. M. Savitskiy, *Fiz. Metal. Metalloved.* **16**, no. 5, 663 (1963) [*English Transl.: Phys. Metals Metallogr.* **16**, no. 5, 22 (1963)].

[36] P. V. Tamarin, G. E. Chuprivkov, and S. S. Shalyt, *Zh. Eksp. Teor. Fiz.* **55**, 1595 (1968) [*English Transl.: Sov. Phys. JETP* **28**, 836 (1969)].

[37] H. J. Born, S. Legvold, and F. H. Spedding, *J. Appl. Phys.* **32**, 2543 (1961).

[38] H. A. Johansen and R. C. Miller, *J. Less-Common Metals* **1**, 331 (1959).

[39] L. R. Edwards, J. Schaefer, and S. Legvold, *Phys. Rev.* **188**, 1173 (1969).

[40] C. Van Baarle, A. J. Cuelenaere, G. J. Roest, and M. K. Young, *Physica* **30**, 244 (1964).

[41] R. E. Fryer, C. C. Lee, V. A. Rowe, and P. A. Schroeder, *Physica* **31**, 1491 (1965).

[42] K. K. Zhilik, *Fiz. Metal. Metalloved.* **23**, no. 1, 187 (1967) [*English Transl.*: *Phys. Metals Metallogr.* **23**, no. 1, 195 (1967)].
[43] J. W. Christian, J. P. Jan, W. B. Pearson, and I. M. Templeton, *Proc. Roy. Soc. (London)* **A245**, 213 (1958).
[44] J. P. Jan, W. B. Pearson, and I. M. Templeton, *Can. J. Phys.* **36**, 627 (1958).
[45] G. M. Maxwell, J. N. Lloyd, and D. V. Keller, Jr., *Rev. Sci. Instrum.* **38**, 1084 (1967).
[46] I. P. Druzhinina, T. M. Vladimirskaya, and A. A. Fraktovnikova, *Izmer. Tekh.* no. 4, 48 (1966) [*English Transl.*: *Meas. Tech.* no. 4, 505 (1966)].
[47] A. R. Mackintosh and L. R. Sill, *J. Phys. Chem. Solids* **24**, 501 (1963).
[48] C. A. Hampel, "Rare Metals Handbook," p. 635. Reinhold, New York, 1961.
[49] I. P. Druzhinina, T. M. Vladimirskaya, and A. A. Fraktovnikova, *Izmer. Tekn.* no. 8, 41 (1966) [*English Transl.*: *Meas. Tech.* no. 8, 1032 (1966)].
[50] V. Raag and H. V. Kowker, *J. Appl. Phys.* **36**, 2045 (1965).
[51] I. Weinberg and C. W. Schultz, *J. Phys. Chem. Solids* **27**, 474 (1966).
[52] J. P. Moore, R. K. Williams, and D. L. McElroy, *Proc. Conf. Thermal Conduct., 7th Gaithersburg*, 1967 NBS Publ. 302 (1968); *Proc. Conf. Thermal Conduct., 8th Purdue University*, 1968 Plenum Press, New York, 1969.
[53] A. R. Edwards, *Phil. Mag.* **8**, 311 (1963).
[54] D. Griffiths and B. R. Coles, *Proc. Phys. Soc.* **82**, 127 (1963).
[55] G. Haase and G. Schneider, *Z Phys.* **144**, 256 (1956).
[56] F. J. Blatt, D. J. Flood, V. A. Rowe, P. A. Schroeder, and J. E. Cox, *Phys. Rev. Lett.* **18**, 395 (1967).
[57] M. J. Laubitz, *Can. J. Phys.* **38**, 887 (1960).
[58] W. Fulkerson, J. P. Moore, and D. L. McElroy, *J. Appl. Phys.* **37**, 2639 (1966).
[59] N. V. Kolomoets and M. V. Vedernikov, *Fiz. Tverd. Tela* **3**, 2735 (1961) [*English Transl.*: *Sov. Phys. Solid State* **3**, 1996 (1962)].
[60] K. Schröder and A. Giannuzzi, *Phys. Status Solidi* **34**, K133 (1969).
[61] P. Radhakrishna and M. Nielsen, *Phys. Status Solidi* **11**, 111 (1965).
[62] D. Greig and J. P. Harrison, *Phil. Mag.* **12**, 71 (1965); T. Farrel and D. Greig, *J. Phys. C: Proc. Phys. Soc.* **3**, 138 (1970).
[63] G. R. Caskey, D. J. Sellmyer, and L. G. Rubin, *Rev. Sci. Instrum.* **40**, 1280 (1969).
[64] I. Nagy and L. Pal, *Phys. Rev. Lett.* **24**, 894 (1970).
[65] P. R. F. Simon, *Proc. Int. Conf. Low Temp. Phys., 9th* (J. G. Daunt, D. O. Edwards, F. J. Milford, and M. Yaqub, eds.), Vol. B, p. 1045. Plenum Press, New York, 1965.
[66] R. Fletcher and D. Greig, *Phys. Lett.* **17**, 6 (1965); *Phil. Mag.* **17**, 21 (1968).
[67] N. H. Hau, Dissertation, Ohio Univ., Athens, Ohio, 1966 (unpublished).
[68] R. P. Huebener, *Phys. Rev.* **146**, 490 (1966).
[69] G. T. Meaden and N. H. Sze, *J. Low-Temp. Phys.* **1**, 567 (1969).
[70] L. R. Sill and S. Legvold, *Phys. Rev.* **137**, A1139 (1965).
[71] O. S. Galkina, E. I. Kondorskii, P. A. Markov, and N. N. Firsov, *Zh. Eksp. Teor. Fiz.* **56**, 1565 (1969) [*English transl.*: *Sov. Phys. JETP* **29**, 840 (1969)].
[72] E. I. Kondorskii, O. S. Galkina, P. A. Markov, and Yu. M. Borovikov, *Zh. Eksp. Teor. Fiz.* **57**, 130 (1969) [*English transl.*: *Sov. Phys. JETP* **30**, 76 (1970)].
[73] L. R. Edwards, Thesis, Iowa State Univ. (1967).
[74] G. T. Meaden, *Proc. Roy. Soc.* **A276**, 553 (1963).
[75] P. Haen and G. T. Meaden, *Cryogenics* **5**, 194 (1965).
[76] J. A. Lee and R. O. A. Hall, *J. Less-Common Metals* **1**, 356 (1959).
[77] P. Costa, *J. Nucl. Mater.* **2**, 75 (1960).
[78] J. A. Lee, J. P. Evans, R. O. A. Hall, and E. King, *J. Phys. Chem. Solids* **11**, 278 (1959).
[79] R. Lallement, *J. Phys. Chem. Solids* **24**, 1617 (1963).

11. Monovalent Metals

The absolute thermoelectric power of *alkali metals* at low temperatures is shown in Fig. 14. The positive hump in the thermopower due to

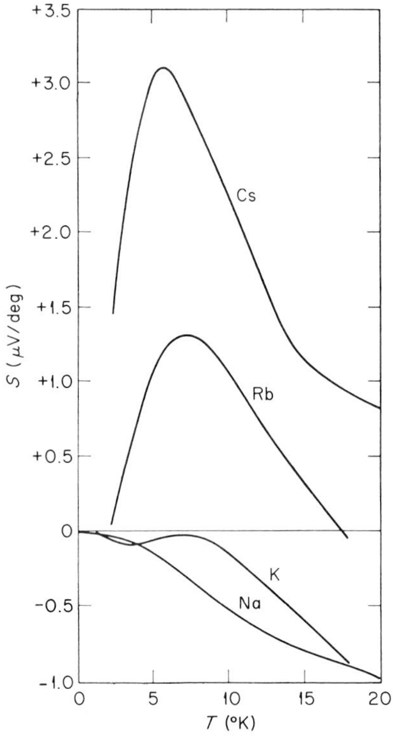

Fig. 14. Thermoelectric power of alkali metals at low temperatures [after D. K. C. MacDonald, "Thermoelectricity: An Introduction to the Principles." Wiley, New York, 1962].

phonon drag is seen to become more pronounced in the heavier alkali metals. From the temperature dependence of the thermoelectric power it appears that in the alkalis the phonon drag component has vanished by 0°C.[79] From thermoelectric measurements near 0°C Dugdale[79] obtained for the quantity $(\partial \ln \rho / \partial \ln E)_{E_F}$ in the alkalis the values listed in Table II. Recent measurements by Kendall[80] of the alkalis near room temperature are in very close agreement with these values. The thermopower of the *noble metals* is shown in Fig. 4 over a wide temperature

[79] J. S. Dugdale, *Science* **134**, 77 (1961).

[80] P. W. Kendall, *Phys. Chem Liquids* **1**, 33 (1968).

range. At low temperatures all noble metals show a pronounced positive phonon drag peak similar to that indicated in Fig. 5 for Au. The phonon drag peak in Ag and Au occurs at about 25°K and in Cu at about 70°K.[9] Generally, in the monovalent metals the peak in S^g appears at $\frac{1}{10}$ to $\frac{2}{10}$ of the Debye temperature. Near room temperature S^g in the noble metals is negligible and S varies proportional to the absolute temperature. The quantity $(\partial \ln \rho / \partial \ln E)_{E_F}$ obtained from the thermopower of the noble metals near room temperature is listed in Table II.

TABLE II. EXPERIMENTAL AND CALCULATED VALUES OF THE DERIVATIVE $(\partial \ln \rho / \partial \ln E)_{E_F}$ FOR THE MONOVALENT METALS

Metal	Experiment[a]	Robinson and Dow[b]	Bortolani and Calandra[c] with r-term	Bortolani and Calandra[c] without r-term	Dickey *et al.*[d]	Martinelli[e]
Li	+6.7	+3.49	+3.9	+5.2	+0.7	
Na	−2.7	−0.65	−0.4	−0.6	−2.4	−2.85
K	−3.8	−2.31	−3.0	−2.3	−3.2	
Rb	−2.3	−2.74	−6.5	−2.7	−3.3	
Cs	−0.2	−2.62	−7.1	−2.6	−0.6	
Cu	+1.6					
Ag	+1.1					
Au	+1.5					

[a] J. S. Dugdale, *Science* **134**, 77 (1961).
[b] J. E. Robinson and J. D. Dow, *Phys. Rev.* **171**, 815 (1968).
[c] V. Bortolani and C. Calandra, *Nuovo Cimento* **58**, 393 (1968).
[d] J. M. Dickey, A. Meyer, and W. H. Young, *Proc. Phys. Soc.* **92**, 460 (1967).
[e] L. Martinelli, *Nuovo Cimento* **70**, 58 (1970).

At sufficiently low temperatures the thermoelectric power of any metal is dominated by very small concentrations of impurities or lattice defects. From Eq. (4.11) we see, that if the resistivity due to phonons, ρ_0, becomes much smaller than the impurity resistivity, $\Delta\rho$, the electron diffusion thermoelectric power is given entirely by the impurity component S_i^e. However, more serious than this ordinary impurity effect is very often the complication at low temperatures due to the "giant" thermoelectric powers of magnetic impurities (Kondo effect). Extremely small concentrations of magnetic impurities (a few ppm) very often seriously disturb thermoelectric measurements in well-annealed specimens of high purity below, say, 10°K. Whereas these effects may not be so important in the alkalis (simply because of the small solubility of foreign metals in the alkalis), they definitely play a role in the noble metals and cause large negative thermopowers below about 10°K.[81] Anderson and Nielsen[82] eliminated the residual Fe content in Au samples almost entirely through annealing in an oxydizing atmosphere. A sample thus purified showed at 3°K a small positive thermopower of about $+5 \times 10^{-3}$ μV/°K instead of the large negative value -13 μV/°K before the annealing. In a similar experiment Worobey *et al.*[83] showed that in pure gold films S^e and S^g remain positive down to 1.4°K. For an extremely pure sample of "natural" Cu Gold *et al.*[36] reported a small positive thermopower at a temperature as low as 3°K.

The positive sign of S^g in all monovalent metals except sodium suggests that, except in Na, U-processes are dominant in the electron phonon interaction. The increase in S^g as we go from Na to Cs has been explained qualitatively by Ziman[51,52,84] through an increasing distortion of the Fermi surface in these metals, thus leading to an enhancement of the relative contribution of U-processes at low temperatures. Bailyn[49] pointed out that the relative contribution of U-processes should increase as we go from Na to Cs independent of a progressive distortion of the Fermi surface, thus leading to an increasingly positive S^g without any Fermi surface considerations.

A theoretical estimate of S^g in the noble metals, attempted by Ziman,[56] produced the wrong sign. Further, Ziman's theory predicted that S^g should vanish less rapidly at higher temperatures than experimentally observed. As pointed out by Blatt,[66] the rapid disappearance of S^g at higher temperatures may be due to the dispersion in the phonon system. Hasegawa and Kasuya[85] recently calculated a positive peak in S^g for Cu at about 70°K by adjusting the relaxation time for phonon–phonon U-processes properly. However, in their calculation S^g changes its sign at about 30°K, which appears to be inconsistent with measurements on extremely pure copper.[36] Because of the extreme sensitivity of the phonon drag thermopower to the details in the model and the approximations, the theoretical calculations of the past—mainly the work of Ziman and Bailyn—yielded at best only qualitative agreement with experiment.

[81] D. K. C. MacDonald, W. B. Pearson, and I. M. Templeton, *Proc. Roy. Soc.* **A266**, 161 (1962).

[82] H. H. Andersen and M. Nielsen, *Phys. Lett.* **6**, 17 (1963); Risö Rep. No. 77, February, 1964.

[83] W. Worobey, P. Lindenfeld, and B. Serin, *Phys. Lett.* **16**, 15 (1965); *in* "Basic Problems in Thin Films" (R. Niedermayer and H. Mayer, eds.), p. 601. Vandenhoeck and Ruprecht, Göttingen, 1966.

[84] J. M. Ziman, *in* "The Fermi Surface" (W. A. Harrison and M. B. Webb, eds.), p. 296. Wiley, New York, 1960.

[85] H. Hasegawa and T. Kasuya, *J. Phys. Soc. Japan* **28**, 75 (1970).

A quantitative theoretical treatment of S^g clearly requires the incorporation of the correct Fermi surface and phonon spectrum together with the proper (anisotropic) electron phonon interaction, a truly formidable task.

As indicated by the values of the quantity $(\partial \ln \rho/\partial \ln E)_{E_F}$ in Table II together with Eq. (2.5), the electron diffusion thermopower in Li and in the noble metals is positive, whereas in the remaining alkalis it is negative. Within the free electron model we expect the value $(\partial \ln \rho/\partial \ln E)_{E_F} = -3$,[3] i.e., a negative electron diffusion thermopower. We see that in Na, K, and Rb, S^e is in reasonable agreement with the free electron model as one would expect. The positive electron diffusion thermopower in Li and in the noble metals has been an embarassing problem for a long time. It was felt for some time that the answer could be found essentially in the distortion of the Fermi surface of these metals, i.e., through phase space considerations.[86–88] However, as pointed out by Ziman,[56,84] in order to account for the positive values of S^e in the noble metals, the Fermi surface distortions would have to be too large and would be inconsistent with the Hall effect data. Ham's calculation[89] showed that in Li the Fermi surface distortion is too small to explain the large positive values for S^e. Similar results were reported recently for Cu.[90,91]

Taylor[92] explored the possibility that anisotropy in the electron relaxation time could lead to a positive electron diffusion thermopower in copper, without success. For explaining S^e in the noble metals, Blatt[93] proposed a mechanism based on electron–electron scattering.

It appears that in the electron diffusion thermopower of the monovalent metals the crucial part is played by the term which contains the dependence of the electron mean free path on electron energy. Recently Robinson[32] demonstrated in a model calculation that the electron diffusion thermopower in all monovalent metals can be reasonably understood within a nearly free electron model if the electron–ion interaction is treated properly. His calculation is based on the concept of pseudoatoms and the diffraction model for describing the electron scattering.

[86] See, e.g., Mott and Jones,[1] p. 312; Wilson,[2] p. 207; Ziman,[5] p. 399.
[87] H. Bross and W. Häcker, *Z. Naturforsch.* **16A**, 622 (1961).
[88] I. V. Abarenkov and M. V. Vedernikov, *Fiz. Tverd. Tela* **8**, 236 (1966) [*English trans.: Sov. Phys.-Solid State* **8**, 186 (1966)].
[89] F. S. Ham, *Phys. Rev.* **128**, 2524 (1962).
[90] H. Hasegawa and T. Kasuya, *J. Phys. Soc. Japan* **25**, 141 (1968).
[91] R. W. Williams and H. L. Davis, *Phys. Lett.* **28A**, 412 (1968).
[92] P. L. Taylor, *Proc. Roy. Soc. (London)* **A275**, 209 (1963).
[93] F. J. Blatt, *Phys. Lett.* **8**, 235 (1964).

In the diffraction theory the electrical resistivity is given by the expression[94]

$$\rho = \frac{3\pi}{\hbar e^2} \frac{\Omega}{v_F^2} \int_0^1 a(K) \cdot |u(K)|^2 \cdot 4 \left(\frac{K}{2k_F}\right)^3 \cdot d\left(\frac{K}{2k_F}\right). \tag{11.1}$$

Here Ω is the volume per atom, v_F the Fermi velocity, K the modulus of the scattering vector, and k_F the electron wave vector at the Fermi surface; $a(K)$ is the static *ion-ion structure factor* or interference function which is measured directly in neutron diffraction experiments. It is defined by

$$a(K) = (1/N) \left\langle \left| \sum_i \exp(i\mathbf{K} \cdot \mathbf{R}_i) \right|^2 \right\rangle_T, \tag{11.2}$$

where N is the total number of ions and the summation is taken over all ion sites $\mathbf{R}_i$ and where the angular brackets indicate the thermal trace at temperature T; $u(K)$ is the pseudopotential matrix element. The electron diffusion thermopower is calculated from Eq. (11.1) with Eq. (2.5). It is customary to express S^e in terms of the quantity ξ

$$S^e = -(\pi^2 k_B^2 T/3eE_F)\xi \tag{11.3}$$

with

$$\xi = -(\partial \ln \rho/\partial \ln E)_{E_F}. \tag{11.4}$$

Within the free electron model, from Eq. (11.1) one finds[95,96]

$$\xi = 3 - 2q - \tfrac{1}{2}r. \tag{11.5}$$

The term q arises from the variation of the upper limit of integration and is given by

$$q = 4k_F^2|u(2k_F)|^2 \cdot a(2k_F) \Big/ \int_0^{2k_F} |u(K)|^2 \cdot a(K) \cdot K^3 \cdot dK. \tag{11.6}$$

The term r comes from the variation of the pseudopotential and is

$$r = k_F \int_0^{2k_F} \left(\frac{\partial}{\partial k} |u(K)|^2\right) \cdot a(K) \cdot K^3 \cdot dK \Big/ \int_0^{2k_F} |u(K)|^2 \cdot a(K) \cdot K^3 \cdot dK. \tag{11.7}$$

Encouraged by the results of the model calculation,[32] Robinson and Dow[97] calculated the quantity ξ for the solid alkali metals at high temperatures using a model which allows for detailed numerical treatment of anisotropy. As shown in Table II, column 3, their main result is to account qualitatively for the large positive electron diffusion thermo-

[94] J. M. Ziman, *Phil. Mag.* **6**, 1013 (1961); *Advan. Phys.* **13**, 89 (1964).
[95] C. C. Bradley, T. E. Faber, E. G. Wilson, and J. M. Ziman, *Phil. Mag.* **7**, 865 (1962).
[96] J. M. Ziman, *Advan. Phys.* **16**, 551 (1967).
[97] J. E. Robinson and J. D. Dow, *Phys. Rev.* **171**, 815 (1968).

power in Li. In their calculation they neglected the explicit energy dependence of the pseudopotential, i.e., the r-term in Eq. (11.5). Subsequently, this point was taken into account by Bortolani and Calandra.[98] In Table II we show their results with and without inclusion of the r-term. The main trend of their results is in agreement with experiment except for Rb and Cs. The failure of their calculation for these metals is attributed by the authors to the influence of the d-bands. A calculation of ξ in the solid alkali metals, using the complete Eq. (11.5), was also performed by Dickey *et al.*[99] Their results (Table II, column 6) are in reasonable agreement with experiment. Martinelli[100] recently calculated the derivative $(\partial \ln \rho/\partial \ln E)_{E_F}$ for Na with the pseudopotential formulation obtaining good agreement with experiment (Table II, column 7). Srivastava and Sharma[101] reported calculations of ξ for the alkali metals near the melting point using Eq. (11.5) and yielding very reasonable values.

In evaluating these theoretical results for the monovalent metals, it is important to note that they are very sensitive to small changes in the pseudopotential, particularly near $2k_F$, where the contribution from the structure factor and from phase space is very large.[97,98]

Rösler[102] treated the electron diffusion thermopower of Li, Na, and K at low temperatures with the usual variational method. He obtained the somewhat surprising result that at very low temperatures the contribution to ξ from the energy dependence of the electron scattering vanishes in all three metals.

Nielsen and Taylor[103] suggested recently that second-order terms in the electron scattering, involving virtual intermediate phonons, may result in a significant contribution to the electron diffusion thermopower at low temperatures.

Since the Fermi energy of a metal changes under pressure, one would expect some relation between the volume coefficient of the electrical resistivity and the quantity $(\partial \ln \rho/\partial \ln E)_{E_F}$, which appears in the electron diffusion thermopower.[79,104] Above the Debye temperature θ the electrical resistivity of a pure metal may be written as

$$\rho_0 = KT/M\theta^2, \tag{11.8}$$

[98] V. Bortolani and C. Calandra, *Nuovo Cimento* **58**, 393 (1968).

[99] J. M. Dickey, A. Meyer, and W. H. Young, *Proc. Phys. Soc.* **92**, 460 (1967).

[100] L. Martinelli, *Nuovo Cimento* **70**, 58 (1970).

[101] S. K. Srivastava and P. K. Sharma, *Solid State Commun.* **7**, 601 (1969).

[102] M. Rösler, *Phys. Status Solidi* **37**, 391 (1970).

[103] P. E. Nielsen and P. L. Taylor, *Phys. Rev. Lett.* **21**, 893 (1968); At. Energy Comm. Tech. Rep. 65 (C00-623-152) 1970.

[104] J. S. Dugdale and J. N. Mundy, *Phil. Mag.* **6**, 1463 (1961).

where K describes the electron–phonon interaction and M is the mass of the metallic ions. The isothermal volume coefficient of the resistivity is then

$$\frac{d\ln\rho_0}{d\ln V} = \frac{d\ln K}{d\ln V} - 2\frac{d\ln\theta}{d\ln V}. \tag{11.9}$$

If we know the Grüneisen parameter, $\gamma = -d\ln\theta/d\ln V$, we can determine $d\ln K/d\ln V$ from the pressure coefficient of the electrical resistivity at high temperatures. The derivative $d\ln K/d\ln V$ describes the change in resistivity with volume that would occur if the phonon system did not change under pressure. If the change in K with volume were due entirely to the change in the Fermi energy with volume, and if the Fermi surface did not distort under pressure, we would have

$$\frac{d\ln K/d\ln V}{d\ln\rho/d\ln E} = \frac{d\ln E_F}{d\ln V}. \tag{11.10}$$

For a spherical Fermi surface this ratio has the value $-\frac{2}{3}$. In Table III

Table III. The Ratio $(d\ln K/d\ln V)/(d\ln\rho/d\ln E)$ Obtained from Measurements at 0°C for the Monovalent Metals[a]

Metal	Li	Na	K	Rb	Cs	Cu	Ag	Au
$\dfrac{d\ln K/d\ln V}{d\ln\rho/d\ln E}$	−0.3	−0.7	−0.8	−0.3	—	−0.6	−0.8	−0.5

[a] After J. S. Dugdale, *Science* **134**, 77 (1961).

the ratio $(d\ln K/d\ln V)/(d\ln\rho/d\ln E)$ as obtained at 0°C from resistance measurements under pressure and from thermoelectric measurements is listed for the monovalent metals. We see that for the monovalent metals the value lies between −0.3 and −0.8. For the two metals with a nearly sperical Fermi surface, K and Na, the value of the ratio is very close to $-\frac{2}{3}$.

12. Polyvalent Metals

Among the polyvalent metals, Mg, Al and Pb show a thermoelectric behavior similar to that expected from a free electron model, both S^e and S^g being negative. This is not surprising in view of the almost spherical Fermi surface in these materials.[105] The peak in S^g for these metals occurs again between $\frac{1}{10}$ and $\frac{2}{10}$ of the Debye temperature. From measurements below 9°K, De Vroomen *et al.*[62] determined S^g in Al assuming Eq. (8.3) for the temperature dependence of the thermopower. At very low temperatures they found $S^g = -1.4 \times 10^{-10} \times T^3$ V/°K, in good agreement with the value $S^g = -1.3 \times 10^{-10} \times T^3$ V/°K obtained by Holwech and Sollien[106] in a similar way. It is interesting to compare these relations with data obtained from an extrapolation procedure starting at *high temperatures* for obtaining S^g. From Huebener's data[107] it is estimated, that at 20°K S^g in Al is -0.6 μV/°K, whereas De Vroomen's relation would yield -1.1 μV/°K at 20°K.

For the polyvalent metals with noncubic symmetry one expects considerable *anisotropy* in their thermoelectric properties. Most of these metals belong to one of the uniaxial systems. For these systems the thermoelectric power in some arbitrary direction relative to the principal axis is given by

$$S(\theta) = S_{\parallel} \cos^2 \theta + S_{\perp} \sin^2 \theta, \tag{12.1}$$

where θ is the angle between the principal axis and the arbitrary direction; $S_{\parallel}$ and $S_{\perp}$ is the termoelectric power measured parallel and perpendicular to the principal axis, respectively. Equation (12.1), also called *Thomson's rule*, is valid only if the heat flow is parallel to the direction of the temperature gradient. However, in an anisotropic material temperature gradients perpendicular to the heat flow are developed because of the anisotropy in thermal conductivity. As shown by Kohler,[108] in this case Eq. (12.1) must be modified, yielding the relation

$$w(\theta)\, S(\theta) = w_{\parallel} S_{\parallel} \cos^2 \theta + w_{\perp} S_{\perp} \sin^2 \theta. \tag{12.2}$$

Here w is the heat resistivity. The expression for $S(\theta)$ in Eqs. (12.1) and (12.2) are often referred to as isothermal and adiabatic thermopower, respectively.

Van Baarle *et al.*[109] measured the thermoelectric power of single

[105] J. B. Ketterson and R. W. Stark, *Phys. Rev.* **156**, 748 (1967); W. A. Harrison, "Pseudopotentials in the Theory of Metals," p. 100. Benjamin, New York, 1966; J. R. Anderson and A. V. Gold, *Phys. Rev.* **139**, A1459 (1965).

[106] L. Holwech and V. Sollien, *Phys. Status Solidi* **34**, 403 (1969).

[107] R. P. Huebener, *Phys. Rev.* **171**, 634 (1968).

[108] M. Kohler, *Ann. Phys.* **40**, 196 (1941).

[109] C. Van Baarle, A. J. Cuelenaere, G. J. Roest, and M. K. Young, *Physica* **30**, 244 (1964).

crystals of tin alloyed with small amounts of In between 3.7 and 9°K. From plots of S/T versus T^2 they found a strongly anisotropic phonon drag component, ranging from $S^g = +2.6 \times 10^{-10} \times T^3$ V/°K along the *c*-direction to $S^g = -3.5 \times 10^{-10} \times T^3$ V/°K perpendicular to it. A qualitative explanation of these results in terms of the zone structure of Sn and a nearly spherical Fermi surface has been given by Klemens *et al.*[110] From the measurements of Fryer *et al.*[111] on polycrystalline Sn it appears that the electron diffusion thermopower in Sn is negative.

Rowe and Schroeder[112] recently studied the thermopower in the hexagonal close-packed metals Mg, Zn, and Cd between 1.2 and 300°K. Their results are shown in Fig. 15. For each metal there is a pronounced peak in S between $\frac{1}{10}$ and $\frac{2}{10}$ of the Debye temperature, which is attributed to the phonon drag component. In addition to these peaks small positive peaks at very low temperatures appear in $S_\perp$ for Mg and in $S_\parallel$ for Zn. Both components S^e and S^g are seen to be highly anisotropic in all three metals. In their experiment Rowe and Schroeder were motivated by the fact, that the well known Fermi surfaces in Mg, Zn, and Cd are topologically similar, however, with significant differences such that a correlation between the phonon drag thermopower and the Fermi surface topology seemed rather likely. In Fig. 16 we show an O.P.W. model for the Fermi surface of magnesium.[105] In terms of this model the phonon drag thermopower of Mg, Zn, and Cd, as shown in Fig. 15, may be understood as follows,[112] keeping in mind the discussion of magnitude and sign of S^g illustrated in Fig. 10. The small negative thermopower in $Zn_\perp$ at very low temperatures may be associated with the small needles in Zn. The absence of this initial negative dip in $Cd_\perp$ and $Mg_\perp$ is consistent with the absence of needles in Cd and with the presence of rather enlarged needles in Mg. The initial positive thermopower in $Zn_\parallel$ may be due to the small monster waists in Zn. Again the absence of this positive contribution in $Cd_\parallel$ and $Mg_\parallel$ is consistent with the absence of monster waists in Cd and with the presence of larger waists in Mg. The positive peaks in $Zn_\perp$ and $Cd_\perp$ may arise from the approximately equal caps in these metals. The small positive peak at lower temperature in $Mg_\perp$ is in agreement with the rather small caps in Mg. A calculation by Tsuji and Kunimune[113] of the electron diffusion thermopower in Cd yielded qualitative agreement with experiment.

Powell[114] measured the thermopower of Ga single crystals between

[110] P. G. Klemens, C. Van Baarle, and F. W. Gorter, *Physica* **30**, 1470 (1964).
[111] R. E. Fryer, C. C. Lee, V. A. Rowe, and P. A. Schroeder, *Physica* **31**, 1491 (1965).
[112] V. A. Rowe and P. A. Schroeder, *J. Phys. Chem. Solids* **31**, 1 (1970).
[113] M. Tsuji and M. Kunimune, *J. Phys. Soc. Japan* **18**, 1569 (1963).
[114] R. F. Powell, *Brit. J. Appl. Phys.* **2**, 1467 (1969).

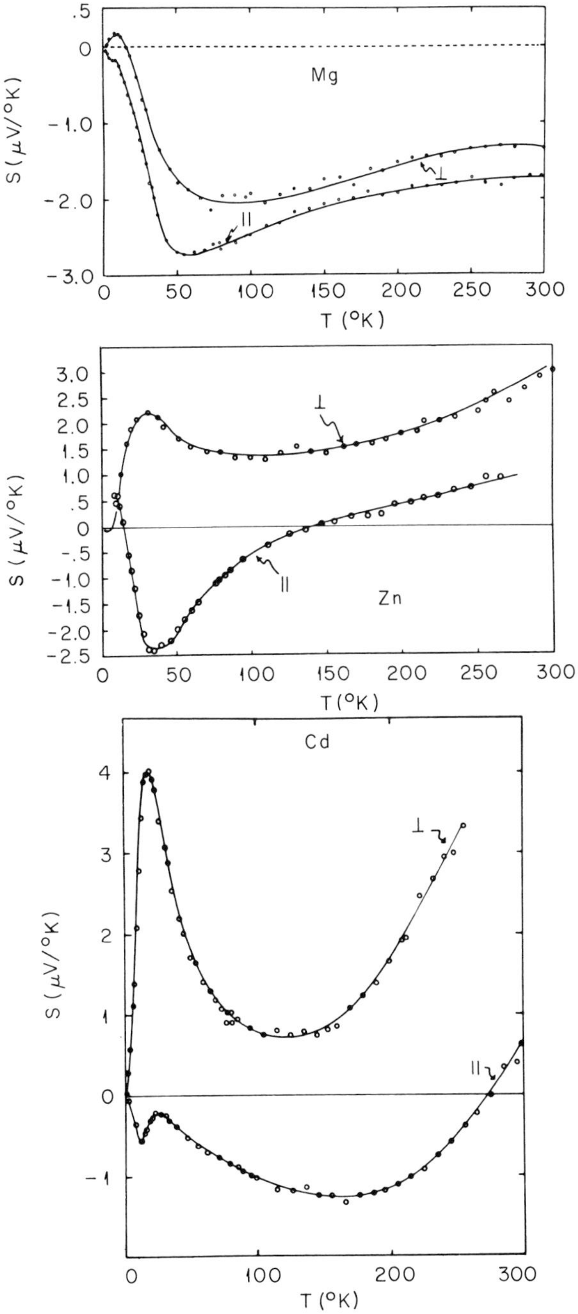

FIG. 15. Thermoelectric power of Mg, Zn, and Cd versus temperature. The symbols || and ⊥ indicate the component parallel and perpendicular to the principal axis, respectively [after V. A. Rowe and P. A. Schroeder, *J. Phys. Chem. Solids* **31**, 1 (1970)].

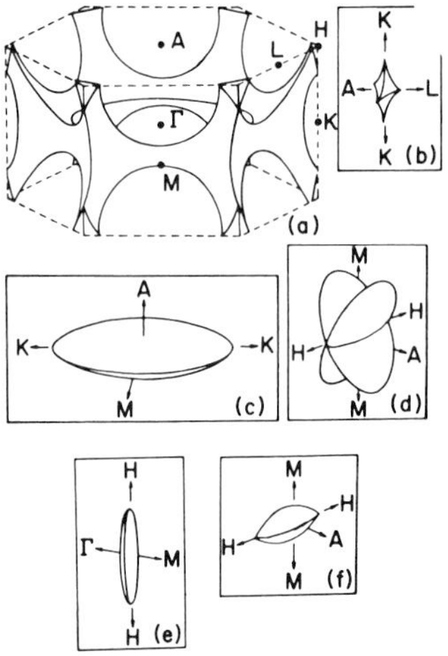

FIG. 16. O. P. W. Model for the Fermi surface of magnesium: (a) monster (holes); (b) caps (holes); (c) lens (electrons); (d) butterflies (electrons); (e) needles (electrons); (f) cigars (electrons) [after J. B. Ketterson and R. W. Stark, *Phys. Rev.* **156**, 748 (1967)].

10 and 200°K and found S to be strongly anisotropic. At about $\frac{1}{10}$ of the Debye temperature, negative phonon drag peaks occur in the [100], [010], and [001] direction. From the data at higher temperatures it appears that S^e is positive in all directions.

13. Transition Metals

Above about 300°K Ni, Pd, and Pt have a relatively large negative thermopower which increases approximately proportional to the absolute temperature. This indicates a rather large electron diffusion component in these materials. At low temperatures these metals show a peak in the thermopower which is attributed to the phonon drag component. In Pd and Pt, S^g is positive (Fig. 4), whereas in Ni, S^g is negative. The large negative values of S^e in Ni, Pd, and Pt can be understood qualitatively from the *model of Mott*[1,115] for the band structure in these metals. According to this model the Fermi energy is located somewhere in the middle of the s band and slightly below the upper band edge of the narrow d band, where the density of states in the d band decreases rapidly with increasing energy. Assuming that the current is carried mainly by the s electrons and that the electrical conductivity is limited by the s–d transitions, the electron mean free path is proportional to the

[115] N. F. Mott, *Proc. Roy. Soc.* **A156**, 368 (1936).

inverse of the density of states in the d band. Since the density of states in the d-band decreases strongly with increasing energy at the Fermi energy, one would expect from Eq. (2.5) large negative values of S^e.

Conversely, the large positive thermopower in Cr, Mo and W at high temperatures can be understood in terms of the d bands just beginning to be filled to a high density of d states.

The positive phonon drag thermopower in Pd and Pt may be understood in terms of phonon induced s–d scattering.[116] In a phonon induced s–d transition the final electron velocity $\mathbf{v}(\mathbf{k}')$ is very small because of the large effective mass in the d-band. Therefore, for such a transition the electron velocity in the direction of the phonon wavevector is generally reduced. In Ni, the situation appears to be more complicated because of the existence of two conduction bands for both spin directions and the possibility of spin mixing.[117] As discussed by Farrell and Greig,[117,118] in Ni at low temperatures we have two separate bands for both spin directions. At low temperatures electron transport is very little affected by s–d scattering since the d-part of the band contributing mainly to the conductivity is completely filled. Therefore, the argument leading to a positive phonon drag thermopower in Pd and Pt does not apply to nickel. At high temperatures both bands contribute to the electron transport because of spin mixing, and Mott's model applies qualitatively to the electron diffusion component of the thermopower. According to Farrell and Greig, the temperature coefficient of S^e in nickel at low temperatures is expected to be different from that at high temperatures.

Recently Vedernikov[78] summarized the experimental results on the thermopower in the transition metals. It appears that metals from the same group in the periodic system show somewhat similar thermoelectric behavior at high temperatures. In most transition metals the dependence of the thermopower on temperature is complicated, showing extrema, bends, and changes in sign. At low temperatures many transition metals show a peak in the thermopower which may be due to the phonon drag effect. Aside from Mott's qualitative model, a theoretical understanding of the thermopower is still lacking for most transition metals. It has been demonstrated in a variety of cases that thermoelectricity in the transition metals clearly shows the influence of *ferromagnetic and antiferromagnetic ordering*, as do the other transport properties.

Aisaka and Shimuzi[119] calculated the electrical and thermal conduc-

[116] R. Fletcher and D. Greig, *Phys. Lett.* **17**, 6 (1965); *Phil. Mag.* **17**, 21 (1968).

[117] D. Greig and J. P. Harrison, *Phil. Mag.* **12**, 71 (1965); T. Farrel and D. Greig, *J. Phys. C (Proc. Phys. Soc.)* **3**, 138 (1970).

[118] T. Farrell and D. Greig, *J. Phys. C (Proc. Phys. Soc.)* **1**, 1359 (1968).

[119] T. Aisaka and M. Shimuzi, *J. Phys. Soc. Japan* **28**, 646 (1970).

tivity and the thermoelectric power at high temperatures for palladium, platinum, rhodium, iridium, molybdenum, and tungsten. They used the Mott model of s–d scattering and the density of states as obtained from low temperature specific heat measurements with the rigid band model. Taking into account the temperature dependence of the Debye temperature, they could explain qualitatively the experimentally observed temperature variations of the thermopower except for Rh and Ir.

In ferromagnetic and antiferromagnetic metals electrons are scattered by spin waves. Analogous to the scattering by phonons resulting in phonon drag effects, the electron–magnon interaction produces *magnon drag*. In magnetic materials a magnon current flows down the temperature gradient, which causes a thermoelectric voltage because of its interaction with the electron system. The theory of magnon drag follows precisely that of phonon drag.[120] Analogous to Eq. (5.1) the magnon drag component, S^m, of the thermopower is approximately given by

$$S^m \approx (C_m/n\mathbf{e})\, \alpha_m\,, \tag{13.1}$$

where C_m is the magnon specific heat and α_m the relative probability for a magnon to interact with a conduction electron. At low temperatures where the magnon–electron interaction is dominant one would expect, from Eq. (13.1), S^m to vary proportionally to $T^{3/2}$ in a ferromagnet and proportional to T^3 in an antiferromagnet.

Following the work of Bailyn,[120] magnon drag has been treated by Gurevich *et al.*,[121] and Roesler.[122] Roesler estimated for iron S^m to be very small compared with S^e. Bhandari and Verma[123] calculated for gadolinium S^m to be only of the order of a few percent of the phonon drag component. Measurements with cobalt between 1.2 and 6°K[124] showed the thermopower to vary proportionally to T and did not indicate a contribution due to magnon drag. Blatt *et al.*[125] measured the thermopower in iron over a wide temperature range and concluded that in Fe magnon drag plays a dominant role. Their conclusion appears somewhat

[120] M. Bailyn, *Phys. Rev.* **126**, 2040 (1962).

[121] L. E. Gurevich and G. M. Nedlin, *Zh. Eksperim. Teor. Fiz.* **45**, 576 (1963); **46**, 1056 (1964) [*English trans.*: *Sov. Phys. JETP* **18**, 396 (1964); **19**, 717 (1964)]; L. E. Gurevich and I. Ya. Korenblit, *Fiz. Tverd. Tela* **6**, 2471 (1964) [*English transl.*: *Sov. Phys. Solid State* **6**, 1960 (1965)].

[122] M. Rösler, *Phys. Status Solidi* **7**, K75 (1964).

[123] C. M. Bhandari and G. S. Verma, *Nuovo Cimento* **60**, 249 (1969).

[124] P. Radhakrishna and M. Nielsen, *Phys. Status Solidi* **11**, 111 (1965).

[125] F. J. Blatt, D. J. Flood, V. A. Rowe, P. A. Schroeder, and J. E. Cox, *Phys. Rev. Lett.* **18**, 395 (1967).

surprising in view of the estimates of S^m.[122,123] Markov[126] recently calculated the electron diffusion thermopower in iron below 200°K, taking into account the electron scattering by spin waves and interband transitions. He finds a large positive peak in the diffusion thermopower at about 100°K, in qualitative agreement with the measurements of Blatt *et al.*[125] Certainly, additional experimental evidence regarding the occurrence of magnon drag would be desirable. Here careful measurements on the variation of S with temperature and on the influence of impurities would seem very useful. However, a meaningful separation of the thermopower into three components may not be so easy.

14. Lanthanides and Actinides

Measurements of the Seebeck effect in the lanthanides and actinides have been performed recently. The thermopower of the lanthanides is complicated by ferromagnetic and antiferromagnetic ordering effects and shows pronounced anisotropy.[127,128] Generally, a change in slope of the $S(T)$ curves appears at the magnetic order–disorder temperature. At high temperatures the electron diffusion component predominates, whereas at low temperatures phonon drag and magnon drag may play a significant role. A separation of the thermopower at low temperatures into the three different components S^e, S^g, and S^m appears to be rather difficult.

The thermopower of the actinides shows a positive peak at low temperatures.[129] Attempts for an understanding of the thermoelectric behavior of the actinides in terms of their electronic band structure have been reported.[130,131]

VI. Alloys

A large number of thermoelectric measurements with alloys have been performed as a function of alloy composition and temperature. Thermoelectricity in alloys clearly lends itself to a wide variety of experimental and theoretical studies. In the following we do not attempt to give a

[126] P. A. Markov, *Fiz. Metal. Metalloved.* **25**, no. 6, 1043 (1968) [*English transl.*: *Phys. Metals Metalogr.* **25**, no. 6, 86 (1968)].
[127] H. J. Born, S. Legvold, and F. H. Spedding, *J. Appl. Phys.* **32**, 2543 (1961).
[128] L. R. Sill and S. Legvold, *Phys. Rev.* **137**, A1139 (1965).
[129] G. T. Meaden, *Proc. Roy. Soc.* **A276**, 553 (1963).
[130] J. A. Lee, J. P. Evans, R. O. A. Hall, and E. King, *J. Phys. Chem. Solids* **11**, 278 (1959).
[131] R. Lallement, *J. Phys. Chem. Solids* **24**, 1617 (1963).

complete description of the work carried out in the past. (This would merely result in something like a strongly magnified version of the list presented in Table I.) Instead, we shall discuss only the typical features of thermoelectricity in alloys, illustrating the main points with a few characteristic examples.

15. Electron Diffusion Thermopower of Alloys

At temperatures which are sufficiently high or sufficiently low such that the phonon drag thermopower is negligible, the thermopower of an alloy is given in good approximation by Eq. (4.6) or (4.11), and the change in the thermopower due to alloying is given approximately by Eq. (4.8). As we have seen in Section 4, measurements of the electron diffusion thermopower in an alloy yield the quantity S_i^e defined in Eq. (4.4). The quantity S_i^e can be obtained from a Nordheim–Gorter plot illustrated in Fig. 9 for copper–zinc alloys at 300°K. Similar plots have been obtained from measurements at constant T near room temperature, e.g., by Crisp *et al.*,[132] Weinberg,[133] and Wright[134] for a series of copper and silver alloys. By plotting the intercepts S_i^e against temperature, Wright[134] demonstrated that S_i^e is proportional to T in agreement with Eq. (4.4). The Nordheim–Gorter relation for binary alloys and its extension to ternary alloys has been verified by Domenicali *et al.*[135]

From Eq. (4.11) we see that at low temperatures, where the electron scattering by the solute ions is dominant ($\Delta\rho \gg \rho_0$), we have $S^e \approx S_i^e$. Here, the electron diffusion thermopower is identical with the characteristic thermopower of the impurity admixture. On the other hand, in a dilute alloy at high temperatures, where the electrons are scattered mainly by phonons ($\Delta\rho \ll \rho_0$), we have $S^e \approx S_0^e$. At intermediate temperatures S^e is a combination of S_0^e and S_i^e as indicated in Eq. (4.11). Once the quantity S_i^e is determined for a single temperature, the diffusion thermopower of the alloy, $S^e(T)$, can be obtained for *all temperatures* using Eq. (4.11) or the more fundamental Eq. (4.6), and assuming S_0^e and S_i^e to be proportional to T. This is demonstrated in Fig. 17 which shows the absolute thermopower of the alloy Au + 4.99 at.% Pt in the temperature range between 4.2 and 300°K. In this alloy the phonon

[132] R. S. Crisp, W. G. Henry, and P. A. Schroeder, *Phil. Mag.* **10**, 553 (1964); R. S. Crisp and W. G. Henry, *ibid.* **11**, 841 (1965).

[133] I. Weinberg, *Phys. Rev.* **157**, 564 (1967).

[134] L. S. Wright, *Can. J. Phys.* **46**, 1711 (1968).

[135] C. A. Domenicali and F. A. Otter, Jr., *Phys. Rev.* **95**, 1134 (1954); C. A. Domenicali, *ibid.* **112**, 1863 (1958).

drag component is nearly completely suppressed because of the relatively high concentration of the platinum admixture. The solid line shows the measured curve of $S^e(T)$, whereas the crosses are calculated using Eq. (4.6) after the coefficients S_0^e/T and S_i^e/T have been determined from the measured values of the thermopower at 200 and 300°K. We see that the determination of S_i^e at a single temperature allows us to reconstruct the function $S^e(T)$ over the whole temperature range.

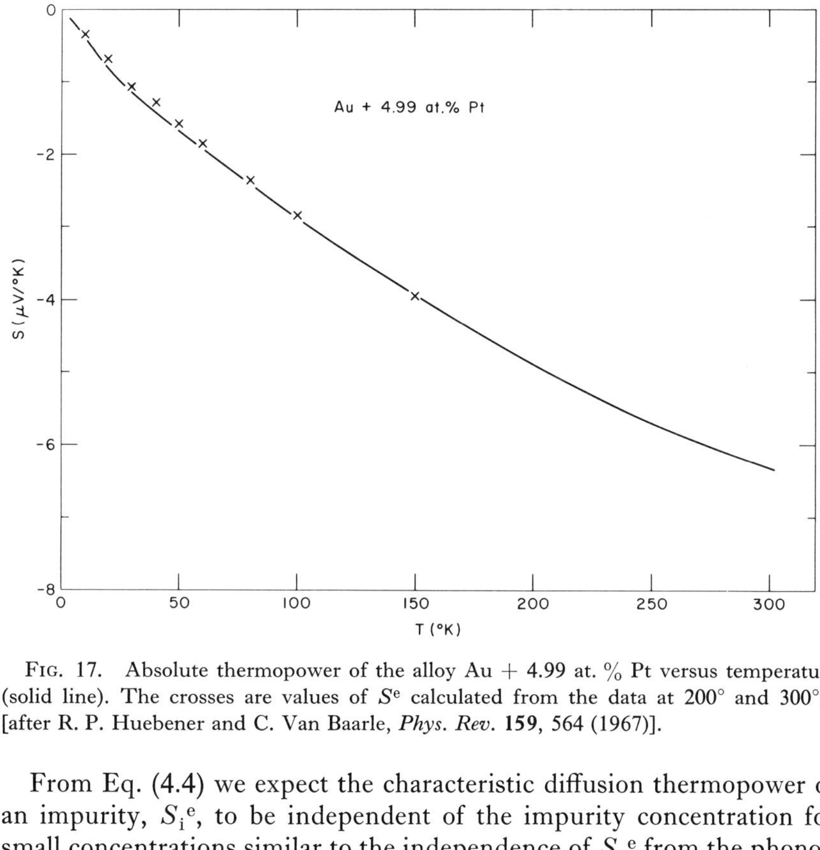

FIG. 17. Absolute thermopower of the alloy Au + 4.99 at. % Pt versus temperature (solid line). The crosses are values of S^e calculated from the data at 200° and 300°K [after R. P. Huebener and C. Van Baarle, *Phys. Rev.* **159**, 564 (1967)].

From Eq. (4.4) we expect the characteristic diffusion thermopower of an impurity, S_i^e, to be independent of the impurity concentration for small concentrations similar to the independence of S_0^e from the phonon concentration in the crystal. This independence of S_i^e of the impurity concentration is often observed experimentally, e.g., in measurements by Van Baarle *et al.*[136] and Guénault[68] on dilute Ag-base alloys.

We have mentioned in Section 4 that the Nordheim–Gorter rule, as contained in Eqs. (4.8) and (4.11), is valid only if the Wiedemann–Franz

[136] C. Van Baarle, F. W. Gorter, and P. Winsemius, *Physica* **35**, 223 (1967).

law and Matthiessen's rule are satisfied. The Wiedemann–Franz law usually holds near and above the Debye temperature and when the electrons are scattered mainly by impurities. However, Matthiessen's rule is usually not obeyed. We write the electrical resistivity of an alloy at temperature T as

$$\rho_{\text{alloy}}(T) = \rho_{\text{solvent}}(T) + \rho_{\text{solute}}(0) + \Delta(T), \tag{15.1}$$

where the deviation from Matthiessen's rule, $\Delta(T)$, describes the temperature dependence of the impurity resistivity. Usually, the ratio $\Delta(T)/\rho_{\text{solute}}(0)$ increases with temperature and can rise to a value of 0.1–0.2 or higher. Further, the impurity resistivity at zero temperature, $\rho_{\text{solute}}(0)$, is often not exactly proportional to the impurity concentration. Deviations from Matthiessen's rule can occur because of changes in the phonon spectrum and changes in the electronic band structure due to alloying. Blatt and Lucke[137] considered the validity criteria for the Nordheim–Gorter relation and showed that it can remain valid in some cases where Matthiessen's rule is violated.

The Nordheim–Gorter rule in its original form provides a scheme for analyzing thermoelectric data taken at constant temperature for different impurity concentrations. In this form it is found very often to be invalid for higher solute concentrations because one or both of the thermoelectric parameters in Eq. (4.11) are varying with solute concentration. More useful for treating thermoelectric data of alloys appears to be a scheme which considers data taken at constant solute concentration and different temperatures. We write the electron diffusion thermopower associated with the electron scattering by phonons and by impurities as

$$S_{\text{ph}}^{\text{e}} = bT \qquad \text{and} \qquad S_{\text{i}}^{\text{e}} = cT, \tag{15.2}$$

respectively, where b and c are constants. From Eq. (4.11) it follows that

$$S^{\text{e}}/T = b + \Delta\rho(\rho_0 + \Delta\rho)^{-1}(c - b). \tag{15.3}$$

Hence a plot of S^{e}/T versus $(\rho_0 + \Delta\rho)^{-1}$ at constant solute concentration should result in a straight line. The conditions for this are more general than those required for the linearity of the Nordheim–Gorter plot, since the linearity is preserved even when b and c vary with solute concentration. A plot of S^{e}/T versus $(\rho_0 + \Delta\rho)^{-1}$ is shown in Fig. 18 for Ag–Pd alloys at different Pd concentrations. The intercept on the vertical axis, i.e., the parameter b in Eq. (15.3), is seen to decrease to a minimum around 6 at.% Pd and to rise again for higher Pd concentra-

[137] F. J. Blatt and W. H. Lucke, *Phil. Mag.* **15**, 649 (1967).

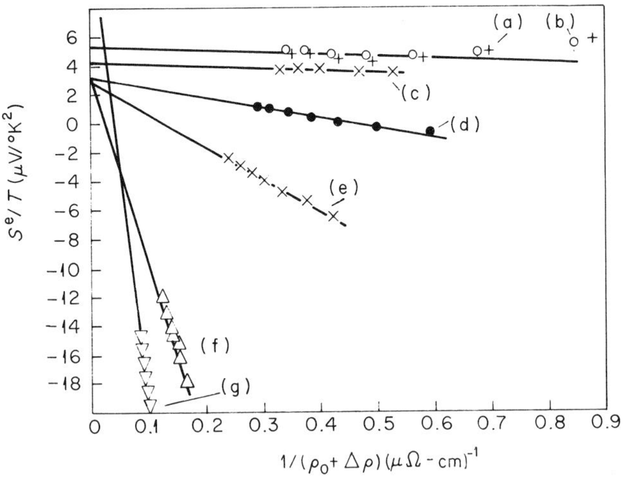

FIG. 18. S^e/T versus $(\rho_0(T) + \Delta\rho)^{-1}$ for Ag–Pd alloys at different Pd concentrations. (a) pure Ag, (b) 500 ppm Pd, (c) 0.12 at. % Pd, (d) 0.53 at. % Pd, (e) 1.28 at. % Pd, (f) 6.3 at. % Pd, (g) 10 at. % Pd. [after P. A. Schroeder, R. Wolf, and J. A. Woolam, *Phys. Rev.* **138**, A105 (1965)].

tions. Similar results were found in Au–Pt alloys.[71] In Table IV the parameters b and c, defined in Eq. (15.2) and obtained from measurements of S^e in Au–Pt alloys, are listed for different Pt concentrations. We see that both parameters clearly vary with the Pt concentration. A variation of the parameter c with solute concentration has also been observed in dilute magnesium alloys.[138] In conclusion we note that the electron diffusion thermopower in binary alloys very often can be described by two independent contributions both of which are proportional to T with the proportionality constants varying with solute concentration.

TABLE IV. THE PARAMETERS b AND c DEFINED IN EQ. (15.2) AND DETERMINED FROM S^e OF Au–Pt ALLOYS, FOR DIFFERENT Pt CONCENTRATIONS[a]

at. % Pt	b (10^{-3} $\mu V/°K^2$)	c (10^{-2} $\mu V/°K^2$)
0	7.05	—
0.11	5.65	−1.13
0.50	4.61	−2.45
1.03	3.26	−2.65
4.99	2.81	−3.37

[a] After R. P. Huebener and C. Van Baarle, *Phys. Rev.* **159**, 564 (1967).

[138] E. I. Salkovitz, A. I. Schindler, and E. W. Kammer, *Phys. Rev.* **107**, 1549 (1957).

The variation of the parameters b and c in Eq. (15.2) with solute concentration can be understood in terms of the change in the electronic band structure with alloying. If we dissolve, e.g., Pd in Ag, the number of conduction electrons per atom will successively decrease because of the rather small electron per atom ratio in Pd and the Fermi surface can be expected to shrink. Thus, in the noble metals, the admixture of the metal next down in the periodic system may change the Fermi surface towards a more spherical shape.[139,140] Conversely, the addition of a metal from a higher column in the periodic system is expected to enhance the contact of the Fermi surface with the Brillouin zone. We shall see further below, that the change in the electronic band structure due to alloying will also strongly affect the phonon drag thermopower.

The influence of alloying on the electron diffusion thermopower is particularly interesting in the *transition metals* Ni, Pd, and Pt. According to Mott's model,[115] in these metals the dominant contribution to S^{e} arises from the energy dependence of n_d, the density of states in the d band, i.e.,

$$S^{e} \approx \frac{\pi^2 k_B{}^2 T}{3|e|} \left(\frac{\partial \ln n_d(E)}{\partial E} \right)_{E_F}. \tag{15.4}$$

A shift of the Fermi level across the d band due to alloying is then expected to influence the diffusion thermopower strongly. Figure 19 shows the absolute thermopower at 273°K of the Pd–Ag system over the whole composition range. By adding Pd to Ag, S is seen to pass from a small positive value to relatively large negative values. Presumably, the sudden decrease of S near 40 at.% Pd indicates the appearance of d band holes. With increasing Pd concentration the Fermi level moves from the upper band edge down into the d band. The increase of S at higher Pd concentrations is due apparently to the fact that the value $|\, \partial \ln n_d(E)/\partial E \,|_{E_F}$ decreases as we go farther away from the upper edge of the d band. By adding Rh to Pd, the d band becomes further depleted, and the electron diffusion thermopower increases to positive values.[140] Hau[141] recently measured the thermopower of the Pd–Au system over the whole composition range. At 300°K he found a dependence of S on composition very similar to that shown in Fig. 19. We note that the phonon drag component in the Pd alloys is negligible at 273°K.

The measurements by Taylor and Coles (Fig. 19) are complicated by the fact that they contain contributions both from electron scattering

[139] P. A. Schroeder, R. Wolf, and J. A. Woollam, *Phys. Rev.* **138**, A105 (1965).
[140] J. C. Taylor and B. R. Coles, *Phys. Rev.* **102**, 27 (1956).
[141] N. H. Hau, Dissertion, Ohio Univ., Athens, Ohio, 1966 (unpublished).

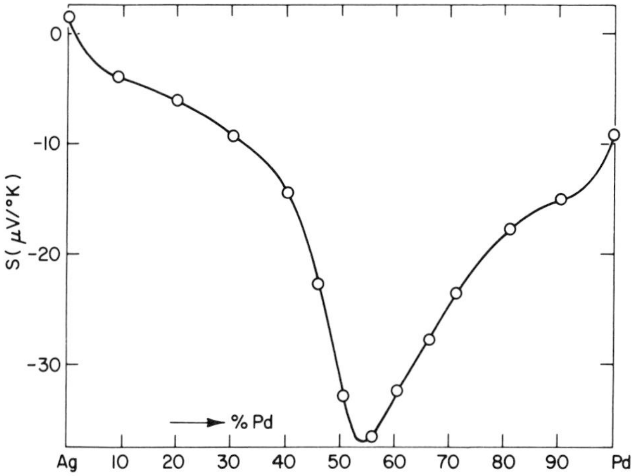

FIG. 19. Absolute thermopower of the Ag–Pd system at 273°K versus the Pd concentration [after J. C. Taylor and B. R. Coles, *Phys. Rev.* **102**, 27 (1956)].

by phonons and by the impurity admixture. Fletcher and Greig[116] measured the electron diffusion thermopower in Pd and Pt alloys at very low temperatures, where only the electron scattering by impurities is important. Their results substantiate the behavior reported by Taylor and Coles for higher temperatures. For the solution of Ag in Pd they found S_i^e to be negative, whereas the admixture of Rh to Pd was found to lead to more positive values of S_i^e. In dilute alloys of Pt with Au and Ir, respectively, they obtained very similar results. Fletcher and Greig found that the sign reversal of S_i^e occurs at the same alloy composition for which one expects a change in sign of the quantity $(\partial \ln n_d(E)/\partial E)_{E_F}$ from electron specific heat data taken with the same alloys. They give a detailed discussion in terms of Mott's analysis of the electron-impurity scattering in transition metals and the rigid band model. A qualitative account of the diffusion thermopower in Pd-rich Pd–Ag and Pd–Rh alloys was also given by Kimura and Shimizu[142] using the Mott model for s–d scattering and rigid bands. Dugdale and Guénault[143] considered the low temperature transport properties of the Pd–Ag alloy series. They pointed out that a rigid band model is strictly not applicable, since on alloying with Ag the d-band holes in Pd are completely filled at approximately 60 at.% Ag, whereas Fermi surface studies of pure Pd indicate that there are only 0.36 holes per atom in the d band. To account for this apparent discrepancy Dugdale and Guénault modified the Mott model

[142] H. Kimura and M. Shimizu, *J. Phys. Soc. Japan* **19**, 1632 (1964).
[143] J. S. Dugdale and A. M. Guénault, *Phil. Mag.* **13**, 503 (1966).

slightly by assuming the *s* and d band to remain unchanged in shape upon alloying, while the s band moves relative to the d band in such a way as to accommodate the additional s electrons. With this model they could explain the impurity resistivity for the alloy series as well as the electron diffusion thermopower in the Pd-rich alloys.

The variation of the electron diffusion thermopower in Ni with alloying has been discussed in terms of the displacement of the Fermi level by several authors.[144]

We have seen from Eq. (4.4) that the quantity S_i^e is determined by the dependence of the electron scattering at the impurity on the electron energy. In a model calculation Domenicali *et al.*[135] investigated the electron diffusion thermopower of binary alloys using Eq. (4.11) with different values for the energy dependence of the impurity resistivity as parameter. Calculations of S_i^e from basic principles have been carried out by a number of authors. Friedel and his collaborators were the first to address themselves to this problem. A calculation[145] for dilute noble metal alloys using a screened Coulomb potential and the Born approximation and assuming a spherical Fermi surface yielded for the quantity $(\partial \ln \Delta\rho / \partial \ln E)_{E_F}$ values between -1 and -2, in qualitative agreement with experiment for nontransition metal impurities. Subsequently the calculation was refined using square well potentials for representing the impurity and a phase shift analysis.[146] In Eq. (4.12) we have expressed the impurity resistivity in terms of the momentum transfer cross section Σ. The phase shift expression for Σ is

$$\Sigma = (4\pi/k^2) \sum_{l=1}^{\infty} l \sin^2(\eta_{l-1} - \eta_l), \tag{15.5}$$

where k is the electron wave vector at the Fermi surface, η_l are the phase shifts of the asymptotic wave function caused by the perturbation; l is the angular momentum quantum number. From Eq. (15.5) we find the derivative

$$\frac{\partial \ln \Sigma}{\partial \ln E} = -1 + \frac{\sum_{l=1}^{\infty} l \sin[2(\eta_{l-1} - \eta_l)] \left(\frac{\partial \eta_{l-1}}{\partial E} - \frac{\partial \eta_l}{\partial E} \right) E}{\sum_{l=1}^{\infty} l \sin^2(\eta_{l-1} - \eta_l)}. \tag{15.6}$$

From this formula together with Eq. (4.13) Friedel *et al.*[146] obtained

[144] N. V. Kolomoets, *Fiz. Tverd. Tela* **8**, 997 (1966) [*English trans.*: *Sov. Phys. Solid State* **8**, 799 (1966)]; V. E. Panin and V. P. Fadin, *Phil. Mag.* **18**, 1301 (1968).

[145] J. Friedel, *J. Phys. Radium* **14**, 561 (1953).

[146] P. De Faget de Casteljau and J. Friedel, *J. Phys. Radium* **17**, 27 (1956).

values for the quantity $(\partial \ln \Delta\rho/\partial \ln E)_{E_F}$ similar to those found in the Born approximation.[145] Fischer[147] calculated S_i^e for dilute alkali metal alloys using screened Coulomb potentials and the Born approximation yielding results in fair agreement with experiment. A calculation of the derivative $(\partial \ln \Delta\rho/\partial \ln E)_{E_F}$ for impurities in monovalent and polyvalent metals using different scattering potentials was also performed by Léonard.[148] His results for monovalent metals are qualitatively in agreement with the experimental data. However, he failed to account for the change in sign of the derivative $(\partial \ln \Delta\rho/\partial \ln E)_{E_F}$ observed in dilute Al and Mg alloys at a critical value of the valency difference. It appears that in these metals S_i^e may be largely influenced by energy band changes.[149]

We note that a relation similar to Eq. (11.10) should exist between the volume coefficient of the impurity resistivity $\Delta\rho$ and the derivative $(\partial \ln \Delta\rho/\partial \ln E)_{E_F}$. This relation is reasonably satisfied for polyvalent impurities in the noble metals taking for the quantity $\partial \ln E_F/\partial \ln V$ the free electron value $-\frac{2}{3}$.[149]

The calculation of the quantity $(\partial \ln \Delta\rho/\partial \ln E)_{E_F}$ requires the choice of the proper scattering potential (square well, screened Coulomb potential, etc.) with two or more adjustable parameters for describing the impurity. The parameters of the potential are then fixed such that the Friedel condition[150]

$$(2/\pi) \sum_l (2l + 1)\, \eta_l = Z \tag{15.7}$$

is satisfied. Here Z is the excess charge which must be screened by the conduction electrons. It appears that the impurity resistivity $\Delta\rho$ is not very sensitive to the detail of the scattering potential as long as the Friedel sum rule (15.7) is obeyed.[3,149] The sensitivity of the derivative $(\partial \ln \Delta\rho/\partial \ln E)_{E_F}$ to the detailed form of the scattering potential has not been tested systematically. It is clear that for an accurate treatment one must include anisotropy in the Fermi surface and in the electron scattering. The derivative $(\partial \ln \Delta\rho/\partial \ln E)_{E_F}$ must then be integrated over the Fermi surface, and it becomes inadequate to use the free electron value $+1.0$ for the term $\partial \ln[n(E)\, v(E)]/\partial \ln E$ in Eq. (4.13).

Whereas for nontransition metal impurities in the noble metals the derivative $(\partial \ln \Delta\rho/\partial \ln E)_{E_F}$ lies between -1 and -2, *transition metal impurities* show much larger values. In Fig. 20 experimental values of the quantity $\Delta x = -(\partial \ln \Delta\rho/\partial \ln E)_{E_F}$ for various substitutional im-

[147] K. Fischer, *Phys. Lett.* **13**, 18 (1964).
[148] P. Léonard, *J. Phys.* **28**, 328 (1967).
[149] J. Friedel, *Can. J. Phys.* **34**, 1190 (1956).
[150] J. Friedel, *Phil. Mag.* **43**, 153 (1952).

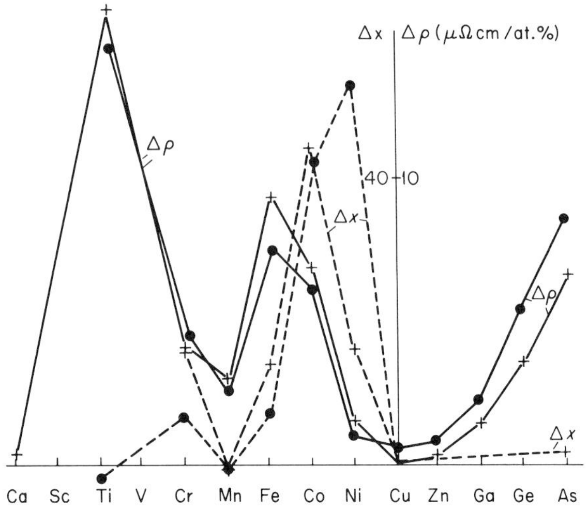

FIG. 20. Residual resistivity $\Delta\rho$ (solid lines) and the quantity $\Delta x = -(\partial \ln \Delta\rho/\partial \ln E)_{E_F}$ (dashed lines) for various substitutional impurities in copper (crosses) and gold (dots) versus the difference in valency [after J. Friedel, *Can. J. Phys.* **34**, 1190 (1956)].

purities in Cu and Au are plotted versus the difference in valency (dashed line). Friedel[149] noted that the curves for Δx are closely related to the analogous curves for the resistivity $\Delta\rho$ also shown in Fig. 20 (solid lines). Roughly speaking, the curve for Δx is the derivative of the curve of $\Delta\rho$ versus the valency difference. This behavior has been explained in principle by Friedel in terms of resonance scattering of the conduction electrons by virtual d states near the Fermi surface. Subsequently, these ideas were developed further, and the occurrence of localized magnetic moments associated with transition metal impirities was studied in detail. A number of excellent reviews on this subject has been given elsewhere.[11–16]

16. PHONON DRAG THERMOPOWER OF ALLOYS

As indicated in Eq. (7.2), the reduction of the phonon drag thermopower due to alloying is determined by τ_i, the relaxation time for phonon scattering by the impurity admixture. Often, a few atomic percent impurities are sufficient to suppress S^g completely. For phonon scattering by point defects in a cubic crystal we have in the limit of low phonon frequencies[151]

$$\frac{1}{\tau_i} = f \cdot c \left\{ \frac{1}{12} \left(\frac{\Delta M}{M} \right)^2 + \left[\frac{1}{\sqrt{6}} \frac{\Delta F}{F} - \sqrt{\left(\frac{2}{3}\right)}\, Q\gamma \frac{\Delta R}{R} \right]^2 \right\} \omega^4. \qquad (16.1)$$

[151] P. G. Klemens, *Proc. Phys. Soc. (London)* **A68**, 1113 (1955).

Here f is a constant, c the mole fraction of the defects, and γ the Grüneisen constant; M is the atomic mass of the crystal, F the force constant of a linkage, R the nearest-neighbor distance, and ΔM, ΔF, and ΔR are the changes at the location of the point defect. The constant Q contains the contribution to the scattering from lattice strains outside the nearest neighbors. In Cu–Al and Cu–Au alloys[152] the suppression of S^g could be explained entirely by the mass difference term in Eq. (16.1). However, other systems show an appreciable or even dominant contribution from the term due to the strain field around the defect.[37,152,153] A more accurate treatment of phonon scattering by point defects [154] leads to an expression for τ_i^{-1} which contains a resonance and damping term in the denominator and which yields the Rayleigh scattering law in the limit $\omega \rightarrow 0$. Such a resonance scattering law has been used for analyzing the change ΔS^g due to lattice vacencies in Pt.[153] Resonant phonon scattering may lead to a strong suppression of S^g at very low temperatures, where Rayleigh scattering is usually rather weak. Ample evidence for resonant phonon scattering by point defects has been obtained from experiments on the thermal conductivity of electrical insulators.[154]

The phonon drag thermopower can also be affected by *changes in the electronic band structure* due to alloying. Alloys of copper with different impurities having a higher valency show a resurgence of S^g at higher impurity concentrations, which appears to depend on the electron per atom ratio, and which may be associated with the approach of the Fermi surface to the (200) faces of the Brillouin zone upon alloying.[132,155] The rapid decrease of S^g in Ag due to alloying with Pd has been explained in part by the withdrawal of the Fermi surface from the zone boundary.[139] The contribution of U-processes to S^g clearly depends on a portion of the Fermi surface being close to a zone boundary.

We have pointed out in Section 9 that *anisotropy in the electron scattering* can yield anomalies in the phonon drag thermopower of dilute alloys. This effect is limited to very low temperatures where the electrons are scattered predominantly by the impurity admixture. An interesting case are dilute Au–Pt alloys,[71] which show a negative phonon drag thermopower below about 25°K (Fig. 21), whereas in pure Au S^g is definitely positive in this temperature range.[83] This sign reversal of S^g due to the Pt admixture has been explained[74] through the strong scattering at the Pt ions of the "d-like" electrons from the concave parts of the Fermi surface in the [110] direction. In this way the positive contribution

[152] I. Weinberg, *Phys. Rev.* **139**, A838 (1965); **146**, 486 (1966).
[153] R. P. Huebener, *Phys. Rev.* **146**, 490 (1966).
[154] A. A. Maradudin, *Solid State Phys.* **18**, 274 (1966).
[155] W. G. Henry and P. A. Schroeder, *Can. J. Phys.* **41**, 1076 (1963).

to S^g from these electrons is reduced, leading to an overall negative phonon drag thermopower. On the other hand, the addition of Pd to Ag causes a slight enhancement of the (positive) S^g in silver,[68] which is consistent with a very small concave region of the Fermi surface in

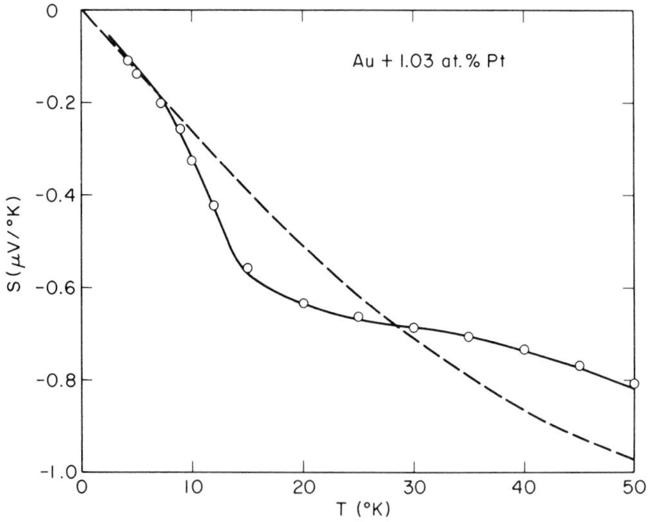

FIG. 21. Absolute thermopower of the alloy Au + 1.03 at. % Pt versus temperature (dots and solid line). The dashed line is the function $S^e(T)$ calculated from the data at 200 and 300°K [after R. P. Huebener and C. Van Baarle, *Phys. Rev.* **159**, 564 (1967)].

the [110] direction. Recently, a strong enhancement of the negative phonon drag thermopower of Al due to alloying has been observed below about 50°K.[107] From these measurements it was concluded that in Al the third band contributes relatively little to the electrical conductivity, in agreement with theoretical estimates.[156] An enhancement of S^g due to alloying was also observed in Ni[72,117] and was explained on the basis of Eq. (5.7) and two bands of opposite spin.

VII. Size Effect

Since electrons and phonons are scattered by the crystal surface, both S^e and S^g will be different from their bulk values, if the smallest sample dimension becomes comparable with or smaller than the bulk mean free paths for electrons and phonons. Measurements of the thermo-

[156] N. W. Ashcroft, *Phys. Kondens. Mater.* **9**, 45 (1969).

electric size effect are conveniently performed with a thermocouple consisting of a very thin foil or film and a relatively thick wire of the same material. The electrical resistivity difference between a foil of the thickness a_1 and a cylindrical wire with the diameter a_2, caused by the electron scattering at the surface, is[157]

$$\Delta\rho = \rho_{\text{foil}} - \rho_{\text{wire}} = \frac{3}{4}\left(\frac{1}{2a_1} - \frac{1}{a_2}\right)\rho_0 l(1 - p) \qquad \text{(VII.1)}$$

Here ρ_0 is the bulk resistivity and l the *bulk* value of the electron mean free path (averaged over the Fermi surface). The parameter p indicates the fraction of the electrons specularly reflected at the surface. In Eq. (VII.1) it is assumed that p is the same in the foil and in the wire and that $l/a_1 \ll 1$ and $l/a_2 \ll 1$. The appearance of the product $\rho_0 l$ in Eq. (VII.1) implies that $\Delta\rho$ is independent of temperature, i.e., that Matthiessen's rule is valid for the scattering at the surface. Deviations from Matthiessen's rule may occur from the fact that with decreasing temperature l becomes too large for Eq. (VII.1) to remain exactly valid. However, an exact treatment[157] indicates that the resistivity increment in a thin foil due to scattering at the surface varies only by about 18% over the range $0 < l/a_1 < 0.5$. Therefore, in many cases the deviations from Matthiessen's rule can be expected to be small.

The change in the *electron diffusion thermopower* due to the crystal surface is readily obtained by inserting Eq. (VII.1) into Eq. (4.8). With the further assumption that the parameter p is independent of the electron energy and with $\Delta\rho \ll \rho_0$ we find

$$\begin{aligned}\Delta S^{\mathrm{e}} &= S^{\mathrm{e}}_{\text{foil}} - S^{\mathrm{e}}_{\text{wire}} \\ &= \left(\frac{1}{2a_1} - \frac{1}{a_2}\right)(1 - p)\frac{\pi^2 k_{\mathrm{B}}^2 T}{4|e|}\, l \left(\frac{\partial \ln l}{\partial E}\right)_{E_{\mathrm{F}}}. \end{aligned} \qquad \text{(VII.2)}$$

We see from Eq. (VII.2) that measurements of the size effect on S^{e} allow determination of the derivative $(\partial \ln l/\partial E)_{E_{\mathrm{F}}}$. Subsequently, the variation of the Fermi surface area with electron energy, $(\partial \ln \mathfrak{S}/\partial E)_{E_{\mathrm{F}}}$, may be estimated with Eq. (2.5). It is clear that here the derivatives $(\partial \ln l/\partial E)_{E_{\mathrm{F}}}$ and $(\partial \ln \mathfrak{S}/\partial E)_{E_{\mathrm{F}}}$ must be understood as average values over the Fermi surface. If the Fermi surface is strongly anisotropic, the separation of S^{e} into the two contributions from $(\partial \ln l/\partial E)_{E_{\mathrm{F}}}$ and $(\partial \ln \mathfrak{S}/\partial E)_{E_{\mathrm{F}}}$ is, of course, not very meaningful.

Studies of the size effect on the *phonon drag thermopower* should yield

[157] E. H. Sondheimer, *Advan. Phys.* 1, 1 (1952).

information on the phonon mean free path in the crystal, as do measurements of the size effect on the heat conductivity in electrical insulators. Taking the phonon mean free path for boundary scattering in a foil of thickness a_1 and in a cylindrical wire of diameter a_2 as[157] $8a_1/3(1-p)$ and $4a_2/3(1-p)$, respectively, and ignoring the dependence of the bulk phonon mean free path, λ_0, on phonon frequency, we find from Eq. (7.2) for the difference in S^g between the foil and the wire[158]

$$\Delta S^g = S^g_{\text{foil}} - S^g_{\text{wire}} = -\frac{3}{4}\left(\frac{1}{2a_1} - \frac{1}{a_2}\right)(1-p)\, S_0{}^g\lambda_0 . \qquad \text{(VII.3)}$$

In Eq. (VII.3) it is assumed that $\lambda_0/a_1 \ll 1$ and $\lambda_0/a_2 \ll 1$. We note that the size effect on S^g is proportional to the *bulk* mean free path λ_0 of the dominant phonon mode.

Measurements of the thermoelectric size effect clearly require well-annealed specimens of high purity in order to detect the influence of the specimen *surface* only. The first experiments of this kind were performed with thin film samples prepared by vacuum evaporation.[159–165] Later, thin foil specimens, obtained by cold rolling and subsequent annealing, were employed.[106,158,166] Recently, further experiments have been carried out with evaporated films of the noble metals.[83,167–169] The experimental results are usually analyzed, assuming *diffuse* scattering of electrons and phonons at the specimen surface ($p = 0$). For not too thin films, the variation of ΔS with the inverse film thickness, indicated in Eqs. (VII.2) and (VII.3) appears to be generally established. However, the experimental results on the quantity $(\partial \ln l/\partial \ln E)_{E_F}$ seem to be somewhat unclear. For Au, between 77°K and room temperature, Huebener[166] obtained for this quantity the value -0.53 ± 0.19, in agreement with the value -0.61 ± 0.20 reported by Worobey *et al.*[83] and the value -0.60 ± 0.04 measured by Lin and Leonard.[169a] On the

[158] R. P. Huebener, *Phys. Rev.* **140**, A1834 (1965).

[159] E. Justi, M. Kohler, and G. Lautz, *Naturwissenschaften* **38**, 475 (1951); *Z. Naturforsch.* **6a**, 456, 544 (1951).

[160] J. Savornin and G. Couchet, *C. R. Acad. Sci. Paris* **234**, 1608 (1952).

[161] J. Savornin and F. Savornin, *J. Phys. Radium* **17**, 283 (1956).

[162] F. Savornin, *Ann. Phys. (Paris)* **5**, 1355 (1960).

[163] F. Savornin, J. Savornin, and A. Donnadieu, *C. R. Acad. Sci. Paris* **254**, 3348 (1962).

[164] L. Reimer, *Z. Naturforsch.* **12a**, 525 (1957).

[165] R. Nossek, *Z. Naturforsch.* **16a**, 1162 (1961).

[166] R. P. Huebener, *Phys. Rev.* **136**, A1740 (1964).

[167] J. Gouault, *J. Phys.* **28**, 931 (1967).

[168] J. Gouault and M. Hubin, *C. R. Acad. Sci. Paris* **B265**, 1478 (1967).

[169] R. K. Angus and I. D. Dalgliesh, *Phys. Lett.* **31A**, 280 (1970).

[169a] S. F. Lin and W. F. Leonard, *J. Appl. Phys.* **42**, 3634 (1971).

other hand, Angus and Dalgliesh[169] found the value +0.9 in Au near room temperature. The positive value of $(\partial \ln l/\partial \ln E)_{E_F}$ reported by the latter authors is supported by their experimental values of +2.3 and +2.7 for Cu and Ag, respectively. Gouault[167] found in Cu the value +2.1. The positive values of $(\partial \ln l/\partial \ln E)_{E_F}$ measured by Angus and Dalgliesh and by Gouault in the noble metals appear somewhat surprising in view of the fact that in the noble metals the positive sign of S^e is likely to be explained by a negative value of the derivative $(\partial \ln l/\partial \ln E)_{E_F}$.[32] For the alkalis K, Rb, and Cs Nossek[165] reported the values 1.7, 2.1, and 2.5, respectively, for the derivative $(\partial \ln l/\partial \ln E)_{E_F}$.

From measurements with thin Pt foils Huebener[158] obtained $(\partial \ln l/\partial E)_{E_F} = 3.4 \pm 0.8(\text{eV})^{-1}$, corresponding to $(\partial \ln l/\partial \ln E)_{E_F} = 30 \pm 7$ with the Fermi energy of 8.9 eV.[170] Recent measurements[171] with evaporated films of Pd and Pt yielded similar results. The large positive derivative $(\partial \ln l/\partial \ln E)_{E_F}$ in these metals can be understood qualitatively with Mott's model for s–d scattering (Eq. (15.4)). Holwech and Sollien[106] studied the low temperature thermopower in thin foils of dilute alloys of Al with Fe and Cu and found a strong energy dependence of the electron scattering by the impurity admixture.

The *size effect on the phonon drag thermopower* has been investigated in Pt.[158] From these experiments the phonon mean free path in Pt has been determined as a function of temperature. A reduction of S^g due to phonon scattering at the specimen surface has also been observed in Au films.[83] Wiebking[172] recently studied the thermopower in thin films prepared by condensation at liquid He temperatures for a series of metals and observed a complete suppression of S^g. Clearly, in his specimens thermoelectricity is dominantly influenced by structural lattice defects.

VIII. Superconductors

17. Motion of the Magnetic Flux Structure

Until a few years ago it was generally believed that, together with the electrical resistivity, thermoelectricity disappears in superconductors below their superconducting transition temperature.[173] However,

[170] J. B. Ketterson, F. M. Mueller, and L. R. Windmiller, *Phys. Rev.* **186**, 656 (1969).
[171] C. Reale, *J. Less Common Metals* **12**, 167 (1969).
[172] H. Wiebking, *Z. Phys.* **232**, 126 (1970).
[173] D. Shoenberg, "Superconductivity," p. 87. Cambridge Univ. Press, London and New York, 1952.

recently resistive voltages and the thermoelectric effects have been observed in superconductors in the intermediate and the mixed state, i.e., in the presence of a magnetic field. These transport phenomena below the superconducting transition temperature are now understood in terms of the motion of the magnetic flux structure under the influence of an electric current or a temperature gradient. In the absence of a magnetic field the earlier notion still appears to be correct, that the electrical resistance and thermoelectricity vanish below the superconducting transition temperature. The nature of the forces on the magnetic structure due to an electric current or a temperature gradient is such that among the thermoelectric phenomena the transverse effects (Nernst and Ettingshausen effect) are usually by far larger than the longitudinal effects (Seebeck and Peltier effect). For this reason we include in this section the discussion of the Nernst and Ettingshausen effect, although in the remainder of this article we have restricted ourselves to the longitudinal phenomena only.

In 1952 Shoenberg[174] noted that the arrangement of the normal domains in the intermediate state of a type I superconductor would be unstable in the presence of an electrical current, since the normal domains would move in a direction perpendicular to the current and the magnetic field. He appears to have been the first to suggest the phenomenon of current induced flux motion in a superconductor. Gorter[175] first proposed flux flow as a mechanism for generating a resistive voltage in a type I superconductor. The concept of flux motion was later extended by Gorter[176] and Anderson[177] to flux lines in a type II superconductor. The first experimental evidence for the appearance of electrical resistance due to flux flow (flux flow resistance) was obtained by Kim *et al.*[178,179] in the mixed state of type II superconductors. Shortly afterwards the Hall effect was observed in the mixed state.[180–182] The results on the flux flow resistivity and the Hall effect in the mixed state have been summarized recently by Kim and Stephen.[183] Flux motion

[174] Shoenberg,[173] p. 113.

[175] C. J. Gorter, *Physica* **23**, 45 (1957).

[176] C. J. Gorter, *Phys. Letters* **1**, 69 (1962); **2**, 26 (1962).

[177] P. W. Anderson, *Phys. Rev. Lett.* **9**, 309 (1962).

[178] Y. B. Kim, C. F. Hempstead, and A. R. Strnad, *Phys. Rev.* **131**, 2486 (1963); **139**, A1163 (1965).

[179] Y. B. Kim, C. F. Hempstead, and A. R. Strnad, *Rev. Mod. Phys.* **36**, 43 (1964).

[180] A. K. Niessen and F. A. Staas, *Phys. Lett.* **15**, 26 (1965).

[181] F. A. Staas, A. K. Niessen, and W. F. Druyvesteyn, *Phys. Lett.* **17**, 231 (1965).

[182] W. A. Reed, E. Fawcett, and Y. B. Kim, *Phys. Rev. Lett.* **14**, 790 (1965).

[183] Y. B. Kim and M. J. Stephen, *in* "Superconductivity" (R. D. Parks, ed.), Vol. 2, p. 1107. Dekker, New York, 1969.

in type I superconductors induced by an electrical current has been treated recently by Solomon.[184]

Following the observation of the electrical resistance and of the Hall effect in the mixed state, Volger[185] suggested the existence of the thermomagnetic effects in the flux flow region. Shortly afterwards the existence of the Peltier effect,[186,187] the Ettingshausen effect,[187,188] and the Nernst effect[189] in the mixed state was experimentally established. Further, the Ettingshausen effect was observed in the intermediate state of a type I superconductor,[188] and the Nernst effect was found in thin films of Pb, Sn, and In.[190]

18. Forces on Flux Lines and Entropy Transport

We describe now phenomenologically the motion of a flux line or a flux tube induced by an electric current or a temperature gradient. We consider a flat superconductor the two broad surfaces of which are located in a xy plane [Fig. 22(a)]. The magnetic field H is oriented in the

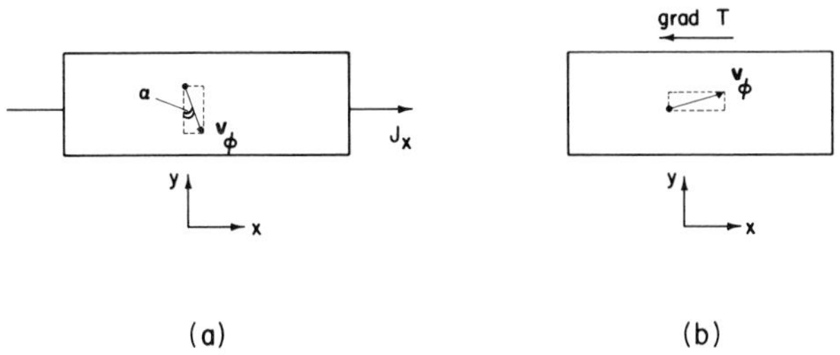

Fig. 22. Sample geometry during flux motion due to an electric current (a) or a temperature gradient (b).

z direction, and an *electric current* with the density $\mathbf{J}_x$ flows in the superconductor in the x direction. In the mixed state, the plate contains a triangular lattice of flux lines each carrying one flux quantum. In the intermediate state the magnetic flux will be distributed in some arrangement of flux tubes which contain more than one and possibly very many

[184] P. R. Solomon, *Phys. Rev.* **179**, 475 (1969).
[185] J. Volger, *in* "Quantum Fluids" (D. F. Brewer, ed.), p. 128. Wiley, New York, 1966.
[186] A. T. Fiory and B. Serin, *Phys. Rev. Lett.* **16**, 308 (1966).
[187] A. T. Fiory and B. Serin, *Phys. Rev. Lett.* **19**, 227 (1967).
[188] F. A. Otter, Jr. and P. R. Solomon, *Phys. Rev. Lett.* **16**, 681 (1966); P. R. Solomon and F. A. Otter, Jr., *Phys. Rev.* **164**, 608 (1967).
[189] J. Lowell, J. S. Munoz and J. B. Sousa, *Phys. Lett.* **24A**, 376 (1967).
[190] R. P. Huebener, *Phys. Lett.* **24A**, 651 (1967); **25A**, 588 (1967).

flux quanta. Under stationary conditions in the presence of the current density $\mathbf{J}_x$, the magnetic flux system moves with the velocity $\mathbf{v}_\varphi$ at the Hall angle α. The equation for the forces acting on the moving vortex line can be written as

$$(\mathbf{J}_x \times \boldsymbol{\varphi}) - f n_s e(\mathbf{v}_\varphi \times \boldsymbol{\varphi}) - \eta \mathbf{v}_\varphi - \mathbf{f}_p = 0. \tag{18.1}$$

The Lorentz force $(\mathbf{J}_x \times \boldsymbol{\varphi})$ is compensated by the Magnus force $f n_s e(\mathbf{v}_\varphi \times \boldsymbol{\varphi})$, the damping force $\eta \mathbf{v}_\varphi$ and the pinning force $\mathbf{f}_p$. Here φ is the flux contained in a vortex line, and n_s the density of superconducting electrons. The constant f indicates the fraction of the Magnus force that is active.[191,192] Except for extremely pure superconductors, f is much smaller than 1; η is a scalar damping factor. The pinning force $\mathbf{f}_p$ is directed opposite to the flux flow velocity $\mathbf{v}_\varphi$. According to Faraday's law, the magnetic flux system moving with velocity $\mathbf{v}_\varphi$ causes the electric field

$$\mathbf{E} = -\text{grad } V = -(\mathbf{v}_\varphi \times \mathbf{B}), \tag{18.2}$$

where B is the magnetic flux density. With the geometry of Fig. 22(a), the component $v_{\varphi y}$ causes the (longitudinal) resistive voltage, whereas the component $v_{\varphi x}$ causes the (transverse) Hall voltage.

The motion of the magnetic flux structure across the specimen is associated with the transport of entropy. If the magnetic flux structure moves as shown in Fig. 22(a), flux lines or flux tubes will be created at the upper and left edges of the sample and will be annihilated at the opposite ends. Since the creation of a flux line in the superconducting environment requires positive entropy, heat energy will be absorbed from the lattice at the upper and left edges of the sample and emitted at the opposite ends. As a result, a transverse and longitudinal temperature gradient will be established, in analogy to the Ettingshausen and Peltier effect.

The heat current density $\mathbf{U}$ coupled with the motion of the flux system is

$$\mathbf{U} = nTS_\varphi \mathbf{v}_\varphi . \tag{18.3}$$

Here n is the flux line density ($n = B/\varphi$), and S_φ the transport entropy per unit length of the flux line or flux tube. Under stationary conditions the heat current density (18.3) is compensated by heat conduction, i.e., we have for the component in the y direction

$$U_y = nTS_\varphi v_{\varphi y} = -\kappa\, \partial T/\partial y. \tag{18.4}$$

[191] A. G. Van Vijfeijken and A. K. Niessen, *Philips Res. Rep.* **20**, 505 (1965); *Phys. Lett.* **16**, 23 (1965).

[192] A. G. Van Vijfeijken, *Philips Res. Rep. Suppl.* No. 8, 1 (1968).

With the relation (18.2) we find

$$\frac{S_\varphi}{\varphi} = \frac{\kappa}{T} \cdot \left| \frac{\partial T}{\partial y} \Big/ \frac{\partial V}{\partial x} \right|. \tag{18.5}$$

We note that the transverse temperature gradient is proportional to the longitudinal electric field, the proportionality constant being determined by S_φ/φ, κ, and T. Relation (18.5) is illustrated in Fig. 23. If $\partial T/\partial y$ and

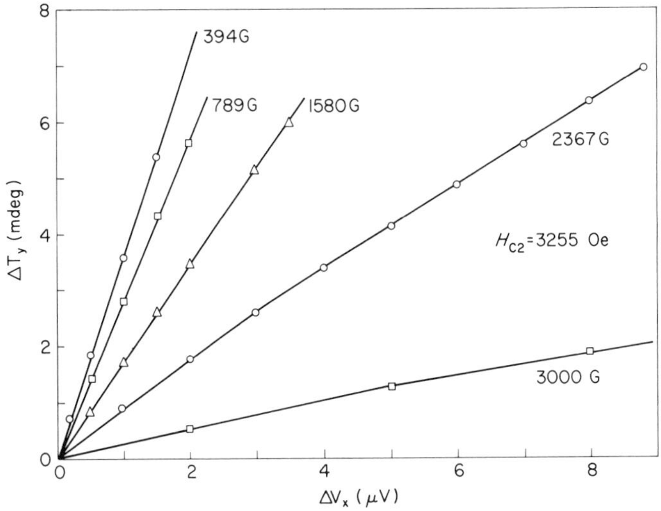

FIG. 23. The transverse temperature difference caused by the Ettingshausen effect versus the longitudinal flux flow voltage for the type II alloy In + 40 at. % Pb at different magnetic fields [after P. R. Solomon and F. A. Otter, Jr., *Phys. Rev.* **164**, 608 (1967)].

$\partial V/\partial x$ is measured in an Ettingshausen experiment, and if the heat conductivity is known, the transport entropy per unit length and per unit flux, S_φ/φ, can be determined.

From Eq. (18.1) we see, that without the Magnus force the component $v_{\varphi x}$ would be zero and the Hall and the Peltier effect would vanish.

Next we consider a flat superconductor with the same orientation as before, however instead of an electric current we apply a *temperature gradient* in the $-x$ direction [Fig. 22(b)]. In a temperature gradient, the forces acting on a moving vortex satisfy the equation

$$-S_\varphi \operatorname{grad} T - fn_{\mathrm{s}}e(\mathbf{v}_\varphi \times \boldsymbol{\varphi}) - \eta \mathbf{v}_\varphi - \mathbf{f}_{\mathrm{p}} = 0. \tag{18.6}$$

The thermal force $-S_\varphi \operatorname{grad} T$ is compensated by the Magnus force, the damping force, and the pinning force. Equation (18.6) is valid only for small values of grad T, such that the superconducting properties of the specimen are approximately constant along the direction of the temperature gradient.[193] The thermal force $-S_\varphi$ grad T in (18.6) is fundamental to all thermal-diffusion phenomena.[21] The fact that the entropy density in the core of a flux line or within a flux tube is higher than in the surrounding superfluid implies that the thermal force $-S_\varphi$ grad T always exists and that it is oriented from the hot end to the cold end of the sample. It can be shown from the Onsager relations, that the quantity S_φ in the expression for the thermal force is identical with the transport entropy S_φ, which is defined by Eq. (18.3).[193]

From Eq. (18.6) we see that, in general, a temperature gradient in $-x$ direction will cause a motion of the flux system in the x- and y-directions. The velocity components $v_{\varphi x}$ and $v_{\varphi y}$ will then lead to a transverse and longitudinal electric field, in analogy to the Nernst and Seebeck effect, respectively. The voltages due to the Nernst and Seebeck effect are expected to increase proportional to the temperature gradient for small values of grad T. This is illustrated for the Nernst effect in Fig. 24. We

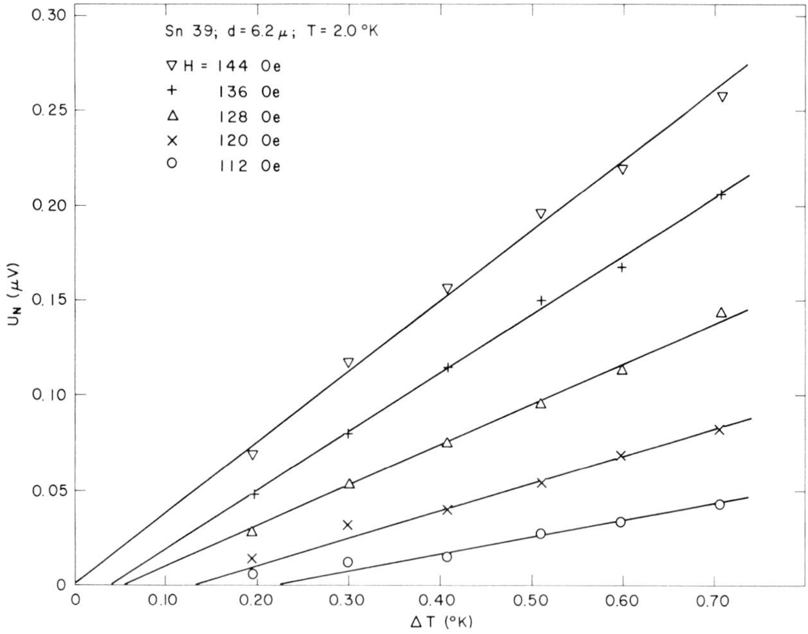

FIG. 24. Nernst voltage versus the longitudinal temperature difference in a tin film with 6.2 μ thickness at 2.0°K and different magnetic fields [after V. A. Rowe and R. P. Huebener, *Phys. Rev.* **185**, 666 (1969)].

[193] R. P. Huebener and A. Seher, *Phys. Rev.* **181**, 701 (1969).

note from Eq. (18.6), that without the Magnus force the component $v_{\varphi y}$ would be zero and the Seebeck effect would vanish.

We have seen that the longitudinal effects (Peltier and Seebeck effect) occur only if the Hall angle is finite, i.e., if the motion of the magnetic flux structure has a finite component perpendicular to the driving force. This is opposite to the situation in a normal metal, where the transverse effects require a finite Hall angle. According to Eqs. (18.1) and (18.6), for very pure superconductors with $f = 1$ and $\eta = 0$ the flux lines move perpendicular to the driving force, and we find only the Hall voltage and the longitudinal thermomagnetic effects. For $\eta \gg fn_s e\varphi$, the flux lines move predominantly in the direction of the driving force, and the transverse thermomagnetic effects will be dominant.

The size of the Magnus force and thereby of the Hall angle depends critically on the microscopic model for the damping mechanism. This question has been treated in a number of theoretical papers,[191,192,194,195] which predicted different values for the Hall angle. Experimentally the situation appears to be far from clear.[196–198] However, the Hall angle is generally small, 10^{-3} or less, except for extremely pure substances.[196–201]

If we neglect the Magnus force and the Hall effect, Eqs. (18.1) and (18.6) are replaced by

$$(\mathbf{J}_x \times \boldsymbol{\varphi}) - \eta \mathbf{v}_\varphi - \mathbf{f}_\mathrm{p} = 0 \tag{18.7}$$

and

$$-S_\varphi \operatorname{grad} T - \eta \mathbf{v}_\varphi - \mathbf{f}_\mathrm{p} = 0. \tag{18.8}$$

From the measurement of the Nernst effect and the flux flow resistivity, the transport entropy S_φ can then be obtained in the following way.[193] From Eq. (18.8), together with Faraday's law, we see that the transverse electric field $\partial V/\partial y$, arising from the Nernst effect, plotted versus $\partial T/\partial x$ should result in a straight line, the slope of which is equal to

$$S_1 = S_\varphi B/\eta. \tag{18.9}$$

[194] J. Bardeen and M. J. Stephen, *Phys. Rev.* **140**, A1197 (1965).
[195] P. Nozières and W. F. Vinen, *Phil. Mag.* **14**, 667 (1966).
[196] A. T. Fiory and B. Serin, *Proc. Int. Conf. Sci. Superconduct. Stanford Univ.*, 1969, *Physica* **55**, 73 (1971).
[197] C. H. Weijsenfeld, *Physica* **45**, 241 (1969).
[198] B. Byrnak and F. B. Rasmussen, *Proc. Int. Conf. Sci. Superconduct. Stanford Univ.*, 1969, *Physica* **55**, 357 (1971).
[199] Y. Muto, K. Noto, M. Hongo, and K. Mori, *Phys. Lett.* **30A**, 480 (1969).
[200] A. T. Fiory and B. Serin, *Phys. Rev. Lett.* **21**, 359 (1968).
[201] B. W. Maxfield, *Solid State Commun.* **5**, 585 (1967).

From Eq. (18.7) we note that a plot of the resistive voltage $\partial V/\partial x$ versus $\mathbf{J}_x$ should yield a straight line with the slope

$$S_2 = \varphi B/\eta. \tag{18.10}$$

The transport entropy per unit length and per unit flux is then found as

$$S_\varphi/\varphi = S_1/S_2\,. \tag{18.11}$$

The scheme to determine the transport entropy from the slopes S_1 and S_2 according to Eq. (18.11) requires that the Lorentz force and the thermal force are considerably larger than the pinning force, and that the flux system is in the linear flux flow regime. The Nernst and Seebeck effect can be complicated by the fact, that sometimes relatively large temperature gradients are necessary to overcome flux pinning thus leading to appreciable variation of the superconducting sample properties along the direction of the temperature gradient. On the other hand, in the measurement of the Ettingshausen and Peltier effect the influence of flux pinning is eliminated, since the temperature gradient can be obtained directly as a function of the flux flow voltage. Further during these experiments the sample is almost at uniform temperature, except for the small temperature variation associated with the Ettingshausen and Peltier effect itself. For these reasons, the Ettingshausen and Peltier effect appear to be more useful experimentally than the Nernst and Seebeck effect. Usually the Nernst and Ettingshausen effect in the mixed or intermediate state are about 2–3 orders of magnitude larger than in normal metals and can be distinguished from the normal contribution without difficulty. For the Peltier effect the normal contribution and the contribution from flux flow can have similar magnitudes below the superconducting transition temperature.[187] The same is to be expected for the Seebeck effect. However, measurements of the Seebeck effect in the mixed or intermediate state have not been reported.

19. Experiments and Comparison with Theory

After the existence of the thermomagnetic effects in the mixed or the intermediate state had been demonstrated experimentally,[186–190] Solomon and Otter[202] and Lowell *et al.*[203] studied both the Nernst and Ettingshausen effect in a dirty type II superconductor (Nb–Mo alloy). Muto

[202] P. R. Solomon and F. A. Otter, Jr., *Proc. Int. Conf. Low Temp. Phys.*, 11*th*, *St. Andrews*, 1968 Vol. II, p. 841.

[203] J. Lowell, J. S. Munoz, and J. B. Sousa, *Proc. Int. Conf. Low Temp. Phys.*, 11*th*, *St. Andrews*, 1968 Vol. II, p. 858; *Phys. Rev.* **183**, 497 (1969).

et al.[204] measured the Ettingshausen effect in a similar alloy. Serin and Fiory[196,205] investigated the Peltier and Ettingshausen effect in Nb and V. Measurements of the Nernst effect were carried out by Huebener and Seher[193] in Nb. The latter authors also studied the Nernst and Ettingshausen effect in thin films of Pb, Sn, and In.[206–210] Vidal *et al.*[211] recently studied the Ettingshausen effect in a type II superconductor by an ac method using a second-sound technique.

It is convenient to express the experimental results on the thermomagnetic effects in superconductors in terms of the transport entropy, S_φ, per flux line or flux tube of unit length. As a function of temperature, S_φ is found to increase with temperature at low temperatures, to pass through a maximum and to vanish at the critical temperature T_C. As a function of magnetic field, S_φ is found to decrease gradually from a maximum value at low fields to zero at the critical field H_C or H_{C2}. This behavior is illustrated in the three-dimensional plot shown in Fig. 25.

For a type I superconductor at low magnetic fields the transport entropy can readily be obtained from the difference in the entropy density of normal and superconducting material[207] and is given by

$$S_\varphi = (H_C(0)/2\pi)(T/T_c^2) \cdot \varphi. \tag{19.1}$$

Here$H_C(0)$ is the critical field at zero temperature. Equation (19.1) is valid only if the flux tubes are well separated from each other and if they are sufficiently large such that the surface energy contribution can be neglected. With increasing magnetic field, the normal regions in a type I superconductor become less and less localized until, at H_C, the whole specimen consists of a single piece of normal material. Therefore, S_φ vanishes as we approach H_C. The maximum in S_φ at intermediate temperatures can be understood qualitatively from Eq. (19.1). At low temperatures S_φ increases with temperature because of the factor T in

[204] Y. Muto, K. Mori, and K. Noto, *Proc. Int. Conf. Sci. Superconduct., Stanford Univ.*, 1969, *Physica* **55**, 362 (1971).

[205] B. Serin and A. T. Fiory, *Proc. Int. Conf. Low Temp. Phys.*, 11*th*, *St. Andrews*, 1968 Vol. II, p. 886.

[206] R. P. Huebener, V. A. Rowe, and A. Seher, *Proc. Int. Conf. Low Temp. Phys.*, 11*th*, *St. Andrews*, 1968 Vol. II, p. 837.

[207] R. P. Huebener and A. Seher, *Phys. Rev.* **181**, 710 (1969).

[208] V. A. Rowe and R. P. Huebener, *Phys. Rev.* **185**, 666 (1969).

[209] R. P. Huebener and V. A. Rowe, *Solid State Commun.* **7**, 1763 (1969).

[210] V. A. Rowe and R. P. Huebener, *Bull. Amer. Phys. Soc.* **15**, 343 (1970).

[211] F. Vidal, Y. Simon, M. Le Ray, and P. Thorel, *Rev. Phys. Appl.* **4**, 50 (1969); Y. Simon and F. Vidal, *Phys. Lett.* **30A**, 109 (1969).

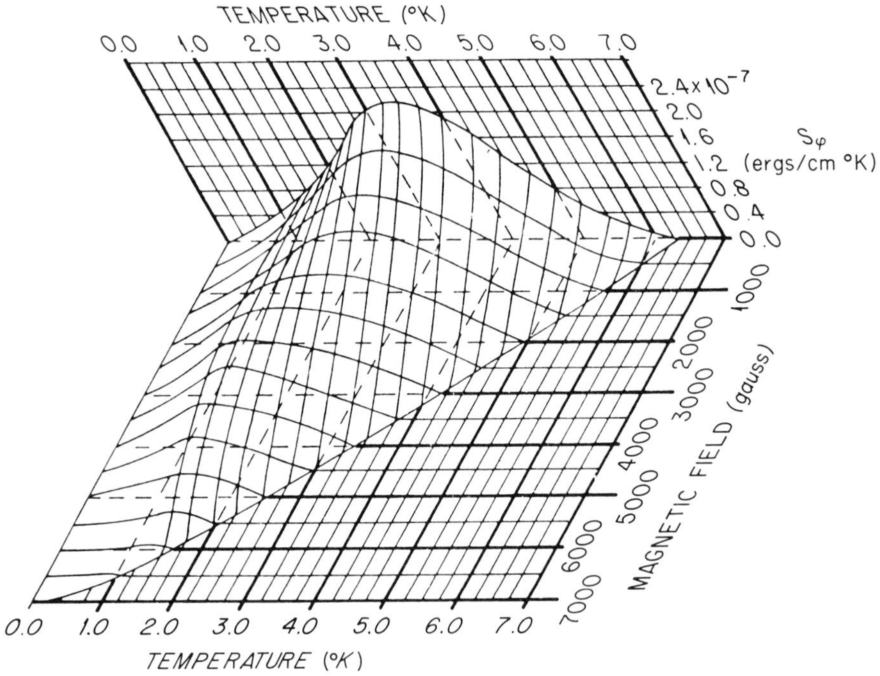

FIG. 25. Three-dimensional plot of the transport entropy S_φ per flux line of unit length versus B and T for the type II alloy In + 40 at. % Pb [after P. R. Solomon and F. A. Otter, Jr., *Phys. Rev.* **164**, 608 (1967)].

Eq. (19.1). As we approach T_C, the area occupied by a flux tube becomes larger, the normal regions are again less and less localized and S_φ vanishes. In thin films of type I superconductors the transport entropy was found to pass through a maximum as a function of film thickness.[207,208,210] This behavior can be understood from the variation of the magnetic flux structure with film thickness. The experimental values of the transport entropy in type I superconductors appear to agree reasonably with Eq. (19.1) at not too high temperatures and magnetic fields. For lead at 4.2°K Eq. (19.1) yields the value $S_\varphi/\varphi = 10.4$ G/°K. Clem[212] suggested the existence of local temperature gradients due to flux tube motion, leading to an expression for the transport entropy of the form given in Eq. (19.1) but multiplied with the correction factor

$$g = \frac{[1 - H/H_C(T)][1 - \kappa_S/\kappa_n]}{1 - [H/H_C(T)][1 - \kappa_S/\kappa_n]} \,. \tag{19.2}$$

[212] J. R. Clem, *Phys. Rev.* **176**, 531 (1968).

Here κ_s and κ_n are the heat conductivity in the superconducting and normal state, respectively. The expression for S_φ corrected in this way reduces to zero at the critical field. However, its value at low magnetic fields and its field dependence appears to be inconsistent with experiment.[210]

For a type II superconductor the transport entropy can readily be calculated again for the case of well separated individual flux lines, i.e., for *small magnetic fields*. Stephen[213] and Van Vijfeijken[214] calculated S_φ using simple thermodynamic arguments. According to Stephen[213] for small magnetic fields the transport entropy per flux line of unit length is given by

$$S_\varphi = -(1/4\pi)(\partial H_{C1}/\partial T) \cdot \varphi_0 , \tag{19.3}$$

where H_{C1} is the lower critical field and φ_0 the flux quantum. The models of Stephen and Van Vijfeijken are both in reasonable agreement with experiment[188,196] well below T_C. However, they fail to reproduce the reduction of S_φ to zero as one approaches T_C. For Nb at 4.2°K Eq. (19.3) yields the value[193] $S_\varphi/\varphi_0 = 13$ G/°K. A phenomenological theory for the thermomagnetic effects in the mixed state has been derived by Ohta.[215]

The first microscopic theory of the thermomagnetic effects in type II superconductors near the upper critical field, H_{C2}, was developed by Caroli and Maki.[216] Subsequently, it was refined and extended.[217–220] The theory is based on the time dependent Ginzburg Landau equations and is a generalization of earlier work by Schmid.[221] Maki's theory correctly predicts that S_φ approaches zero at $T = 0$ and $T = T_C$ and that S_φ goes to zero at H_{C2}. For the magnetic field dependence of S_φ near the upper critical field, Maki[218] obtained the following expression in the dirty limit, defined by $l \ll \xi_0$, where l is the electron mean free path and ξ_0 the BCS coherence length,

$$-H_{C2}\left(\frac{\partial S_\varphi(t)}{\partial B}\right)_{H_{C2}} = \frac{\hbar c H_{C2}(t)\, L_D(t)}{4eT[1.16(2\kappa_2^2(t) - 1) + 1]} \tag{19.4a}$$

$$= \frac{2(2\pi)^2\, \hbar k_B}{e^2} \frac{\sigma\kappa_1^2(0)\, \Phi(0)}{[1.16(2\kappa_2^2(t) - 1) + 1]} . \tag{19.4b}$$

213 M. J. Stephen, *Phys. Rev. Lett.* **16**, 801 (1966).
214 A. G. Van Vijfeijken, *Phys. Lett.* **23**, 65 (1966).
215 T. Ohta, *Jap. J. Appl. Phys.* **6**, 645 (1967).
216 C. Caroli and K. Maki, *Phys. Rev.* **164**, 591 (1967).
217 K. Maki, *Phys. Rev. Lett.* **21**, 1755 (1968).
218 K. Maki, *J. Low Temp. Phys.* **1**, 45 (1969).
219 K. Maki, *Progr. Theoret. Phys.* **41**, 902 (1969).
220 K. Maki, *Proc. Int. Conf. Sci. Superconduct., Stanford Univ.*, 1969, *Physica* **55**, 124 (1971).
221 A. Schmid, *Phys. Kondens. Mater.* **5**, 302 (1966).

Here, t is the reduced temperature T/T_C, σ the dc conductivity, and $\kappa_1(0)$ the Ginzburg Landau parameter at zero temperature; κ_2 is the temperature dependent parameter introduced by Maki[222]; $L_D(t)$ and $\Phi(t)$ are universal functions. The expression given in Eq. (19.4) is in excellent agreement with the experimental results for Nb–Mo alloys.[203,204,223] The refined Caroli–Maki theory for the pure limit,[219] defined by $l \gg \xi_0$, shows only qualitative agreement with experiment. A summary of the results for pure type II superconductors has been given recently by Serin.[196]

Maki[224] extended his theory of vortex motion in superconductors to the intermediate state of type I material. However, the experimental results on the transport entropy[210] appear to be inconsistent with his model.

ACKNOWLEDGMENT

The author benefitted greatly from numerous conversations with John E. Robinson, who freely gave his time for advice and discussions.

[222] K. Maki, *Physics* **1**, 21, 127 (1964).
[223] F. A. Otter, Jr. and P. R. Solomon, *Proc. Int. Conf. Sci. Superconduct., Stanford Univ.*, 1969, *Physica* **55**, 351 (1971).
[224] K. Maki, *Progr. Theoret. Phys.* **42**, 448 (1969).

Dynamical Diffraction Theory by Optical Potential Methods

P. H. DEDERICHS

Institut für Festkörperforschung der Kernforschungsanlage Jülich
Jülich, West Germany

I. Introduction

Two theories are widely used to describe the intensities observed in electron or X-ray diffraction by crystals. The "kinematical theory" treats the crystal as perturbation and is therefore valid only for sufficiently small crystals. For larger crystals one has to take into account the multiple scattering of the incident wave. This problem, which is simplified substantially by the periodicity of the crystal, has been dealt with first by Darwin in 1914.[1] More fundamentally, the problem has been treated in a series of papers by Ewald in 1916 and 1917[2] and later on by von Laue.[3] These papers form the basis of the so-called "dynamical theory," which has been extended further by Bethe[4] for the case of electron diffraction.

The dynamical theory for X-ray diffraction is summarized in the books of von Laue,[5] Zachariasen,[6] and James[7] as well as in two more recent review articles of Batterman and Cole[8] and James.[9] For electron diffraction we refer to the books of von Laue,[10] Heidenreich,[11] and Hirsch *et al.*[12]

In the last twenty years there has been a renewed and steadily increasing interest in dynamical diffraction of X-rays and electrons, which is due partly to the availability of large, perfect crystals and to the development of the electron microscope. New branches have evolved, such as low-energy-electron-diffraction (LEED), channeling of high energy electrons, and positrons or dynamical scattering of Mössbauer quanta. Also, the dynamical theory has made considerable progress. For instance, starting with the papers of Molière[13] and Yoshioka,[14] the theory

[1] C. G. Darwin, *Phil. Mag.* **27**, 315, 675 (1914).

[2] P. P. Ewald, *Ann. Phys.* **49**, 1, 117 (1916); **54**, 519 (1917).

[3] M. von Laue, *Ergeb. Exakt. Naturwiss.* **10**, 133 (1931).

[4] H. Bethe, *Ann. Phys.* **87**, 55 (1928).

[5] M. von Laue, "Röntgenstrahl-Interferenzen." Akademie-Verlag, Frankfurt, 1960.

[6] W. H. Zachariasen, "Theory of X-Ray Diffraction in Crystals." Dover, New York, 1945.

[7] R. W. James, "The Optical Principles of the Diffraction of X-Rays." G. Bell and Sons, London, 1950.

[8] B. W. Batterman and H. Cole, *Rev. Mod. Phys.* **36**, 681 (1964).

[9] R. W. James, *Solid State Phys.* **15**, 55 (1963).

[10] M. von Laue, "Materiewellen und ihre Interferenzen." Akademie-Verlag, Leipzig, 1944.

[11] R. D. Heidenreich, "Fundamentals of Transmission Electron Microscopy." Interscience, New York, 1964.

[12] P. W. Hirsch, A. Howie, R. B. Nicholson, D. W. Pashley, and M. J. Whelan, "Electron Microscopy of Thin Crystals." Butterworths, London, 1965.

[13] G. Molière, *Ann. Phys.* **35**, 272, 297 (1939).

[14] H. Yoshioka, *J. Phys. Soc. Japan* **12**, 618 (1957).

for the elastic or "coherent" wave could be sufficiently generalized to take into account the effects of inelastic waves, thermal motion, or statistical defects.

First, in Sections 2 and 3, we shall review the conventional form of the dynamical theory. We shall present the basic principles in a short, but self-contained way and shall emphasize some new methods such as the band structure for complex wave vectors and the t-matrix method. Moreover, we develop at the same time the theory for electron, neutron and X-ray diffraction by a finite crystal, i.e., the matching of the wave fields, the calculations of diffracted intensities, the discussions of Laue and Bragg cases, etc. These topics have been dealt with by a number of review articles.[5–12,14a] Second, in Sections 4–7, we shall review the theory for the coherent wave, which is known as the optical potential method in nuclear physics. We shall explicitly derive the corrections to the potential coefficients due to inelastic waves, thermal motion, and statistically distributed defects, both for electron and X-ray diffraction.

Since this is a purely theoretical article, we shall not give long lists of tables of atomic form factors, wavelength, etc. Nevertheless, we feel obliged to give the reader who is not familiar with dynamical diffraction an idea of the order of magnitude of the most important quantities. Therefore, the following table gives some typical values of the energy E, etc. for the cases of neutron, X-ray, electron, and low-energy-electron diffraction, which are abbreviated by the symbols n, X, e, and LEED, respectively. The most important quantity for diffraction is the extinction length, which is essentially the thickness of the crystal for which the kinematical theory breaks down. For neutrons and LEED, the extinction lengths differ by a factor 10^5, meaning that the dynamical theory is absolutely necessary for LEED, but that most experiments with neutrons and X-rays are well described by the kinematical theory. For electrons and X-rays the absorption length $1/\mu$, given in the fourth line of the following table, is roughly a factor 10 larger than the extinction length. Here, neutrons are an exception since they are not absorbed for all practical purposes. The last quantity $1/\Delta\mu$ is the absorption length for the case when a Bragg reflection is excited (anomalous transmission). For X-rays, the absorption is then reduced by a factor ≈ 30, known as the Bormann effect, whereas the absorption of electrons is only slightly reduced.

[14a] A more detailed treatment of the subjects in Sections 2 and 3 of these articles, including also discussions of boundary conditions, Laue and Bragg cases, etc., is given in P. H. Dederichs, KFA-JÜL-Report, Jül.-797-FF 1971, which is available from Zentralbibliothek der Kernforschungsanlage, 517 Jülich, Germany.

Energy, E	n (10 meV)	X (10 keV)	e (100 keV)	LEED (100 eV)
Wavelength, λ	1 Å	1 Å	0.05 Å	1 Å
Extinction length, d_{ext}	10^5 Å	10^4 Å	10^2–10^3 Å	5 Å
Absorption length, $1/\mu$	10^8 Å	10^5 Å	10^3–10^4 Å	10 Å
$1/\Delta\mu$	$>10^8$ Å	$30 \cdot 10^5$ Å	$3 \cdot (10^3$–$10^4)$ Å	10 Å

II. Diffraction of Electrons

In this part we shall first review the basic facts and theorems of the band theory as far as they are important for diffraction. For more detail, the reader is referred to the literature.[15] In particular, we shall discuss the band structure for complex **k** vectors. Furthermore, we shall consider the two-beam and some simple multiple beam cases.

1. Bloch Waves

The motion of an electron in a crystal is governed by the periodic lattice potential $V(\mathbf{r}) = V(\mathbf{r} + \mathbf{R})$, which is invariant with respect to lattice translations $\mathbf{R}$. $V(\mathbf{r})$ can be split up into the contributions $v(\mathbf{r} - \mathbf{R}_n)$ of the different unit cells at $\mathbf{R}_n$.

$$V(\mathbf{r}) = \sum_n v(\mathbf{r} - \mathbf{R}_n) = \sum_n \left\{ \frac{-Ze^2}{|\mathbf{r} - \mathbf{R}_n|} + \int_{V_c} \frac{e^2\rho(\mathbf{r}')}{|\mathbf{r} - \mathbf{R}_n - \mathbf{r}'|} \, d\mathbf{r}' \right\}. \quad (1.1)$$

Here $\rho(\mathbf{r})$ is the electron density in the first cell (V_c = volume of the unit cell). Because of electric neutrality the volume integral of $\rho(\mathbf{r})$ over one cell is equal to Z.

Because of the periodicity, the potential can be expanded in a Fourier series over reciprocal lattice vectors $\mathbf{h}$:

$$V(\mathbf{r}) = \sum_{\mathbf{h}} V_{\mathbf{h}} \, e^{i\mathbf{hr}} \qquad \text{with} \qquad V_{\mathbf{h}} = (1/V_c) \int_{V_c} e^{-i\mathbf{hr}} V(\mathbf{r}) \, d\mathbf{r}. \quad (1.2)$$

[15] See, e.g., W. Brauer, "Einführung in die Elektrontheorie der Metalle." Vieweg, Braunschweig, 1967. J. Callaway, "Energy Band Theory." Academic Press, New York, 1964. P. T. Landsberg, Ed., "Solid State Theory." Wiley-Interscience, London, 1969.

The coefficients $V_{\mathbf{h}}$ are evaluated by using (1.1) and transforming the sum over the different cells into an integral over the whole space.

$$V_{\mathbf{h}} = -(4\pi e^2/V_c h^2)(Z - f_{\mathbf{h}}) \quad \text{with} \quad f_{\mathbf{h}} = \int_{V_c} d\mathbf{r}\, e^{-i\mathbf{hr}}\rho(\mathbf{r}). \tag{1.3}$$

Since $f_{\mathbf{h}}$, the "atomic scattering factor for X-rays," is always smaller or equal to $f_0 = Z$, all coefficients $V_{\mathbf{h}}$ are negative. For large h only the interaction with the nucleus remains, whereas $V_{\mathbf{h}}$ approaches a constant for small $\mathbf{h}$. By expanding $f_{\mathbf{h}}$ in powers of $\mathbf{h}$, we obtain for cubic symmetry

$$V_0 = -(2\pi e^2 Z/3V_c)\langle \mathbf{r}^2 \rangle, \tag{1.4}$$

where $\langle\ \rangle$ means the average over the electron density of an atom. Since $V(\mathbf{r})$ is real, we have according to (2.2)

$$V_{\mathbf{h}} = V^*_{-\mathbf{h}}\,. \tag{1.5}$$

Furthermore, if S is a symmetry operation of the lattice, e.g., inversion, then $V(\mathbf{r}) = V(S\mathbf{r})$. For the coefficients $V_{\mathbf{h}}$, this means

$$V_{\mathbf{h}} = V_{S\mathbf{h}}, \qquad \text{e.g., } V_{\mathbf{h}} = V_{-\mathbf{h}} \quad \text{for inversion.} \tag{1.6}$$

The modifications for $V_{\mathbf{h}}$ for nonprimitive lattices are obvious. If $f_{\mathbf{h}}^{\mu}$ is the atomic scattering factor for X-rays of atom μ with the position $\mathbf{R}^{\mu}$ in the unit cell, we have

$$V_{\mathbf{h}} = -(4\pi e^2/V_c h^2) \sum_{\mu} (Z^{\mu} - f_{\mathbf{h}}^{\mu})\, e^{-i\mathbf{hR}^{\mu}}. \tag{1.7}$$

Due to the periodicity of $V(\mathbf{r})$, the Hamiltonian commutes with the translation operator $T_{\mathbf{R}} = e^{i\mathbf{pR}/\hbar}$. Therefore, the appropriate eigenfunctions are Bloch waves $\phi_{\mathbf{k}}(\mathbf{r})$, being simultaneously eigenfunctions of $T_{\mathbf{R}}$ with the eigenvalues $e^{i\mathbf{kR}}$. From this translation symmetry, it follows that the Bloch wave $\phi_{\mathbf{k}}(\mathbf{r})$ is essentially a plane wave modulated by a periodic function $u_{\mathbf{k}}(\mathbf{r})$, the so-called Bloch function.

$$\phi_{\mathbf{k}}(\mathbf{r}) = e^{i\mathbf{kr}} u_{\mathbf{k}}(\mathbf{r}) \qquad \text{with} \qquad u_{\mathbf{k}}(\mathbf{r}) = u_{\mathbf{k}}(\mathbf{r} + \mathbf{R}). \tag{1.8}$$

Because of the periodicity, $u_{\mathbf{k}}(\mathbf{r})$ can be expanded in plane waves $e^{i\mathbf{hr}}$ analogously to (1.2). Therefore, we have

$$\phi_{\mathbf{k}}(\mathbf{r}) = \sum_{\mathbf{h}} C_{\mathbf{h}}(\mathbf{k})\, e^{i(\mathbf{k}+\mathbf{h})\mathbf{r}}. \tag{1.9}$$

To determine the coefficients $C_{\mathbf{h}}$, we introduce this ansatz into the

Schrödinger equation and obtain the following infinite system of homogeneous equations:

$$\{E_{\mathbf{k}} - (\hbar^2/2m)(\mathbf{k} + \mathbf{h})^2\}\, C_{\mathbf{h}}(\mathbf{k}) = \sum_{\mathbf{h}'} V_{\mathbf{h}-\mathbf{h}'} C_{\mathbf{h}'}(\mathbf{k}). \tag{1.10}$$

For each $\mathbf{k}$ this system has solutions for a infinite number of band energies $E_\nu(\mathbf{k})$ ($\nu = 1, 2, \ldots$) for which the determinant vanishes:

$$\det \left\| \left(E_{\mathbf{k}} - \frac{\hbar^2(\mathbf{k} + \mathbf{h})^2}{2m} \right) \delta_{\mathbf{h},\mathbf{h}'} - V_{\mathbf{h}-\mathbf{h}'} \right\| = 0. \tag{1.11}$$

The Bloch waves $\phi_{\mathbf{k},\nu}(\mathbf{r})$ determined in this way form a complete and orthogonal set of functions. Further, for each $\mathbf{k}$ the Bloch functions $u_{\mathbf{k},\nu}(\mathbf{r})$ are orthonormal and complete in the space of all periodic functions. The orthonormalization condition leads to

$$(1/V_{\mathrm{c}}) \int_{V_{\mathrm{c}}} d\mathbf{r}\, u^*_{\mathbf{k},\nu'}(\mathbf{r})\, u_{\mathbf{k},\nu}(\mathbf{r}) = \delta_{\nu,\nu'} = \sum_{\mathbf{h}} C_{\mathbf{h}}{}^*(\mathbf{k}, \nu')\, C_{\mathbf{h}}(\mathbf{k}, \nu), \tag{1.12}$$

whereas completeness means

$$\delta_{\mathbf{h},\mathbf{h}'} = \sum_{\nu} C^*_{\mathbf{h}'}(\mathbf{k}, \nu)\, C_{\mathbf{h}}(\mathbf{k}, \nu). \tag{1.13}$$

Equations (1.12) and (1.13) mean that the matrix $M_{\nu,\mathbf{h}} = C_{\mathbf{h}}(\mathbf{k}, \nu)$ is unitary ($MM^\dagger = 1 = M^\dagger M$). M represents the transformation from the plane waves $e^{i\mathbf{hr}}$, i.e., the Bloch functions for $V = 0$ to the Bloch functions $u_{\mathbf{k},\nu}(\mathbf{r})$ for the potential $V(\mathbf{r})$.

An important quantity is the current density $\mathbf{j}_{\mathbf{k},\nu}(\mathbf{r})$ of a Bloch wave $\phi_{\mathbf{k},\nu}$, which is periodical in space due to the translation properties of $\phi_{\mathbf{k},\nu}$. By averaging over a unit cell and using (1.9), we obtain for the mean current density

$$\langle \mathbf{j}_{\mathbf{k}\nu}(\mathbf{r}) \rangle_{V_{\mathrm{c}}} = (\hbar/m) \sum_{\mathbf{h}} (\mathbf{k} + \mathbf{h}) |C_{\mathbf{h}}(\mathbf{k}, \nu)|^2. \tag{1.14}$$

Thus the currents of the different plane waves add incoherently. This result can be simplified further by using the Schrödinger equation in the form (1.10). Multiplying by $C_{\mathbf{h}}$, summing over $\mathbf{h}$, and differentiating with respect to $\hbar\mathbf{k}$, we obtain after some calculation

$$\langle \mathbf{j}_{\mathbf{k},\nu} \rangle_{V_{\mathrm{c}}} = \frac{1}{\hbar} \frac{\partial E_\nu(\mathbf{k})}{\partial \mathbf{k}}. \tag{1.15}$$

Therefore, the current is always perpendicular to the two-dimensional dispersion surface $E_\nu(\mathbf{k}) = \text{constant}$.

2. Complex k, E and Symmetries of $E_\nu(\mathbf{k})$

Up to now, we have tacitly assumed that the eigenvalues $\mathbf{k}$ and E are real. However, this has not necessarily to be so, because the determinant (1.11) has formal solutions also for complex $\mathbf{k}$ and E. For instance, the eigensolutions $\phi_{\mathbf{k}}$ for complex $\mathbf{k}$ are damped waves, decreasing in one direction and increasing in the opposite. But they are also eigenfunctions of $T_{\mathbf{R}}$ and H; the only, but very important, difference is that a scalar product cannot be defined due to the exponential increase.

In an infinite ideal crystal and for stationary problems, such eigenfunctions for complex $\mathbf{k}$ and E are only of pathological interest since all the Bloch waves for real $\mathbf{k}$ and E form already a complete set. However, scattering of electrons by a finite crystal produces "damped" Bloch waves (with real E) quite naturally, as will be seen in Section 3. Furthermore, for a finite crystal there are no divergence difficulties for the scalar product because the volume of integration is finite. Similarly, by considering time-dependent problems, e.g., initial value problems, complex E's can occur as decay constants. Therefore, in the following we shall allow E and k to have complex values and shall consider the symmetries of the function $E_\nu(\mathbf{k})$ in the complex E, $\mathbf{k}$ space.

First, we see from the determinant (1.11) by replacing $\mathbf{k}$ by $\mathbf{k} + \mathbf{h}''$ and then introducing $\mathbf{h} + \mathbf{h}''$ and $\mathbf{h}' + \mathbf{h}''$ as new summation indices, that for $\mathbf{k} + \mathbf{h}$, we have the same manifold of allowed energies E_1, E_2,... as for the vector $\mathbf{k}$. Therefore,

$$E_\nu(\mathbf{k}) = E_\nu(\mathbf{k} + \mathbf{h}), \tag{2.1}$$

as long as $\mathbf{k}$ and $\mathbf{k} + \mathbf{h}$ refer to the same band (i.e., belong to the same Riemann's sheet; see the following).

Further, if S is again a symmetry operation of the crystal, then $\mathbf{h}$ and $S\mathbf{h}$ are reciprocal lattice vectors. Using (1.6) and substituting $S\mathbf{h}$ and $S\mathbf{h}'$ as new summation indices in (1.11), we obtain

$$E_\nu(\mathbf{k}) = E_\nu(S\mathbf{k}), \tag{2.2}$$

showing that E has the same symmetry as the lattice. For instance, for the inversion this means $E_\nu(\mathbf{k}) = E_\nu(-\mathbf{k})$.

Further, since $V(\mathbf{r})$ is a local potential, the matrix element of $V(\mathbf{r})$ in the determinant depends only on the difference $\mathbf{h} - \mathbf{h}'$. Because the first term in (1.11) is symmetrical in $\mathbf{h}$ and $\mathbf{h}'$, we have

$$E_\nu(\mathbf{k}) = E_\nu(-\mathbf{k}) \tag{2.3}$$

independent of the existence of the inversion as a symmetry operation, which is seen by interchanging $\mathbf{h}$ and $\mathbf{h}'$ in (1.11).

The reality of the potential $V(\mathbf{r})$ leads to a further symmetry relation. By forming the complex conjugate of the determinant and using (1.5) we find

$$E(\mathbf{k}) = E^*(\mathbf{k}^*), \tag{2.4}$$

where $E(\mathbf{k})$ and $E^*(\mathbf{k})$ generally, but not necessarily refer to the same band ν. This result is especially important for the subsequent discussion.

Looking back at the symmetries (2.1)–(2.4) for $E_\nu(\mathbf{k})$, we note that similar relations are valid for the eigenvalues $e^{i\mathbf{kR}}$ of the translation operator $T_\mathbf{R}$. Namely we have

$$e^{i\mathbf{kR}} = e^{i(\mathbf{k}+\mathbf{h})\mathbf{R}} = e^{i(S\mathbf{k})(S\mathbf{R})} = e^{i(-\mathbf{k})(-\mathbf{R})} = (e^{-i\mathbf{k}^*\mathbf{R}})^*. \tag{2.5}$$

Therefore, in the absence of degeneracies, the corresponding Bloch waves are equal apart from phase factors.

$$\phi_{\mathbf{k},\nu}(\mathbf{r}) \sim \phi_{\mathbf{k}+\mathbf{h},\nu}(\mathbf{r}) \sim \phi_{S\mathbf{k},\nu}(S\mathbf{r}) \sim \phi_{-\mathbf{k},\nu}(-\mathbf{r}) \sim \phi^*_{\mathbf{k}^*,\nu}(-\mathbf{r}). \tag{2.6}$$

For special choices of the phase factor we refer to the literature.[15]

So far we have considered only purely elastic scattering by the potential $V(\mathbf{r})$. However, this is a poor approach because inelastic effects often are very important and cannot be neglected. In Part IV of this review we will see that some of the effects of inelastic scattering can be taken into account in a relatively simple way by considering only the *coherent wave.* For this coherent wave inelastic effects lead to an apparent absorption, which can be described by a socalled "*optical potential.*" As it is shown in Part IV, the "optical potential" is no longer such a simple local potential as $V(\mathbf{r})$. For instance, it is nonlocal, i.e., an integral operator

$$U\phi(\mathbf{r}) = \int d\mathbf{r}'\, U(\mathbf{r}, \mathbf{r}')\, \phi(\mathbf{r}'). \tag{2.7}$$

Moreover, it is non-Hermitean: $U \neq U^\dagger$. Most important, however, is the fact that in an infinite crystal the optical potential U is periodic.

$$U(\mathbf{r}, \mathbf{r}') = U(\mathbf{r} + \mathbf{R}, \mathbf{r}' + \mathbf{R}). \tag{2.8}$$

Therefore, Bloch's theorem, etc., remains valid. For instance, for the dispersion condition (1.11), we get for such a potential

$$\det\|\{E - (\hbar^2/2m)(\mathbf{k} + \mathbf{h})^2\}\, \delta_{\mathbf{h},\mathbf{h}'} - U_{\mathbf{k}+\mathbf{h},\mathbf{k}+\mathbf{h}'}\| = 0, \tag{2.9}$$

with

$$U_{\mathbf{k}+\mathbf{h},\mathbf{k}+\mathbf{h}'} = (1/V_c) \int_{V_c} d\mathbf{r} \int_{-\infty}^{\infty} d\mathbf{r}'\, e^{-i(\mathbf{k}+\mathbf{h})\mathbf{r}}\, U(\mathbf{r}, \mathbf{r}')\, e^{i(\mathbf{k}+\mathbf{h}')\mathbf{r}'}. \tag{2.10}$$

Here, the regions of integration for $\mathbf{r}$ and $\mathbf{r}'$ can also be interchanged due to the periodicity of U.

We shall discuss now the symmetries of $E_\nu(\mathbf{k})$ for the optical potential U. First we note that the periodicity (2.1) of $E_\nu(\mathbf{k})$ in $\mathbf{k}$ space is unchanged being a direct consequence of the periodicity of the potential. Similarly, Eq. (2.2) because of the symmetry operation S does not change either. However, the relations (2.3) and (2.4) will be modified.

According to (2.10), we get the following identities for the matrix elements of U:

$$U_{\mathbf{k}+\mathbf{h},\mathbf{k}+\mathbf{h}'} = (\hat{U})_{-\mathbf{k}-\mathbf{h}',-\mathbf{k}-\mathbf{h}} = (U^\dagger)_{\mathbf{k}+\mathbf{h}',\mathbf{k}+\mathbf{h}}\,, \tag{2.11}$$

where $\hat{U}$ is the transpose of U, and $U^\dagger$ is the Hermitian adjoint. Making the analogous substitutions as in (2.3) and (2.4), we obtain equations connecting the band structure of the potentials U, $\hat{U}$, and $U^\dagger$:

$$E_\nu^{\{U\}}(\mathbf{k}) = E_\nu^{\{\hat{U}\}}(-\mathbf{k}) = \{E_\nu^{\{U^\dagger\}}(\mathbf{k}^*)\}^*. \tag{2.12}$$

For the special cases $U = \hat{U}$ or $U = U^\dagger$, we then have symmetry relations for $E_\nu(\mathbf{k})$:

$$E_\nu(\mathbf{k}) = \begin{cases} E_\nu(-\mathbf{k}) & \text{for} \quad U = \hat{U} & (2.13\text{a}) \\ E_\nu{}^*(\mathbf{k}^*) & \text{for} \quad U = U^\dagger & (2.13\text{b}) \\ E_\nu{}^*(-\mathbf{k}^*) & \text{for} \quad U^\dagger = \hat{U} \quad (U(\mathbf{r}, \mathbf{r}')\ \text{real}). & (2.13\text{c}) \end{cases}$$

As will be shown in Part IV, the optical potential U is always symmetrical ($U = \hat{U}$). Therefore, the inversion symmetry $E_\nu(\mathbf{k}) = E_\nu(-\mathbf{k})$ is always valid.

The Bloch waves for complex $\mathbf{k}$ will generally not be orthogonal simply because the integrals diverge for complex $\mathbf{k}$. However, one can still derive an orthogonality condition connecting the eigenfunctions of U and of $U^\dagger$ by the following identity:

$$\begin{aligned} 0 &= (\phi_{\mathbf{k}^*,\nu'}^{\{U^\dagger\}}, H\phi_{\mathbf{k},\nu}^{\{U\}}) - (H^\dagger\phi_{\mathbf{k}^*,\nu'}^{\{U^\dagger\}}, \phi_{\mathbf{k},\nu}^{\{U\}}) \\ &= \{E_\nu^{\{U\}}(\mathbf{k}) - (E_{\nu'}^{\{U^\dagger\}}(\mathbf{k}^*))^*\}(\phi_{\mathbf{k}^*,\nu'}^{\{U^\dagger\}}, \phi_{\mathbf{k},\nu}^{\{U\}}). \end{aligned} \tag{2.14}$$

Because of (2.12), the scalar product has to vanish for $\nu \neq \nu'$. Therefore, we obtain for the Bloch function $u_{\mathbf{k},\nu}(\mathbf{r})$ (the exponential factors $e^{i\mathbf{k}\mathbf{R}}$ in (2.14) cancel each other so that there is no divergence!):

$$(1/V_c)\int_{V_c} d\mathbf{r}\, u^{*\{U^\dagger\}}_{\mathbf{k}^*,\nu'}(\mathbf{r})\, u^{\{U\}}_{\mathbf{k},\nu}(\mathbf{r}) = \delta_{\nu,\nu'}\,. \tag{2.15}$$

This is a generalization of (1.12) for non-Hermitean potentials and complex **k**. In particular, for Hermitean potentials with complex **k** we have

$$(1/V_c)\int_{V_c} d\mathbf{r}\, u^{*}_{\mathbf{k}^*,\nu'}(\mathbf{r})\, u_{\mathbf{k},\nu}(\mathbf{r}) = \delta_{\nu,\nu'}\,. \tag{2.16}$$

Thus, one can see that due to these orthogonality relations and to the divergence of the scalar product for different **k** and **k**′, one has to be somewhat careful by operating with "damped" Bloch waves. However, many relations and methods familiar from "normal" Bloch waves can be used for "damped" Bloch waves as well.

3. Band Structure for Complex k (One Dimension)

For the scattering of electrons by a crystal, the energy E is real, of course, and essentially given by the incident plane wave. If we assume that the crystal potential is strictly periodic inside the crystal and is simply cut off at the crystal surface, then for the representation of the wave function in the crystal, we need only those Bloch waves which, whether oscillating or damped, all have the same energy E.

Moreover, we consider only the scattering at crystals being infinite in two directions (x and y) and limited in the other (z), i.e., we consider only the scattering by a crystal filling the half space $z \geqslant 0$ or by a crystal slab filling the space with $0 \leqslant z \leqslant t$. For these crystals, the periodicity of the crystal potential in x and y directions is not perturbed by the crystal surfaces because the potential in the vacuum $V(\mathbf{r}) = 0$ fulfills every periodicity condition. Therefore, the potential in the whole space has the xy periodicity of the crystal potential. Accordingly, the xy-components $\mathscr{K}$ of the incident plane wave $\mathbf{K} = (\mathscr{K}, K_z)$ are good quantum numbers, meaning that all allowed Bloch waves in the crystal must have the same reduced xy-component $\mathscr{K}$, which is also real.

Therefore, only the z-component k_z of the Bloch vector $\mathbf{k} = (\mathscr{K}, k_z)$ can be complex and we are interested in the band structure $E_\nu(k_z)$ as a complex function E of the complex z-component k_z, with $\mathbf{k}_{x,y} = \mathscr{K}$ being given and real. In particular, we have to know those complex k_z values that are compatible with real energies $E_\nu(k_z) = E$. These k_z values lie on lines in the two-dimensional k_z-plane named "real lines" by Heine.[16] On his work the following section is based.

First we want to discuss the *one-dimensional band structure*. We start with the symmetry relations for $E_\nu(k)$, which are for a local and real potential according to (2.3), (2.4), and (2.1):

$$E(k) = E(-k) = E^*(k^*) = E(k + n2\pi/a) \qquad \text{for} \quad n = 0, \pm 1, \pm 2, \ldots . \tag{3.1}$$

One sees immediately that for $k = k^*$, we have $E(k) = E^*(k)$, i.e., the real axis $k = k^*$ is a real line $E = E^*$. Moreover, the lines $k = n\pi/a + ik''$ with real k'' can be real lines, too, as they are perpendicular to the real axis. From (3.1), we have

$$E(n\pi/a + ik'') = E(-n\pi/a - ik'') = E^*(-n\pi/a + ik'') = E^*(n\pi/a + ik''). \tag{3.2}$$

Therefore E is either real, if the energies on both sides refer to the same band ($E_\nu = E_\nu^*$), or there are two bands with complex conjugate energies ($E_\nu = E_{\nu'}^*$).

As will be shown later, these are already all the real lines in the one-dimensional case, namely the real axis, the imaginary axis, and the Brillouin zone boundary $k = +\pi/a + ik''$. Furthermore, we have on these lines $E_\nu(k) = E_\nu(-k) = E_\nu(k^*)$. Therefore, the whole band structure is specularly symmetrical with respect to the real and the imaginary axis.

The Schrödinger equation in one dimension with a periodic potential $V(x) = V(x + a)$ is an ordinary second-order differential equation having two linear independent solutions for each energy. From this it follows simply that the function $\mu(E) = \cos ka$ is an entire function of the complex variable E as has been shown by Kohn.[17] Hence the inverse $E(\mu)$ of the entire function $\mu(E)$ is an analytic function of μ except where $d\mu/dE = 0$. In the vicinity of such a point μ_γ we have

$$\mu = \mu_\gamma + \tfrac{1}{2}\mu_\gamma''(E - E_\nu)^2 \qquad \text{or} \qquad E = E_\gamma + ((2/\mu_\gamma'')(\mu - \mu_\gamma))^{1/2}.$$

As a function of k this means

$$E = E_\gamma + c((k - k_\gamma))^{1/2}. \tag{3.3}$$

Therefore E is an analytic function of k in the whole k-plane with the exception of the branch points k_γ.

[16] V. Heine, *Proc. Phys. Soc.* **81**, 300 (1963); *Surface Sci.* **2**, 1 (1964).
[17] W. Kohn, *Phys. Rev.* **115**, 809 (1959).

Previously, we have seen that the real axis also is a real line and that at the points $k_0 = n\pi/a$, real lines leave the real axis going into the complex plane. At what points k_0 on the real axis can this occur generally? In order to see this we expand $E(k)$ and $E(k^*)$ in the vicinity of k_0.

$$\begin{aligned} E(k) &= E(k_0) + E'(k_0)\,\delta k + \tfrac{1}{2}E''(k_0)(\delta k)^2 + \cdots \qquad \text{with} \quad \delta k = k - k_0 \\ E(k^*) &= E(k_0) + E'(k_0)\,\delta k^* + \tfrac{1}{2}E''(k_0)(\delta k^*)^2 + \cdots . \end{aligned} \tag{3.4}$$

Because for a real line we have $E(k) = E(k^*)$ and because k should be complex, we find

$$E'(k_0) = 0 \qquad \text{and} \qquad \delta k = i\,\delta k'' \qquad \text{with real} \quad \delta k''. \tag{3.5}$$

Therefore, real lines can leave the real axis only at extrema of $E_\nu(k)$ and only at right angles (Fig. 1). Moreover, such an extremum k_0 is a

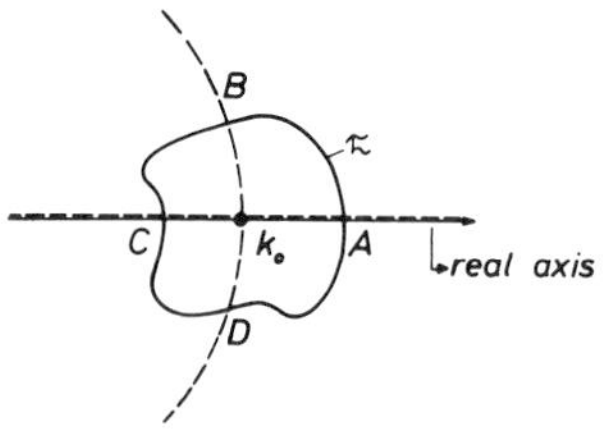

FIG. 1. Real lines (dashed) in the vicinity of a saddle point k_0 and contour $\mathscr{C}$ surrounding k_0.

saddle point, i.e., either a maximum on the real axis ($(\delta k)^2 \geqslant 0$) and a minimum on the perpendicular real line with

$$(\delta k)^2 = -(\delta k'')^2 \leqslant 0 \qquad \text{or vice versa.}$$

In the one-dimensional case, we have extrema at the positions $k_0 = 0$ and $k_0 = +\pi/a$ (or $n\pi/a$) which follows directly from $E_\nu(k) = E_\nu(-k)$ and $E_\nu(\pi/a + k) = E_\nu(\pi/a - k)$. No other extrema can occur because this would automatically lead to more than two linear independent solutions, which is a contradiction (Fig. 2). Moreover the real lines, leaving the real axis at $k_0 = n\pi/a$, are straight lines, as has been shown by (3.2).

The behavior of the real lines in the vicinity of a saddle point also can be studied by the following contour integral around a contour $\mathscr{C}$ surrounding k_0 close enough to include no branch point (Fig. 17). By a well-known theorem we have

$$I = \int_{\mathscr{C}} dk\, d/dk\{\ln f(k)\} = 2\pi i(Z - P), \tag{3.6}$$

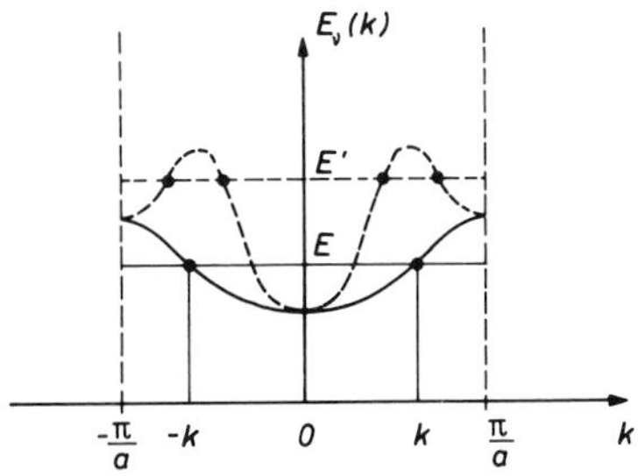

FIG. 2. Band structure in one dimension. For each energy E there are two linearly independent solutions k and $-k$ (full lines). Dashed lines, giving four independent solutions, are impossible.

where Z and P are the number of zeros and poles, respectively, enclosed by the contour and counted according to their multiplicity. Putting $f(k) = E(k) - E(k_0)$ we have due to the saddle point at k_0, $Z = 2$, and $P = 0$ and consequently

$$I = 4\pi i = \{\ln|E(k) - E(k_0)| + i \arg(E(k) - E(k_0))\}_{\mathscr{C}}. \tag{3.7}$$

Here, the bracket with the index $\mathscr{C}$ means the change of the bracket by going around the contour. Because the ln gives no contribution, the argument of $\{E(k) - E(k_0)\}$ increases from 0 to 4π on going around the contour. Therefore, there are four points k on the contour with real $E(k)$, namely the k values belonging to $\arg\{E(k) - E(k_0)\} = 0, \pi, 2\pi$, and 3π, which are the crossing points of the real lines with the contour (points ABCD in Fig. 1). From this simple theorem it follows directly that real lines cannot simply terminate since we have the result $I = 4\pi i$ for every closed contour surrounding k_0. For the same reason they cannot branch. Moreover, the energy varies monotonically along the real line except at the saddle points k_0 since every point $dE/dk = 0$ has to be a crossing point of real lines.

Because the arguments (3.6) and (3.7) always hold as long the contour does not enclose a branch point k_γ, there are only two possibilities for the real line in the complex plane (line BD in Fig. 1). Either it reaches a branch point k_γ or it does not. If it reaches a branch point then it behaves in the vicinity of k_γ as (3.3), namely, the real line runs around the branch point into another Riemann sheet of the complex k-plane, from where it will loop back to the real axis, running on the same line in k-space, but on the other edge of the branch cut in the next Riemann sheet. All along the line, the energy varies monotonically until the next saddle point k_0' on the real axis in the next Riemann sheet is reached, where the line crosses the real line on the real axis.

On the other hand, if the real line encloses no branch point, then the line has to run to infinity, while the energy always varies monotonically. However, then the k-value gets extremely large, $\hbar^2 k^2/2m \gg V(r)$, and the band structure can be calculated by neglecting $V(r)$. But for free electrons we have

$$E(k) = (k + n2\pi/a)^2 (\hbar^2/2m), \qquad n = 0, \pm 1, \pm 2, \ldots, \tag{3.8a}$$

or with $k = k' + ik''$ and $-\pi/a < k' \leqslant \pi/a$, we obtain

$$(2m/\hbar)^2 E_\nu(k) = (k' + n2\pi/a)^2 + 2ik'' \cdot (k' + n2\pi/a) - k''^2 \tag{3.8b}$$

and only obtain a real line for $k' = 0$ and $n = 0$. This is the imaginary axis along which E decreases to $-\infty$ for increasing $|k''|$. Therefore, the imaginary axis is the only real line running to infinity, all other ones reach a branch point and loop back to the real axis.

In conclusion, in the one-dimensional case we can summarize the result as follows: The whole band structure is specularly symmetric to the real and the imaginary axis. For large negative energies we have purely imaginary k-values (Figs. 3 and 4a, point A). By moving along the imaginary axis to $k = 0$, the energy increases to the lowest values of the first band (point B). Then moving along the real axis the energy,

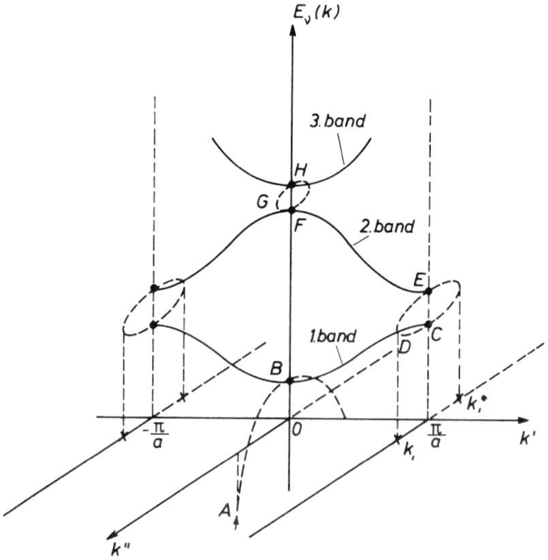

FIG. 3. Complex band structure in one dimension ($k = k' + ik''$) (●, saddle points; ×, branch points; solid line, k real; dashed line, k complex).

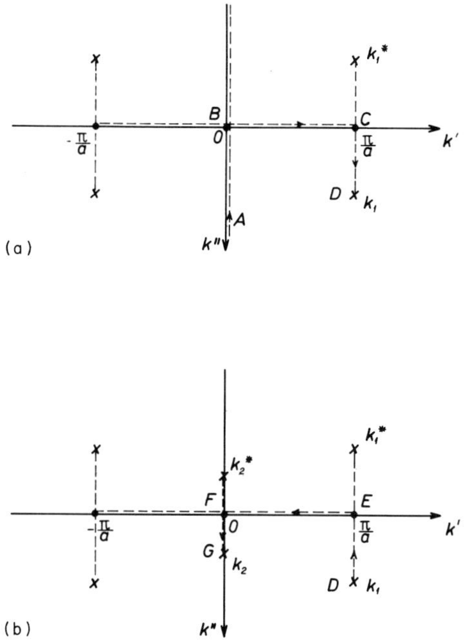

FIG. 4. (a) Real lines in first Riemann sheet (dashed) (●, saddle points; ×, branch points). (b) Real lines in second Riemann sheet (dashed) (●, saddle points; ×, branch points).

continuously increasing, assumes all the values of the first allowed band until one reaches the saddle point C at the Brillouin zone boundary. Here the real line again enters the complex plane and moves to the branch point k_1, the energy increasing to D. There the real line leaves the first Riemann sheet and moves in the second one back from k_1 to the real axis, where the energy reaches the bottom of the second band (point E). From here on the energy assumes all the values of the second allowed energy band, whereas the real lines run to $k = 0$ in the second sheet (F). By circumventing the branch point $k_2(G)$, the lines run into the third Riemann sheet and to the third band on the real axis (H), etc.

4. BAND STRUCTURE FOR COMPLEX k (THREE DIMENSIONS AND $V \neq V^\dagger$)

For the three-dimensional case, many results of the one-dimensional band structure remain valid. First we want to define the appropriate basis vectors of the lattice for the scattering at a given crystal surface lying in the xy-plane. We choose the two shortest (nonparallel) translation vectors $\mathbf{a}_1$ and $\mathbf{a}_2$ in the crystal surface. Then the total potential in the crystal and in the vacuum is invariant under a surface translation $\mathscr{R}^n = n_1\mathbf{a}_1 + n_2\mathbf{a}_2$. Furthermore, the third basis vector $\mathbf{a}_3$ is perpendicular to both $\mathbf{a}_1$ and $\mathbf{a}_2$ and gives the shortest periodicity in the z-direction. As an example, we have plotted in Fig. 5 the basis vectors $\mathbf{a}_1$,

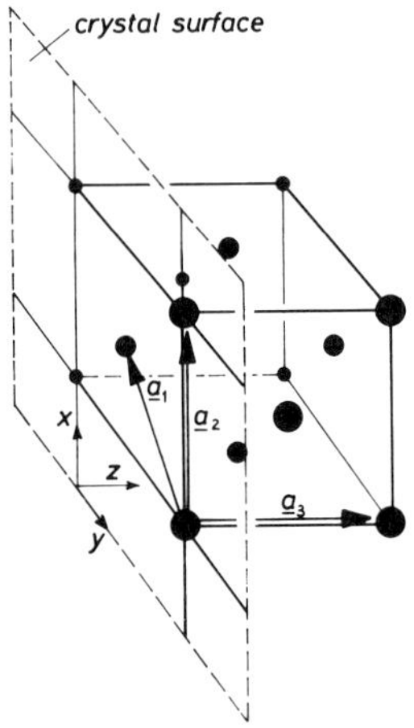

FIG. 5. Choice of the basis vectors for a (100) surface of a fcc crystal ($\mathbf{a}_1$, $\mathbf{a}_2$, in crystal boundary; $\mathbf{a}_3$, perpendicular to crystal boundary). *Note*: In this diagram and all following, underlined letters refer to boldfaced letters in text.

$\mathbf{a}_2$, and $\mathbf{a}_3$ for the (100) surface of a fcc crystal. This is a nonprimitive description of a primitive lattice, with each unit cell contining two atoms. Analogously, $\mathbf{b}_1$ and $\mathbf{b}_2$ are the reciprocal lattice vectors of the surface net, lying in the xy-plane, and $\mathbf{b}_3$ points in z-direction.

Then, according to the last section, the reduced xy-component $\mathscr{K}$ (in the surface mesh $(\mathbf{b}_1, \mathbf{b}_2)$) of the Bloch vector $\mathbf{k} = (\mathscr{K}, k_z)$ is determined by the incident plane wave and real. Assuming the surface plane to be a reflection plane, we have for the energy $E(\mathscr{K}, k_z)$ as a function of k_z for a given real $\mathscr{K}$ the following symmetries:

$$E(k_z) = E(-k_z) = E^*(k_z^*) = E(k_z + n\pi/a_3). \tag{4.1}$$

Therefore, the real axis $k_z = k_z^*$ is again a real line. Furthermore, the whole band structure is specularly symmetrical with respect to the real and the imaginary axis. Blount[18] has shown that $E(k_z)$ is an analytic function of k_z everywhere in the complex plane with the exceptions of

[18] E. I. Blount, *Solid State Phys.* **13**, 306 (1962).

branch points of the type (3.3), which are the only singularities. However, in disagreement to the one-dimensional case, extrema do not only occur at $k_z = 0$ or $\pm\pi/a_3$.

Analogously to (3.4) a real line can leave the real axis only at saddle points k_0 and at right angles (Fig. 1). However, these real lines do not necessarily have to be straight lines. They only are straight due to the inversion symmetry for $k_0 = 0$ and $k_0 = \pm\pi/a_3$. Furthermore, the real lines cannot terminate and cannot branch as can be shown by the *Z–P* theorem (3.6) and (3.7). In principle they can cross, for instance in k_0 on the real axis. But a crossing point for complex k_z would be highly accidental. For according to (37) we have $\partial E/\partial k_z = 0$ at a crossing point. As Herring[19] points out, this would be vanishingly probable because the slightest perturbation would destroy E being real and $\partial E/\partial k_z = 0$ at the same point (except for the crossing points on the real axis, being a real line for symmetry reasons). Therefore, along the real lines the energy varies monotonically, except at the saddle points on the real axis.

Again there are only two possibilities for a real line leaving the real axis at a saddle point. Either it can enclose a branch point of the type (3.3), enter there into a new Riemann sheet and loop in this sheet back to another saddle point on the real axis approaching the adjacent allowed energy band, or it has to run to infinity with monotonically varying energy. Because the k_z vector on these lines gets arbitrarily large, the occurrence of these lines can already be seen in the free electrons approximation. For each reciprocal lattice vector $\mathbf{h} = (\boldsymbol{\hbar}, h_z)$ with xy-component $\boldsymbol{\hbar}$ we get

$$\begin{aligned}(2m/\hbar^2)\,E(k_z) &= (\mathbf{k} + \mathbf{h})^2 \\ &= (\mathscr{K} + \boldsymbol{\hbar})^2 + (k_z + h_z)^2 \\ &= (\mathscr{K} + \boldsymbol{\hbar})^2 + (k_z' + h_z)^2 + 2ik_z''(k_z' + h_z') - k_z''^2, \quad (4.2)\end{aligned}$$

where we need only the section of the parabola $(k_z + h_z)^2$ which falls into the "first Brillouin zone" $-\pi/a_3 < k_z' \leqslant \pi/a_3$. We see that for $k_z' = -h_z$, we get a real line running to $k'' \to \pm\infty$, i.e., a real line for all reciprocal lattice vectors $\mathbf{h}$ with $-\pi/a_3 < h_z \leqslant \pi/a_3$. Therefore, in three dimensions there are ∞^2 real lines running to infinity, contrary to the one-dimensional case with one line only.

Qualitatively, we have a situation as shown in Fig. 6. We start with the parabola of $\mathbf{h}_{1,0} = (\boldsymbol{\hbar}_1, 0)$ at the energy $(\hbar^2/2m)(\mathscr{K} + \boldsymbol{\hbar}_1)^2$. The reversed parabola $(\mathscr{K} + \boldsymbol{\hbar}_1)^2 - k_z''^2$ belongs to a real line running to infinity. If

[19] C. Herring, *Phys. Rev.* **52**, 365 (1937).

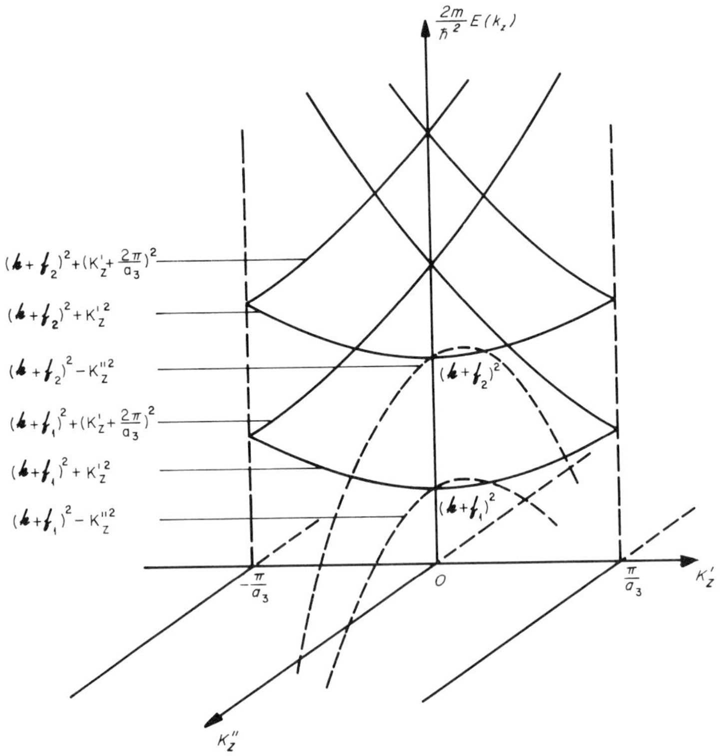

FIG. 6. Free electron band structure in three dimensions (solid line, $k = k.'$ real; dashed line, k complex).

we add to this the parabolas belonging to the reciprocal lattice vectors $\mathbf{h}_{1,n} = (\boldsymbol{h}_1, n2\pi/a_3)$ then we have a typical free electron band structure in one dimension. However, there are a lot of other parabolas in three dimensions, for instance the one of $\mathbf{h}_{2,0} = (\boldsymbol{h}, 0)$ and the sections of the parabolas $\mathbf{h}_{2,0} = (\boldsymbol{h}_2, n2\pi/a)$, which again look like another typical one-dimensional band structure with an additional real line going to infinity. Thus we have had to superimpose an infinite number of one-dimensional band structure with one real "infinite" line each.

Now, by switching on the potential $V(\mathbf{r})$, the degeneracies at the crossing of the different parabolas is removed and the bands split up (Fig. 7). The extrema (saddle points!) on the adjacent bands are connected by a loop because of a real line going into the complex plane around a branch point. Whereas the real lines for $k' = 0$ and $k' = \pm\pi/a$ are straight lines for symmetry reasons, the "additional" real lines

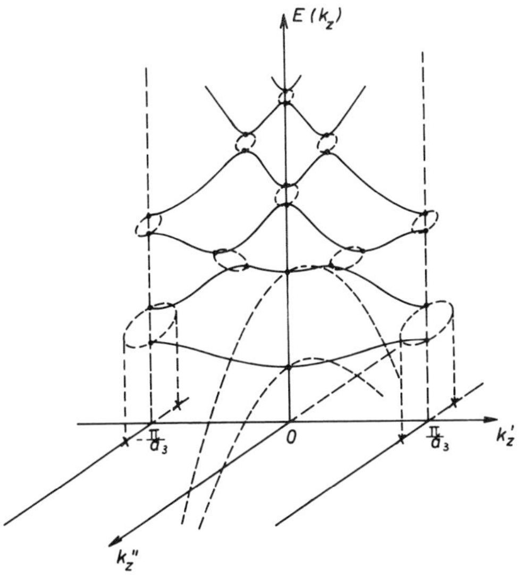

FIG. 7. Band structure in three dimensions (solid lines, bands with real k; dashed lines, bands with complex k; ●, saddle points).

inside the Brillouin zone are not straight. Moreover one sees that for a given energy one gets only a finite number of "allowed" Bloch waves with real **k** (in Fig. 7 at most 4), but always an infinite number of "damped" Bloch waves with complex k_z. The whole band structure is symmetrical with respect to the real and imaginary axis.

Non-Hermitean Potentials

The basic assumptions for the foregoing discussion of the band structure are the symmetry relations (3.1) and (4.1), especially the equation $E_\nu(k_z) = E_\nu^*(k_z^*)$, which results in the real axis being a real line. However, the optical potential is in general a non-Hermitean potential, for which this symmetry relation is not valid. Instead, we have the relation (2.12), namely $E_\nu^{\{U\}}(k_z) = (E_\nu^{\{U\dagger\}}(k_z^*))^*$. Consequently, the real lines for the potential U and the ones for $U^\dagger$ lie symmetrically to each other with respect to the real axis, which is no longer a real line (Fig. 8) (except again for $U = U^\dagger$, where both lines coincide on the real axis). Therefore, all the general predictions for the band structure are no longer valid. If, however, the optical potential is only weakly non-Hermitean, we can study its band structure by means of perturbation theory, starting from a Hermitean potential, as will be shown in the following.

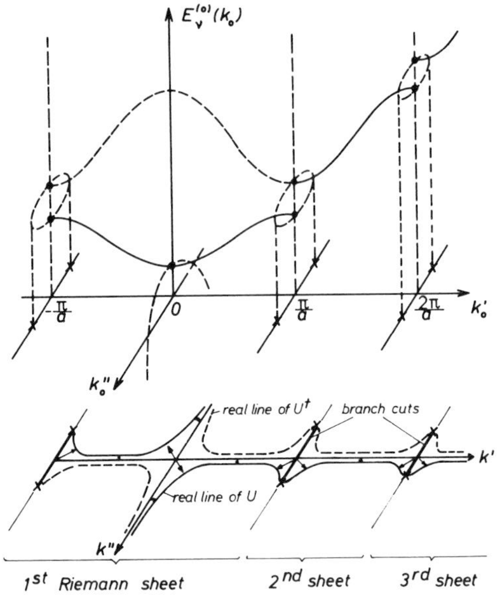

FIG. 8. Real lines for non-Hermitian potentials (●, saddle points of U'; ×, branch points).

Splitting the potential U onto a Hermitean part U' and an anti-Hermitean part iU'', we have

$$U = U' + iU'' \qquad \text{with} \qquad U' = (U + U^\dagger)/2 = U'^\dagger; \qquad U'' = (U - U^\dagger)/2i = U''^\dagger \tag{4.3}$$

we treat U'' as a perturbation. The eigenfunctions $\phi_{\mathbf{k},\nu}^{(0)}$ for the Hermitean potential U' are defined by

$$(H_0 + U')\,\phi_{\mathbf{k},\nu}^{(0)} = E_\nu^{(0)}(\mathbf{k})\,\phi_{\mathbf{k},\nu}^{(0)}. \tag{4.4}$$

To obtain the Bloch wave $\phi_{\mathbf{k},\nu}(\mathbf{r})$ for the potential U for a given energy E and a given xy-component $\mathscr{K}$ of $\mathbf{k} = \{\mathscr{K}, k_z\}$, we make the ansatz for $\phi_{\mathbf{k},\nu}$:

$$\phi_{\mathbf{k},\nu}(\mathbf{r}) = \sum_{\nu'} C_{\nu'}\phi_{\mathbf{k},\nu'}^{(0)}(\mathbf{r}). \tag{4.5}$$

This is really an ansatz for the Bloch function $u_{\mathbf{k},\nu}$ in terms of the $u_{\mathbf{k},\nu}^{(0)}$. building a complete system for real $\mathbf{k}$ (1.13).

Encouraged by the orthogonality (2.16), we assume also completeness for complex $\mathbf{k}$. Then we have for the coefficients C_ν

$$(E_\nu^{(0)}(\mathbf{k}) - E)\, C_\nu + i \sum_{\nu'} U''_{\nu\nu'}(\mathbf{k})\, C_{\nu'} = 0$$

with (4.6)

$$U''_{\nu\nu'}(\mathbf{k}) = (1/V_c) \int_{V_c} d\mathbf{r} \int_{-\infty}^{\infty} d\mathbf{r}\; e^{-i\mathbf{k}\mathbf{r}} u_{\mathbf{k}^*,\nu}^{(0)*}(\mathbf{r}) \cdot U''(\mathbf{r}, \mathbf{r}')\, u_{\mathbf{k},\nu'}^{(0)}(\mathbf{r}')\, e^{i\mathbf{k}\mathbf{r}'}.$$

Now in zeroth order we have $E = E_\nu(\mathbf{k}_0)$ with $\mathbf{k}_0 = (\mathscr{K}, k_{0z})$; in first order $\mathbf{k} = \mathbf{k}_0 + \delta\mathbf{k}$ is determined by

$$E_\nu^{(0)}(\mathbf{k}_0 + \delta\mathbf{k}) - E_\nu^{(0)}(\mathbf{k}_0) = -iU''_{\nu\nu}(\mathbf{k}_0) \tag{4.7}$$

Expanding up to linear terms in $\delta\mathbf{k} = (0, \delta k_z)$, we have

$$\delta k_z = -i \frac{U''_{\nu\nu}(\mathbf{k}_0)}{\partial E_\nu^{(0)}(\mathbf{k}_0)/\partial k_{0z}}, \tag{4.8}$$

and by using the plane wave expansion (1.9),

$$\delta k_z = -i \frac{1}{\partial E_\nu^{(0)}(\mathbf{k}_0)/\partial k_{0z}} \sum_{\mathbf{h},\mathbf{h}'} C_{\mathbf{h}}^*(\mathbf{k}_0, \nu)\, C_{\mathbf{h}'}(\mathbf{k}_0, \nu) \cdot U''_{\mathbf{h}-\mathbf{h}'}, \tag{4.9}$$

where for a local potential $U''(\mathbf{r}, \mathbf{r}') = U''(\mathbf{r})\, \delta(\mathbf{r} - \mathbf{r}')$ the coefficients $U''_{\mathbf{h}}$ are given by (1.2). It is interesting to see that δk_z is a periodic function of $\mathbf{k}_0$. Further for real $\mathbf{k}_0$ in the allowed energy bands δk_z is purely imaginary representing the absorption of the Bloch wave. Moreover it is inversely proportional to the z-component of the group velocity which is plausible from a classical point of view. For $\partial E/\partial k_{0z} \approx 0$, Eq. (4.8) is no longer valid and in (4.5) we have to take the quadratic terms in δk_z into account. Due to dispersion this leads to finite values of δk_z even at the extrema $\partial E/\partial k_{z0} = 0$ (instead of according to (4.8)).

$$\delta k_z = \frac{1 \pm i}{\sqrt{2}} \left\{ \left| \frac{U''_{\nu\nu}(\mathbf{k}_0)}{\frac{1}{2}\partial^2 E_\nu^{(0)}(\mathbf{k}_0)/\partial k_{0z}^2} \right| \right\}^{1/2}. \tag{4.10}$$

Therefore, δk_z, being proportional to $(U'')^{1/2}$ at the extrema, is especially large and the absorption is very effective at these points. For the wave-function we obtain to first order

$$\phi_{\mathbf{k},\nu}(\mathbf{r}) = e^{i(\mathbf{k}_0+\delta\mathbf{k})\mathbf{r}} \left\{ u_{\mathbf{k}_0+\delta\mathbf{k},\nu}(\mathbf{r}) + \sum_{\nu' \neq \nu} \frac{iU''_{\nu\nu'}(k_0)}{E_\nu^{(0)}(\mathbf{k}_0) - E_{\nu'}^{(0)}(\mathbf{k}_0)}\, u_{\mathbf{k}_0,\nu'}^{(0)}(\mathbf{r}) \right\}, \tag{4.11}$$

where the most important term for real $\mathbf{k}_0$ is the damping factor $e^{i\delta k_z z}$ (δk_z imaginary).

Near the branching points k_γ, where the energies of different bands are equal, we have to apply the degenerate perturbation theory. The results are somewhat lengthy and will not be given here.

In Fig. 8 we have plotted the real lines for the non-Hermitean potential U according to the present perturbation theory. For simplicity we have chosen the linear case. Furthermore, $U = \hat{U}$ is assumed to be symmetric, leading to the inversion symmetry (2.13a) in the complex k_z-plane. The upper figure shows $E_\nu^{(0)}(k_0)$ in the expanded zone scheme, whereas the lower shows the real lines in the different sheets. The real lines of the potential $U^\dagger$ (dashed lines) are obtained from the real lines of U by reflection at the real axis. In addition, the crossing points disappeared, being very sensitive to the perturbation (4.10).

5. The Two-Beam Case[5–12]

For electron diffraction the calculations of the Bloch waves and the band structures are simplified very much since the energy is large as compared to the potential $V(\mathbf{r})$. Therefore, one can apply perturbation theory. Starting with the Schrödinger equation (1.10), we have

$$(K_0^2 - (\mathbf{k} + \mathbf{h})^2)\, C_\mathbf{h} = \sum_{\mathbf{h}'(\neq \mathbf{h})} v_{\mathbf{h}-\mathbf{h}'} C \tag{5.1}$$

with

$$v_\mathbf{h} = (2m/\hbar^2)\, V_\mathbf{h}\,, \qquad K_0^2 = K^2 - v_0 = (2m/\hbar^2)(E - V_0).$$

If no Bragg reflection is excited, i.e., if $|K_0^2 - (\mathbf{k} + \mathbf{h})^2| \gg |v_\mathbf{h}|$ for all $\mathbf{h}$, then we have $K_0^2 = k^2$ and $C_\mathbf{h} \cong \delta_{\mathbf{h},0}$. Then the Bloch wave $\phi_{\mathbf{k},\nu}(\mathbf{r})$ is represented by the single plane wave $e^{i\mathbf{kr}}$.

However, if $\mathbf{k}^2 \cong (\mathbf{k} + \mathbf{h})^2 \cong K^2$, a Bragg reflection is excited and we get at least two strong beams $\mathbf{k}$ and $\mathbf{k} + \mathbf{h}$. In this section, we shall discuss the so-called "two-beam case." For more details and especially for a thorough discussion of the Laue and Bragg cases we refer to the literature.[5–12] In the next section we discuss some important multiple-beam cases.

For two strong plane waves $\mathbf{k}$ and $\mathbf{k} + \mathbf{h}$, Eq. (5.1) gives

$$\begin{aligned} (K_0^2 - \mathbf{k}^2)\, C_0 &= v_- C\;, \\ (K_0^2 - (\mathbf{k} + \mathbf{h})^2)\, C_\mathbf{h} &= v\, C_0\,. \end{aligned} \tag{5.2}$$

By setting the determinant equal to zero, we obtain the dispersion equation

$$(K_0^2 - \mathbf{k}^2) \cdot (K_0^2 - (\mathbf{k} + \mathbf{h})^2) = v_\mathbf{h} \cdot v_{-\mathbf{h}}\,. \tag{5.3}$$

The allowed **k** vectors for a given energy lie on a dispersion surface consisting of two branches (Fig. 9). For $v_{\mathbf{h}} = 0$, it degenerates into two spheres with radius K_0, one around the reciprocal lattice vector **h** and the other around the origin. For $v_{\mathbf{h}} \neq 0$ the spheres split up at the intersecting line, where the Bragg condition $\mathbf{k}^2 = (\mathbf{k} + \mathbf{h})^2$ is fulfilled,

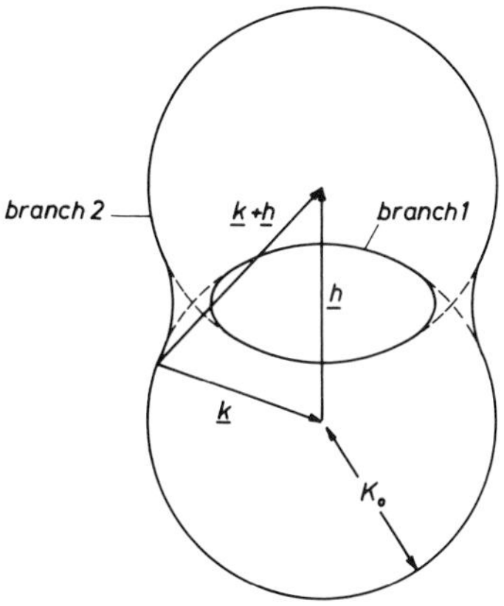

FIG. 9. Dispersion surface for the two-beam case.

and the outer branch 2 completely surrounds the inner branch 1. Exactly in the Bragg condition we have for $\mathbf{k} = \mathbf{k}_B : \mathbf{k}_B{}^2 = (\mathbf{k}_B + \mathbf{h})^2$. Near the Bragg spot $\mathbf{k}_B$ we obtain from (5.3) by setting $\mathbf{k} = \mathbf{k}_B + \delta\mathbf{k}$ and neglecting third- and fourth-order terms in $\delta\mathbf{k}$:

$$4(\mathbf{k}_B \cdot \delta\mathbf{k})((\mathbf{k}_B + \mathbf{h}) \cdot \delta\mathbf{k}) = v_{\mathbf{h}} \cdot v_{-\mathbf{h}} \,. \tag{5.4}$$

Neglecting the higher order terms in $\delta\mathbf{k}$ for $v_{\mathbf{h}} = 0$ is equivalent to replacing the spheres in Fig. 9 by the tangential planes in the point $\mathbf{k}_B$ (Fig. 10).

In this approximation, the dispersion surfaces are hyperbolas, the asymptotes of which are the tangential planes of the spheres. The smallest separation of the two branches is

$$\Delta k = |v_{\mathbf{h}}|/K_0 \cos \theta_B \,. \tag{5.5}$$

In the Laue case of reflection[5–12] the two Bloch waves on the opposite branches of the dispersion surface are excited with equal intensity. The small difference Δk of the **k**-vectors then gives rise to characteristic modulations of the total intensity. Depending on the thickness of the crystal, the total intentsity oscillates between the transmitted and

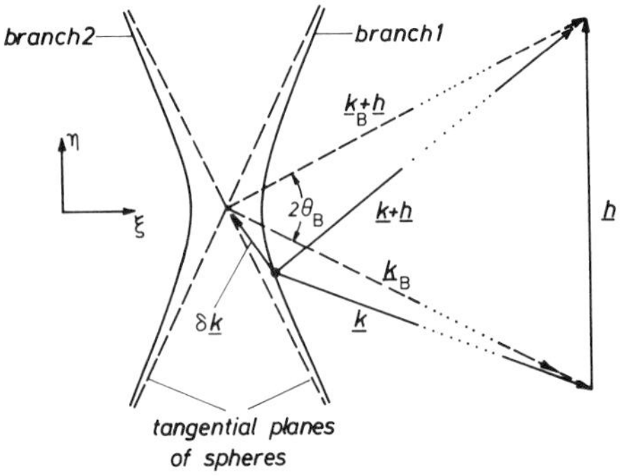

FIG. 10. Dispersion surface near the Bragg spot $\mathbf{k}_B$.

reflected beam back and forth (pendulum solution). The period of the oscillations $d_{ext} = 2\pi/\Delta k$ is called the extinction length.

By introducing $K_0{}^2$ from (5.3) into Eq. (5.2), we can calculate the coefficients C_0 and $C_{\mathbf{h}}$. By using the normalization (1.12), we obtain

$$C_0 = \frac{1}{\sqrt{2}}\{1 \pm W(1 + W^2)^{1/2}\}^{1/2}$$
$$C_{\mathbf{h}} = \pm\operatorname{sgn}(v_{\mathbf{h}})\frac{1}{\sqrt{2}}\{1 \pm W(1 + W^2)^{1/2}\}^{1/2} \tag{5.6}$$

with

$$W = (\mathbf{k} + \tfrac{1}{2}\mathbf{h}) \cdot \mathbf{h}/|v_{\mathbf{h}}| = \delta\mathbf{k} \cdot \mathbf{h}/|v_{\mathbf{h}}|.$$

Here and in the following, the upper sign refers to the inner branch 1, the lower to the outer branch 2.

For $|W| \to \infty$, i.e., if the Bragg condition is not fulfilled, we get for $\mathbf{k}$ vectors on the sphere around the origin, $C_0 = 1$, $C_{\mathbf{h}} = 0$, and on the sphere around $\mathbf{h}$, $C_0 = 0$ and $C_{\mathbf{h}} = \pm 1$ (Fig. 9).

For the Bragg condition we have $W = 0$. The two Bloch waves $\phi_{\mathbf{k}}(\mathbf{r})$ are, in this case,

$$\phi^{\mathrm{I}}(\mathbf{r}) = \frac{1}{\sqrt{2}} e^{i\mathbf{kr}}(1 + e^{i\mathbf{hr}}) = \sqrt{2}\, e^{i(\mathbf{k}+\frac{1}{2}\mathbf{h})\mathbf{r}} \cdot \cos(\mathbf{h} \cdot \mathbf{r}/2) \tag{5.7}$$

and

$$\phi^{\mathrm{II}}(\mathbf{r}) = -\sqrt{2}\, i e^{i(\mathbf{k}+\frac{1}{2}\mathbf{h})\mathbf{r}} \sin(\mathbf{h} \cdot \mathbf{r}/2). \tag{5.8}$$

Since $v_{\mathbf{h}} < 0$, the Bloch wave ϕ^{I} lies on the outer branch and ϕ^{II} on the inner one. This would be reversed for $v_{\mathbf{h}} > 0$ (X-ray diffraction).

Characteristic for both waves are the modulation functions $\cos(\mathbf{h} \cdot \mathbf{r}/2)$ and $\sin(\mathbf{h} \cdot \mathbf{r}/2)$. Therefore ϕ^{I} is always maximal at the atomic positions on the reflecting planes and vanishes in the middle between the planes (Fig. 11). Contrarily, ϕ^{II} has nodal planes at the reflecting planes and is

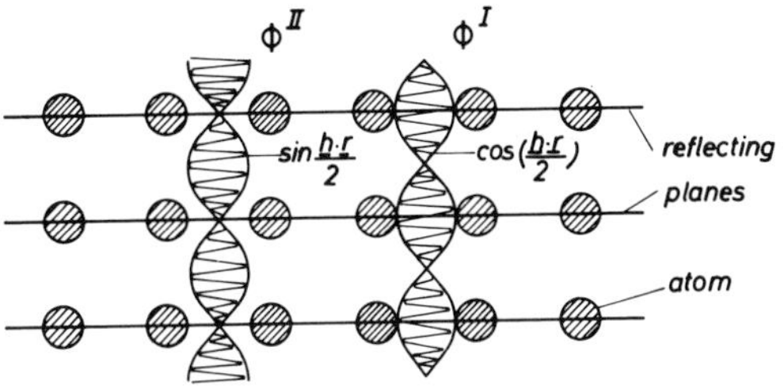

FIG. 11. Modulation factors of ϕ^{I} and ϕ^{II}.

maximal between these planes. Whereas both waves have the same energy, the wave on the outer branch has more kinetic energy due to the fact that $\mathbf{k}$ and $\mathbf{k} + \mathbf{h}$ are larger than on the inner branch. This can also be seen at the form of the wave function ϕ^{I} and ϕ^{II}. For example, the wave ϕ^{I}, being concentrated at the atomic positions, has a larger (but negative) potential energy then ϕ^{II}, and consequently a larger kinetic energy.

It is noteworthy that the Bloch functions on the different branches 1 and 2 of the dispersion surface, but for the same parameter W, are orthogonal. Furthermore, we have the following relations (see also Section 7)

$$\begin{aligned} C_0^1 C_0^2 + C_{\mathbf{h}}^1 C_{\mathbf{h}}^2 &= 0 = C_0^1 C_{\mathbf{h}}^1 + C_0^2 C_{\mathbf{h}}^2, \\ C_0^1 C_0^1 + C_0^2 C_0^2 &= 1 = C_{\mathbf{h}}^1 C_{\mathbf{h}}^1 + C_{\mathbf{h}}^2 C_{\mathbf{h}}^2. \end{aligned} \tag{5.9}$$

According to (1.15) the average current is always perpendicular to the dispersion surface. In the two-beam case it is given by

$$\langle \mathbf{j}_{\mathbf{k}}(\mathbf{r}) \rangle_{V_c} = (\hbar/m)\{\mathbf{k}_B + \mathbf{h}\,\tfrac{1}{2}(1 \pm W/(1 + W^2)^{1/2})\}. \tag{5.10}$$

In the vicinity of the Bragg condition, the direction of the current changes by an angle of $2\theta_B$, namely from the direction $\mathbf{k} \simeq \mathbf{k}_B$ on the sphere around the origin to the direction $\mathbf{k} + \mathbf{h} \simeq \mathbf{k}_B + \mathbf{h}$ on the sphere around $\mathbf{h}$ (Fig. 10). Exactly in the Bragg condition ($W = 0$) the current is parallel to the reflecting planes.

For the absorbing crystal, i.e., for a non-Hermitean potential, the $\mathbf{k}$ vectors are complex. It is clear already from Fig. 11 that the absorption of the Bloch waves ϕ^{I} and ϕ^{II} must be vastly different because the "absorption power" will be concentrated at the atoms. We get a higher than normal absorption for ϕ^{I} and an anomalously low absorption for the sin waves ϕ^{II} representing the "anomalous transmission" effect. According to (4.9) the imaginary part δk_z of $\mathbf{k}$ can be calculated for both waves by assuming an imaginary potential $iU''(\mathbf{r})$. In the symmetrical Laue case, the reflecting planes are perpendicular to the surface and we obtain for the absorption $\mu(W) = 2|\delta k_z|$ of the waves I and II:

$$\mu^{\mathrm{I,II}}(W) = \mu_0 \pm \frac{1}{(1 + W^2)^{1/2}} \mu_{\mathbf{h}} \tag{5.11}$$

with

$$\mu_0 = \frac{2mU_0''}{\hbar^2 K \cos\theta}; \qquad \mu_{\mathbf{h}} = \frac{2mU_{\mathbf{h}}''}{\hbar^2 K \cos\theta},$$

where μ_0 is the normal absorption coefficient of a plane wave. $\mu(W)$ is plotted in Fig. 12. For $|W| \gg 1$, i.e., outside the Bragg condition, we get

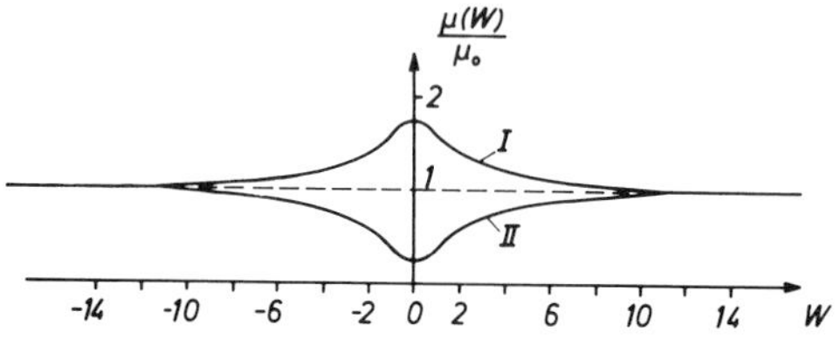

FIG. 12. Absorption for Bloch wave I and II (for $\mu_{\mathbf{h}} = \frac{3}{4}\mu_0$).

the normal absorption, μ_0 whereas with the Bragg condition, the absorption of wave II is $\Delta\mu_{\mathbf{h}} = \mu_0 - \mu_{\mathbf{h}}$. If the absorption is concentrated at the centers of the atoms, then $U_0'' = U_{\mathbf{h}}''$ and the absorption for wave field II vanishes. This is plausible from Fig. 11 because the sin function vanishes at the positions of the centers of the atom. By writing $U''(\mathbf{r})$ analogously to the real potential (1.2) as $U''(\mathbf{r}) = \sum_{\mathbf{R}} u''(\mathbf{r} - \mathbf{R})$ and by expanding $\mu_{\mathbf{h}}$ in powers of $\mathbf{h}$, we obtain for the anomalous absorption coefficient $\Delta\mu_{\mathbf{h}}$, assuming radial symmetry for $U''(\mathbf{r})$,

$$\Delta\mu_{\mathbf{h}} = \mu_0 - \mu_{\mathbf{h}} \cong \mu_0 \tfrac{1}{6}\langle \mathbf{r}^2 \rangle\, \mathbf{h}^2 \tag{5.12}$$

with

$$\langle \mathbf{r}^2 \rangle = \int \mathbf{r}^2 u''(\mathbf{r})\, d\mathbf{r} \Big/ \int u''(\mathbf{r})\, d\mathbf{r}.$$

Accordingly, $\Delta\mu$ varies as $\mathbf{h}^2$ and is proportional to the 2nd moment of the imaginary potential of the atoms. However due to the large spatial extent of the outer orbitals, this is not a good approximation for electron diffraction.

6. Some Multiple-Beam Cases

Multiple-beam cases are very important in electron diffraction due to the small wavelength and the relatively strong interaction. However, unlike the two-beam case, the multiple-beam cases can no longer be solved analytically. For instance, to obtain the dispersion relation in the three-beam case one has to solve a cubic equation. Nevertheless, some simple analytical results can be given for special multiple-beam cases, from which a number of properties can be derived.

First we will discuss qualitatively the effect of the so-called *systematic reflections*.[12,20–22] These are the secondary reflections **2h**, **3h**,... and $-$**h**, $-$**2h**,... lying on the same line as the reciprocal lattice point **h** corresponding to the strong reflection $(\mathbf{k}+\mathbf{h})$ of the two-beam case (Fig. 13). Because the radius K_0 of the Ewald sphere is appreciably

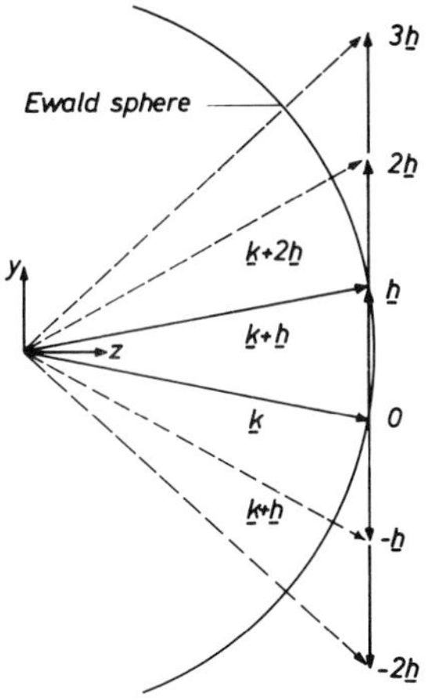

Fig. 13. Systematic reflections **2h**, **3h**, and $-$**h**, $-$**2h**.

larger than h, the reciprocal lattice points $n\mathbf{h}$ lie relatively close to the Ewald sphere and are always excited to some extent. Their influence can be determined by perturbation theory,[4] at least qualitatively.

[20] A. Howie, *Phil. Mag.* **14**, 223 (1966).
[21] M. J. Goringe, A. Howie, and M. J. Whelan, *Phil. Mag.* **14**, 217 (1966).
[22] R. Serneels and R. Gevers, *Phys. Status Solidi* **33**, 703 (1969); **45**, 493 (1971).

Going back to Eq. (5.1) for $C_{\mathbf{h}}$, we assume that in addition to the two strong beams $\mathbf{k}$ and $\mathbf{k} + \mathbf{h}$ we have a number of weak beams $\mathbf{k} + \mathbf{g}$ with $\mathbf{g} = 0, \mathbf{h}$, for which we get approximately

$$C_{\mathbf{g}} \simeq \frac{1}{K_0{}^2 - (\mathbf{k} + \mathbf{g})^2} (v_{\mathbf{g}} C_0 + v_{\mathbf{g}-\mathbf{h}} C_{\mathbf{h}}). \tag{6.1}$$

On the right-hand side, we have neglected the coefficients $C_{\mathbf{g}}$ of the other weak beams which are assumed to be small. Going back with this result for the $C_{\mathbf{g}}$'s into the exact equations for C_0 and $C_{\mathbf{h}}$, we obtain two equations similar to the two-beam case (5.2), but with coefficients $v^B_{\mathbf{h},\mathbf{h}'}$, instead of $v_{\mathbf{h}-\mathbf{h}'}$ ("Bethe potentials").

$$v^B_{\mathbf{h},\mathbf{h}'} = v_{\mathbf{h}-\mathbf{h}'} + \sum_{\mathbf{g}(\neq 0,\mathbf{h})} \frac{v_{\mathbf{h}-\mathbf{g}} v_{\mathbf{g}-\mathbf{h}}}{K_0{}^2 - (\mathbf{k} + \mathbf{g})^2} \,. \tag{6.2}$$

It is seen that the reciprocal lattice points $\mathbf{g}$ lying inside the Ewald sphere ($K_0{}^2 > (\mathbf{k} + \mathbf{g})^2$) give rise to a repulsive potential correction whereas the outer ones give an attractive contribution.

Applying this to the systematic reflections $\mathbf{g} = n\mathbf{h}$ ($n = 2, 3,\ldots, -1, 2,\ldots$) of a lowest order reflection $\mathbf{h}$, for $v_{\mathbf{h}} = v_{-\mathbf{h}}$ we have $v^B_{\mathbf{h},\mathbf{h}} = v^B_{0,0}$ and $v^B_{\mathbf{h},0} = v^B_{0,\mathbf{h}}$. Furthermore, we replace $\mathbf{k}$ in (6.2) by the vector $\mathbf{k}_B$ satisfying the exact Bragg condition leading to $K_0{}^2 - (\mathbf{k} + n\mathbf{h})^2 \simeq -n(n-1)\,\mathbf{h}^2$. Then we obtain for the branch separation, $\Delta k = 2\pi/d_{\text{ext}}$,

$$\Delta k = \frac{|v^B_{\mathbf{h},0}|}{K_0 \cos\theta_B} = \frac{1}{K_0 \cos\theta_B} \left\{ |v_{\mathbf{h}}| + \frac{1}{h^2} \left(v_{2\mathbf{h}} v_{\mathbf{h}} + \tfrac{1}{3} v_{3\mathbf{h}} v_{2\mathbf{h}} + \cdots \right) \right\} \tag{6.3}$$

The first term in the bracket represents the two-beam expression (5.5). Therefore, the extinction distance decreases due to the systematic reflections.

We may also calculate the influence of the systematic reflections upon the absorption. According to Eqs. (4.9) and (5.11) the absorption of a Bloch wave $\mathbf{k}$ for a simple imaginary potential $U''(\mathbf{r})$ may be written as

$$\mu = \sum_{\mathbf{h},\mathbf{h}'} C_{\mathbf{h}}{}^* C_{\mathbf{h}'} \mu_{\mathbf{h}-\mathbf{h}'} \,. \tag{6.4}$$

For instance, we obtain the Bragg condition, for the absorption of the wave field II (Fig. 11) exactly in the Bragg condition by taking into account only those terms which are linear in the "systematic" coefficients $C_{n\mathbf{h}}$:

$$\Delta\mu = \mu_0 - \mu_{\mathbf{h}} + (\mu_{\mathbf{h}} - \mu_{2\mathbf{h}}) \cdot \frac{-v_{\mathbf{h}} + v_{2\mathbf{h}}}{h^2} + (\mu_{2\mathbf{h}} - \mu_{3\mathbf{h}}) \cdot \frac{-v_{2\mathbf{h}} + v_{3\mathbf{h}}}{3h^2} + \cdots . \tag{6.5}$$

Therefore, the systematic reflections diminish the anomalous transmission effect ($v_{\mathbf{h}} < 0$!).

Now, we want to discuss situations where, for symmetry reasons, there are more than two strong waves. A first example is the *three-beam case* shown in Fig. 14a, for which there are three strong waves **k**, **k** + **h**, and **k** + **h**′. In Fig. 14a we have taken $|\mathbf{h}| = |\mathbf{h}'|$. Moreover, we assume

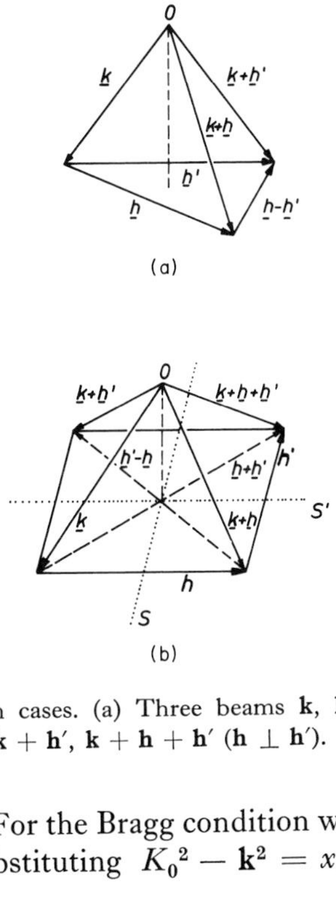

FIG. 14. Multiple beam cases. (a) Three beams **k**, **k** + **h**, **k** + **h**′ ($|\mathbf{h}| = |\mathbf{h}'|$). (b) Four beams **k**, **k** + **h**, **k** + **h**′, **k** + **h** + **h**′ ($\mathbf{h} \perp \mathbf{h}'$).

that $v_{\mathbf{h}} = v_{\mathbf{h}'} = v_{-\mathbf{h}}$. For the Bragg condition we have $\mathbf{k}^2 = (\mathbf{k} + \mathbf{h})^2 = (\mathbf{k} + \mathbf{h}')^2$ and by substituting $K_0{}^2 - \mathbf{k}^2 = x$, we obtain the matrix equation

$$\begin{pmatrix} x & -v_{\mathbf{h}} & -v_{\mathbf{h}-\mathbf{h}'} \\ -v_{\mathbf{h}} & x & -v_{\mathbf{h}} \\ -v_{\mathbf{h}-\mathbf{h}'} & -v_{\mathbf{h}} & x \end{pmatrix} \cdot \begin{pmatrix} C_{\mathbf{h}} \\ C_0 \\ C_{\mathbf{h}'} \end{pmatrix} = 0. \tag{6.6}$$

Due to the symmetry of the problem we have one antisymmetrical solution with $C_0 = 0$ and $C_{\mathbf{h}'} = -C_{\mathbf{h}}$, and with the x value

$$x^a = -v_{\mathbf{h}-\mathbf{h}'} \, . \tag{6.7}$$

Furthermore, we have two symmetrical solutions with $C_{\mathbf{h}} = C_{\mathbf{h}'}$, the x values of which are

$$x^s_{1,2} = \tfrac{1}{2}\{v_{\mathbf{h}-\mathbf{h}'} \pm ((v_{\mathbf{h}-\mathbf{h}'})^2 + 8v_{\mathbf{h}}^2)^{1/2}\}. \tag{6.8}$$

In addition,

$$C_{\mathbf{h}}/C_0 = x/2v_{\mathbf{h}}\,.$$

If the reflection $\mathbf{h} - \mathbf{h}'$ is forbidden ($v_{\mathbf{h}-\mathbf{h}'} = 0$), we have an especially interesting case, where there is no direct coupling between the coefficient $C_{\mathbf{h}}$ and $C_{\mathbf{h}'}$. Nevertheless, by starting with a strong plane wave $\mathbf{k} + \mathbf{h}$ (instead of $\mathbf{k}$ as before) we get a strong wave $\mathbf{k} + \mathbf{h}'$, which is due to the indirect coupling via the plane wave $\mathbf{k}$. This effect has long been known as *Umweganregung*.

As a second example we discuss the symmetrical *four-beam case* shown in Fig. 14b. The reciprocal lattice vectors $\mathbf{h}$ and $\mathbf{h}'$ form a rectangle with $\mathbf{h} + \mathbf{h}'$ and $\mathbf{h}' - \mathbf{h}$ as diagonals. Further we assume $v_{\mathbf{h}} = v_{-\mathbf{h}}$, $v_{\mathbf{h}'} = v_{-\mathbf{h}'}$, and similarly $v_{\mathbf{h}+\mathbf{h}'} = v_{\mathbf{h}-\mathbf{h}'}$. The problem is simplified by considering that Fig. 14b is symmetrical with respect to reflections on the plane going through the origin and the line S and on the plane through O and S′. Therefore, if the Bragg conditions $\mathbf{k}^2 = (\mathbf{k} + \mathbf{h})^2 = (\mathbf{k} + \mathbf{h}')^2 = (\mathbf{k} + \mathbf{h} + \mathbf{h}')^2$ are exactly fulfilled, the eigenfunctions can be chosen as simultaneous eigenfunctions to the reflections S and S′. They can therefore be characterized as being symmetrical or antisymmetrical with respect to both reflections. Of special interest is the totally antisymmetrical wave with $C_0 = -C_{\mathbf{h}} = -C_{\mathbf{h}'} = C_{\mathbf{h}+\mathbf{h}'} = \frac{1}{2}$, which is given by

$$\phi_{\mathbf{k}}{}^a(\mathbf{r}) = e^{i\mathbf{kr}} e^{(i/2)(\mathbf{h}+\mathbf{h}')\mathbf{r}} \sin(\mathbf{hr}/2) \cdot \sin(\mathbf{h}'\mathbf{r}/2). \tag{6.9}$$

Since $\phi^a(\mathbf{r})$ has two sine modulation factors, it vanishes on both reflecting planes $\mathbf{h}$ and $\mathbf{h}'$ and even quadratically at the atomic positions on the intersecting lines of these planes. Therefore, this Bloch wave has an expecially weak absorption, even weaker than the wave ϕ^{I} of the two-beam case. We have

$$\Delta\mu = \mu_0 - \mu_{\mathbf{h}} - \mu_{\mathbf{h}'} + \mu_{\mathbf{h}+\mathbf{h}'}\,. \tag{6.10}$$

By expanding $\mu_{\mathbf{h}}$ into powers of $\mathbf{h}$ analogously to (5.12), even the quadratic terms in $\mathbf{h}$ vanish $\{(\mathbf{h}^2 + \mathbf{h}'^2) = (\mathbf{h} + \mathbf{h}')^2\}$ and the expansion starts with h^4 only. This is due to the fact that ,unlike the two-beam case, the wave field vanishes quadratically at the atomic positions.

7. Orthogonality of Bloch Functions on the Dispersion Surface

In Section 1 we have seen that all Bloch functions $u_{\mathbf{k},\nu}$ with the same $\mathbf{k}$ vectors but different band energies $E_\nu(\mathbf{k})$ are orthogonal. However under special conditions one can derive a complementary orthogonality relation, namely for Bloch functions on the different branches of the dispersion surfaces, i.e., for different $\mathbf{k}$ vectors, but for the same energy.

Starting with the Schrödinger equation (5.1), we may write for $\mathbf{k}$,

$$\mathbf{k} = \mathbf{K} + \delta\mathbf{k} \qquad \text{with} \qquad \delta k \ll K.$$

Here $\mathbf{K}$ can be thought as the wave vector in the vacuum ($\mathbf{K}^2 = \mathbf{k}^2$) and $\delta\mathbf{k}$ describes the derivation of the dispersion surface from the free electron dispersion surface. For a given direction $\mathbf{n}$ of $\delta\mathbf{k} = \delta k\mathbf{n}$ we have to the first order in $\delta\mathbf{k}$,

$$(\mathbf{k} + \mathbf{h})^2 \cong (\mathbf{K} + \mathbf{h})^2 + 2(\mathbf{K} + \mathbf{h}) \cdot \delta\mathbf{k} \cong (\mathbf{K} + \mathbf{h})^2 + 2K \cos \theta_\mathbf{h}\, \delta k, \quad (7.1)$$

with

$$\cos \theta_\mathbf{h} = \{\mathbf{K} + \mathbf{h}/|\mathbf{K} + \mathbf{h}|\} \cdot \mathbf{n}.$$

Therefore, the allowed δk values for a given direction $\mathbf{n}$ are determined

$$\sum_{\mathbf{h}'} M_{\mathbf{hh}'} C_{\mathbf{h}'} = \delta k \cos \theta_\mathbf{h} C_\mathbf{h} \quad (7.2)$$

with

$$M_{\mathbf{hh}'} = (1/2K)\{(K^2 - (\mathbf{K} + \mathbf{h})^2\, \delta_{\mathbf{h},\mathbf{h}'} - v_{\mathbf{h}-\mathbf{h}'}\}.$$

Introducing modified coefficients $\tilde{C}_\mathbf{h} = (\cos \theta_\mathbf{h})^{1/2}\, C_\mathbf{h}$, this can be transformed into an eigenvalue equation:

$$\sum_{\mathbf{h}'} \tilde{M}_{\mathbf{hh}'} \tilde{C}_{\mathbf{h}'} = \delta k\, \tilde{C}_\mathbf{h},$$

with (7.3)

$$\tilde{M}_{\mathbf{hh}'} = \{1/(\cos \theta_\mathbf{h} \cdot \cos \theta_{\mathbf{h}'})^{1/2}\}\, M_{\mathbf{hh}'}.$$

For real potentials $V(\mathbf{r})$, M is a Hermitean matrix ($M = M^\dagger$). Therefore, $\tilde{M}$ is also Hermitean if all $(\cos \theta_\mathbf{h})^{1/2}$ are real, i.e., all $\cos \theta_\mathbf{h} > 0$. Consequently, the eigenvalues $\delta k^{(\nu)}$ lying on the different branches ν of the dispersion surface are real and the Bloch waves are undamped. Further the eigensolutions $\tilde{C}_\mathbf{h}(\nu)$ form a complete and orthogonal set.

$$\sum_{\mathbf{h}} \tilde{C}_\mathbf{h}{}^*(\nu')\, \tilde{C}_\mathbf{h}(\nu) = \delta_{\nu\nu'}\,; \qquad \sum_{\nu} \tilde{C}_\mathbf{h}^*(\nu)\, \tilde{C}_\mathbf{h}(\nu) = \delta_{\mathbf{hh}'}\,. \quad (7.4)$$

However, if either $M \neq M^\dagger$ ($V \neq V^\dagger$) or if $\cos\theta_\mathbf{h} < 0$ for at least one $\mathbf{h}$ then the eigenvalues $\delta k^{(\nu)}$ are complex and the Bloch waves are damped. Further the eigensolutions $\tilde{C}_\mathbf{h}(\nu)$ are no longer orthogonal. Instead, we obtain a relation similar to (2.15) for the eigensolutions of $\tilde{M}$ and $\tilde{M}^\dagger$.

Since Eq. (7.3) refers to the modified coefficients $\tilde{C}_\mathbf{h} = (\cos\theta_\mathbf{h})^{1/2}\, C_\mathbf{h}$, the Bloch functions $u_\mathbf{k}(\mathbf{r})$ on the different branches of the dispersion surface are not orthogonal in general. They are orthogonal only for the special case when all $\cos\theta_\mathbf{h}$ are equal, since then $\tilde{C}_\mathbf{h}$ and $C_\mathbf{h}$ are identical apart from a normalization constant, and the tilde in Eq. (7.4) can be left off. For instance, all $\cos\theta_\mathbf{h}$ are equal, if all reciprocal lattice vectors $\mathbf{h}$ lie in one plane perpendicular to $\mathbf{n}$ (compare Eq. (5.9) of the two-beam case). Moreover, this is also the case for high energy electrons, when $K \gg h$.

III. Diffraction of Low-Energy Electrons, Neutrons, and X-Rays

In diffraction experiments with low-energy electrons the energy is typically of the order of 50 eV and therefore comparable with the mean potential V_0 or the potential of a single atom. Such low energies give rise to many complications. One of these is due to the fact that even the interaction with an isolated atom can no longer be calculated by Born's approximation but has to be treated exactly. With respect to this, the situation is even worse for thermal neutron scattering where the extremely short range interaction with the nucleus is of the order of several tens of mega electron volts, compared with the neutron energy of 0.025 eV. Here the interaction potential normally is replaced by Fermi's pseudopotential.[23] However, this procedure is restricted to the first Born approximation. A dynamical theory has to reconsider this problem, which was investigated first by Goldberger and Seitz.[24] In the first section we will show that the difficulty due to the strong single-particle interaction can be overcome by the introduction of single-scattering matrices.

8. Multiple Scattering with Single-Scattering Matrices[25]

We shall start with the Lippmann–Schwinger equation, which is in operator form

$$\psi = \varphi + G_0 V \psi \qquad \text{where} \qquad G_0 = 1/(E + i\epsilon - H_0). \tag{8.1}$$

[23] E. Fermi, *Ric. Sci.* **7**, Pt. 2, 13 (1936).
[24] M. L. Goldberger and F. Seitz, *Phys. Rev.* **71**, 294 (1947).
[25] M. L. Goldberger and K. M. Watson, "Collision Theory." Wiley, New York, 1964.

Here φ stands for the incident wave $\varphi(\mathbf{r}) = e^{i\mathbf{kr}}$ and $G_0(\mathbf{r}, \mathbf{r}')$ is the free Green's function:

$$G_0(\mathbf{r}, \mathbf{r}') = -\frac{2m}{\hbar^2 4\pi} \frac{e^{iK|r-\mathbf{r}'|}}{|\mathbf{r} - \mathbf{r}'|}, \qquad \text{with} \qquad E = \frac{\hbar^2 K^2}{2m}. \tag{8.2}$$

Equation (8.1) also can be written as

$$\psi = \varphi + G_0 T\varphi, \tag{8.3}$$

where the transition or scattering matrix T is given by

$$T = V\{1/(1 - G_0 V)\} = V + VG_0 T. \tag{8.4}$$

Considering the scattering by many centers, the potential V is a sum of the single-center contributions $v_n(\mathbf{r})$.

$$V(\mathbf{r}) = \sum_n v_n(\mathbf{r}) = \sum_n v(\mathbf{r} - \mathbf{R}^n) \tag{8.5}$$

if all centers are equal. Now the scattering by the potential v_n alone can be described by the single-scattering matrix t_n.

$$t_n = v_n\{1/(1 - G_0 v_n)\} = v_n + v_n G_0 t_n . \tag{8.6}$$

We have $T = t_n$, if only the center n is present ($V = v_n$). In analogy to the incident wave in (8.3) we introduce an "effective incident wave" φ_n for the atom n by writing

$$\psi = \varphi_n + G_0 t_n \varphi_n . \tag{8.7}$$

By multiplying (8.7) with v_n and using (8.6), we obtain

$$v_n \psi = t_n \varphi_n . \tag{8.8}$$

Introducing this into the Lippmann–Schwinger equation (8.1), the wave function ψ can be expressed in terms of the effective fields φ_m:

$$\psi = \varphi + G_0 \sum_m t_m \varphi_m . \tag{8.9}$$

Moreover, by comparing this with the defining equation (8.7) the φ_n have to be solutions of the coupled equations

$$\varphi_n = \varphi + \sum_{m(\neq n)} G_0 t_m \varphi_m . \tag{8.10}$$

Therefore, the effective incident field φ_n for the center n consists of the incident plane wave φ plus the scattered waves $G_0 t_m \varphi_m = G_0 v_m \psi$ from the other centers $m \neq n$, as is illustrated in Fig. 15. These equations have the advantage of clearly separating the scattering properties of the single centers, given by t_m, from the multiple-scattering properties of

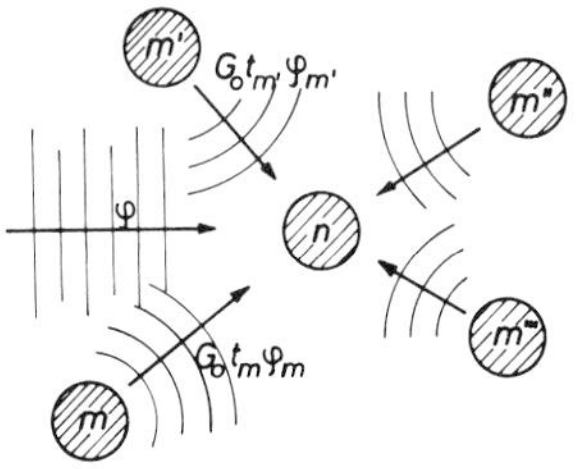

FIG. 15. Contributions to the effective incident wave φ_n.

the whole system. However, we have paid for this by getting a system of coupled equations (8.10) instead of the single equation (8.1).

From (8.9), we obtain a modified approximation by replacing the effective field φ_m by the incident field φ.

$$\psi \cong \varphi + G_0 \sum_m t_m \varphi. \tag{8.11}$$

This approximation, known as pseudokinematical theory in LEED, goes over into the usual Born approximation (kinematical theory) if the t_m are replaced by v_m. Compared to this, the approximation has the advantage that the single-scattering process is treated exactly, whereas the multiple scattering by different centers is still neglected.

In an *infinite crystal* the eigenfunctions $\phi(\mathbf{r})$ for an energy E obey the homogeneous equation

$$\phi = G_0 V \phi. \tag{8.12}$$

Here, one can as well replace G_0 by the advanced Green's function or by the principal value Green's function, i.e.,

$$1/(E - i\epsilon - H_0) \qquad \text{or} \qquad P(1/(E - H_0)). \tag{8.13}$$

By choosing the eigenfunctions ϕ as Bloch waves, we have

$$\phi_{\mathbf{k}}(\mathbf{r} + \mathbf{R}^m) = e^{i\mathbf{k}\mathbf{R}^m} \phi_{\mathbf{k}}(\mathbf{r}). \tag{8.14}$$

Using this and the periodicity of $V(\mathbf{r}) = \sum_n v(\mathbf{r} - \mathbf{R}^n)$, Eq. (8.12) can be written in the interesting form

$$\phi_{\mathbf{k}}(\mathbf{r}) = \int_{-\infty}^{+\infty} d\mathbf{r}'\, G(\mathbf{r}, \mathbf{r}')\, v(\mathbf{r}')\, \phi_{\mathbf{k}}(\mathbf{r}'). \tag{8.15}$$

Here the integral contains only the single potential $v(\mathbf{r})$. Furthermore, the Green's function G, called the complete Greenian by Ziman,[26] is

$$G(\mathbf{r}, \mathbf{r}') = \sum_n G_0(\mathbf{r} - \mathbf{r}' + \mathbf{R}^n)\, e^{-i\mathbf{k}\mathbf{R}^n}. \tag{8.16}$$

It depends explicitly on $\mathbf{k}$ and not only on E as G_0 does. With respect to $\mathbf{r}$, it has the same translation property (8.14) as $\phi_{\mathbf{k}}$.

By introducing the t_n matrices of Eq. (8.6) and the effective fields φ_n as in (8.7), we may also write Eq. (8.12) in the form

$$\phi = G_0 \sum_n t_n \varphi_n \qquad \text{and} \qquad \varphi_n = G_0 \sum_{m(\neq n)} t_m \varphi_m, \tag{8.17}$$

which is quite analogous to the multiple-scattering equations (8.9) and (8.10). Furthermore, we can see from (8.17) that the quasiperiodicity (8.14) also leads to a periodicity condition for the corresponding effective field φ_n, namely

$$\varphi_{n+m}(\mathbf{r} + \mathbf{R}^m) = e^{i\mathbf{k}\mathbf{R}^m} \varphi_n(\mathbf{r}). \tag{8.18}$$

Therefore, all effective waves φ_n can be reduced to a single one, for instance φ_0. By doing this we obtain from Eq. (8.14) in the $\mathbf{r}$-representation

$$\phi(\mathbf{r}) = \int d\mathbf{r}'\, d\mathbf{r}''\, G(\mathbf{r}, \mathbf{r}')\, t(\mathbf{r}', \mathbf{r}'')\, \varphi_0(r''), \tag{8.19}$$

$$\varphi_0(\mathbf{r}) = \int d\mathbf{r}'\, d\mathbf{r}'\, G'(\mathbf{r}, \mathbf{r}')\, t(\mathbf{r}', \mathbf{r}'')\, \varphi_0(\mathbf{r}''), \tag{8.20}$$

with

$$G'(\mathbf{r}, \mathbf{r}') = G - G_0 = \sum_{n\neq 0} G_0(\mathbf{r} - \mathbf{r}' + \mathbf{R}^n)\, e^{-i\mathbf{k}\mathbf{R}^n}. \tag{8.21}$$

These equations can also be obtained directly from (8.15) by substituting $v(\mathbf{r}')\, \phi(\mathbf{r}')$ by $t\varphi_0$.

[26] See, e.g., J. M. Ziman, "On the Band Structure Problem," *in* "Theory of Condensed Matter, Lectures Presented at an International Course, Trieste, 1967. IAEA, Vienna, 1968.

9. Scattering by Muffin-Tin Potentials

Now we apply the multiple-scattering equations of the last section to a system of spherically symmetric, but nonoverlapping potentials. Then the potential $v(\mathbf{r})$ of a single center is

$$v(\mathbf{r}) = \begin{cases} v(|\mathbf{r}|) & \text{for} \quad r < r_s, \\ 0 & \text{for} \quad r > r_s. \end{cases} \tag{9.1}$$

In order to avoid overlap between the potential of the different atoms, r_s has to be smaller than $d_{nn}/2$, where d_{nn} is the nearest-neighbor distance. Such potentials, known as "muffin-tin potentials" have been used extensively for band-structure calculations.[27] By substituting $\mathbf{r} \to \mathbf{R}^n + \mathbf{r}$ in $\varphi_n(\mathbf{r})$, we obtain from Eq. (8.10)

$$\varphi_n(\mathbf{R}^n + \mathbf{r}) = e^{i\mathbf{K}\mathbf{R}^n + i\mathbf{K}\mathbf{r}} + \int d\mathbf{r}'\, d\mathbf{r}'' \sum_{m(\neq n)} G_0(\mathbf{r} - \mathbf{r}' + \mathbf{R}^n - \mathbf{R}^m)$$

$$\times\, t(\mathbf{r}', \mathbf{r}'')\, \varphi_m(\mathbf{R}^m + \mathbf{r}''). \tag{9.2}$$

For the free Green's function one verifies directly that

$$(-(\hbar^2/2m)\, \partial_{\mathbf{r}}^{\,2} - \hbar^2 K^2/2m)\, G_0(\mathbf{r} - \mathbf{r}' + \mathbf{R}^n - \mathbf{R}^m) = -\delta(\mathbf{r} - \mathbf{r}' + \mathbf{R}^n - \mathbf{R}^m). \tag{9.3}$$

The same equation holds if $\partial_{\mathbf{r}}$ is replaced by $\partial_{\mathbf{r}'}$. For $r \leqslant r_s < d_{nn}/2$ and $r' \leqslant r_s$ the source term in (9.2) vanishes and G_0 satisfies the potential-free Schrödinger equation in these regions. The same applies to the incident wave $e^{i\mathbf{K}\mathbf{r}}$, of course. Then it follows from (9.2) that $\varphi_n(\mathbf{R}^n + \mathbf{r})$ also satisfies the potential-free Schrödinger equation for $r \leqslant r_s$. Therefore, by expanding $\varphi_n(\mathbf{R}^n + \mathbf{r})$ (or G_0 and $e^{i\mathbf{K}\mathbf{r}}$) into spherical harmonics $Y_{lm}(\mathbf{r}/r)$, the radial functions $R_l(r)$ have to be solutions of the radial Schrödinger equation without potential. Since this is a differential equation of second order, there are two linearly independent solutions for a given energy, namely the spherical Bessel functions $j_l(Kr)$ and $n_l(Kr)$.[28] Whereas $j_l(Kr)$ is finite everywhere, $n_l(Kr)$ is singular in the origin and has to be dropped. Thus the following expansion for φ_n is valid for $r \leqslant r_s$:

$$\varphi_n(\mathbf{R}^n + \mathbf{r}) = \sum_L i^l \varphi_L{}^n\, j_l(Kr)\, Y_L(\mathbf{r}/r). \tag{9.4}$$

[27] See, e.g., J. M. Ziman, "Principles of the Theory of Solids." Cambridge University Press, Cambridge, 1965.

[28] W. Magnus and F. Oberhettinger, "Formeln und Sätze für die speziellen Funktionen der mathematischen Physik." Springer-Verlag, Berlin, 1948.

Here the index $L = (l, m)$ stands for the angular momentum index l and for the magnetic index m. The $\varphi_L{}^n$ are unknown coefficients, and the spherical harmonics are orthonormalized:

$$\int d\Omega\, Y_{L'}^*(\mathbf{r}/r)\, Y_L(\mathbf{r}/r) = \delta_{L,L'} = \delta_{l,l'}\delta_{m,m'}\,. \tag{9.5}$$

Similarly, we get a souble expansion for $G_0(\mathbf{r}, \mathbf{r}')$ for $r \leqslant r_s$ and $r' \leqslant r_s$.

$$G_0(\mathbf{r} - \mathbf{r}' + \mathbf{R}^n - \mathbf{R}^m) = (2m/\hbar^2) \sum_{L,L'} i^{l-l'} G_{L,L'}^{n-m} \times j_l(Kr)\, j_{l'}(Kr')\, Y_L(\mathbf{r}/r)\, Y_{L'}{}^*(\mathbf{r}/r). \tag{9.6}$$

The corresponding expansion for $e^{i\mathbf{Kr}}$ is[28]

$$e^{i\mathbf{Kr}} = \sum_L i^l 4\pi Y_L{}^*(\mathbf{K}/K)\, j_l(kr)\, Y_L(\mathbf{r}/r). \tag{9.7}$$

Furthermore, due to the rotation invariance of $v(r)$, the scattering matrix $t(\mathbf{r}, \mathbf{r}')$ depends only on r, r' and the angle θ between r and r'. Therefore, an expansion of the form

$$\begin{aligned} t(\mathbf{r}, \mathbf{r}') &= \sum_l (2l + 1)/4\pi\; t_l(r, r')\, P_l(\cos\theta) \\ &= \sum_L t_l(r, r')\, Y_L(\mathbf{r}/r)\, Y_L{}^*(\mathbf{r}'/r) \end{aligned} \tag{9.8}$$

holds, where for the last line we have used the addition theorem for spherical harmonics.[28] We now introduce the expansions (9.4)–(9.8) in Eq. (9.2) and by using the orthogonality of the Y_L's we obtain

$$\varphi_L{}^n = e^{i\mathbf{KR}\,n} \cdot 4\pi Y_L{}^*(\mathbf{K}/K) + \sum_{m(\neq n);L'} G_{L,L'}^{n-m}\, \tau_{l'}\, \varphi_{L'}^m\,. \tag{9.9}$$

Here τ_l is given by

$$\begin{aligned} \tau_l &= \int_0^\infty r'^2\, dr'\, r''^2\, dr''\, j_l(Kr')\, t_l(r', r'')\, j_l(Kr'') \\ &= -(1/K)\, e^{i\delta_l} \sin\delta_l \end{aligned} \tag{9.10}$$

as can be shown by partial wave analysis. δ_l is the phase shift of the lth partial wave. Thus we have reduced the solution of the integral equation (9.2) to the solution of the algebraic equations (9.9). This was only possible because the single potentials do not overlap. Otherwise the expansions (9.4) and (9.6) are not valid. The division into single-center properties and properties of the whole system is still apparent in (9.9). All the information we need about the single potential is contained in τ_l or in the phase shifts δ_l, whereas the coefficients $G_{L,L'}^{n-m}$ are determined by the structure of the system alone. Furthermore, we should point out that by knowing the coefficients $\varphi_L{}^n$ and the effective fields $\varphi_n(\mathbf{r})$ we can obtain the wavefunction ψ from (8.9). The solution of the algebraic equations (9.9) for a crystal slab or half crystal is still a formidable problem because of the infinite number of atoms and angular momenta involved. Of some help is here the xy periodicity, which reduces all effective fields of the same atomic layer in the xy-plane to a single field for each layer. Namely, in analogy to (8.18) we have for a "plane" translation $\mathscr{R}^m$

$$\varphi_{n+m}(\mathbf{r} + \mathscr{R}^m) = \exp[i\mathscr{K}\mathscr{R}^m]\, \varphi_n(\mathbf{r}). \tag{9.11}$$

Therefore, the number of unknown $\varphi_L{}^n$ in (9.9) is the product of the number of atomic layers times the number of angular momenta considered. A particularly simple but also enlightening problem is the case of a monolayer of atoms scattering isotropically ($L = 0$). For this case, we have just one constant, say $\varphi_0{}^0$. The result,[29,30] shows interesting resonances, which are due to quasi-localized surface states, as well as certain threshold effects connected with "surface waves."

For a real crystal one may either try to solve the equations connecting the different monolayers. This method, proposed by Beeby,[31] has been successfully used for LEED.[32] On the other hand, we may try to solve the equations for an infinite crystal, as will be shown. Then we have to match the allowed Bloch waves and allowed plane waves at the crystal surfaces. This method has been used in LEED also.[33-35]

To obtain the Bloch wave $\phi_{\mathbf{k}}(\mathbf{r})$ (Eq. (8.19)) for an infinite crystal we have to solve Eq. (8.20) for the effective field φ_0. Again, for muffin-tin potentials, $\varphi_0(\mathbf{r})$ and the Green's function $G'(\mathbf{r}, \mathbf{r}')$ satisfy for $r, r' \leqslant r_s$ the potential- and source-free Schrödinger equation. Therefore we obtain expansions of the form

[29] E. G. McRae, *J. Chem. Phys.* **45**, 3258 (1966).
[30] K. Kambe, *Z. Naturforsch.* **22a**, 322 (1967).
[31] J. L. Beeby, *J. Phys. C.* **1**, 82 (1968).
[32] C. B. Duke and C. W. Tucker, *Surface Sci.* **15**, 231 (1969); *Phys. Rev. Lett.* **23**, 1163 (1969).
[33] D. S. Boudreaux and V. Heine, *Surface Sci.* **8**, 426 (1967).
[34] J. B. Pendry, *J. Phys. C.* **2**, 2273 (1969).
[35] G. Capart, *Surface Sci.* (1971) in press.

$$\varphi_0(\mathbf{r}) = \sum_L i^l \varphi_L j_l(Kr)\, Y_L(\mathbf{r}/r) \qquad \text{for} \quad r \leqslant r_s\,, \tag{9.12}$$

$$G'(\mathbf{r}, \mathbf{r}') = \sum_{L,L'} i^{l-l'} G_{LL'} j_l(Kr) j_{l'}(Kr') \times Y_L(\mathbf{r}/r)\, Y^*_{L'}(\mathbf{r}'/r) \qquad \text{for} \quad r, r' \leqslant r_s\,. \tag{9.13}$$

Introducing these relations into (8.20) and using the expansion for $t(\mathbf{r}, \mathbf{r}')$ (Eq. (9.8)), we obtain the homogeneous equations

$$\varphi_L = \sum_{L'} G_{LL'} \tau_{l'} \varphi_{L'}\,. \tag{9.14}$$

They have solutions only if the dispersion condition

$$\det|G_{LL'}\tau_{l'} - \delta_{L,L'}| = 0 \tag{9.15}$$

is satisfied connecting the allowed $\mathbf{k}$ and E values. This is the t-matrix version[36] of the Korringa–Kohn–Rostocker method (KKR method) for band structure calculation.[37] In practice, the evaluation of the "structure constants" $G_{LL'}$ represents most of the work.

For the case of s scattering only, the determinant simply gives $G_{00}\tau_0 = 1$. This case is essentially equivalent to the treatment in the following section.

10. Diffraction of Neutrons

In this section we want to apply the multiple-scattering equations to the diffraction of neutrons by an ideal crystal. However, let us first discuss the *scattering by a single nucleus*. Because the wavelength ($\approx$1 Å) is very much larger than the radius r_0 of the nucleus ($\approx 10^{-13}$ cm), we have only s scattering. For the evaluation of the scattering amplitude, we can set the energy $E = 0$. The case of higher energies is discussed later on. Then we get from the Schrödinger equation for $E = 0$,

$$(\partial_r{}^2 - v(r))\, r\psi(r) = 0, \tag{10.1}$$

$$\psi(r) = 1 - \frac{a_0}{r} \qquad \text{or} \qquad r\psi(r) = r - a_0 \qquad \text{for} \quad r \geqslant r_0\,. \tag{10.2}$$

The real constant a_0 is the scattering length for $E = 0$, which is connected with the cross section by $\sigma = 4\pi a_0{}^2$. Graphically, the scattering

[36] J. M. Ziman, *Proc. Phys. Soc.* **86**, 337 (1965).

[37] W. Kohn and N. Rostoker, *Phys. Rev.* **94**, 1111 (1954). F. S. Ham and B. Segall, *Phys. Rev.* **124**, 1786 (1961).

length can be obtained by the intersection of straight line $r - a_0$ with the r-axis (Fig. 16). For an attractive potential ($v(r) < 0$), the curvature of $r\psi$ is negative according to (10.1) if $r\psi$ is positive, and vice versa. Furthermore, $r\psi$ vanishes at the origin. For a relatively weak but negative potential we may therefore obtain the curve of Fig. 16a, leading to a negative scattering length. If the potential strength increases, then a_0

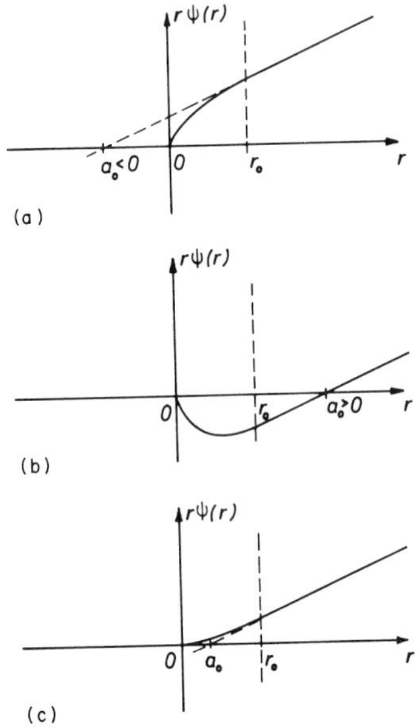

FIG. 16. Scattering length a_0 (a) for an attractive potential without bound state ($a_0 < 0$); (b) for an attractive potential with one bound state ($a_0 > 0$); (c) for a repulsive potential ($0 \leqslant a_0 \leqslant r_0$).

goes to $-\infty$ and the cross section diverges. This is due to the fact that there exists a bound state with zero energy in this case. Namely for a bound state with energy $E = -\hbar^2\varkappa^2/2m$, the wave function for $r \geqslant r_0$ is $r\psi(r) \sim e^{-\varkappa r}$, leading to a constant for $\varkappa = 0$. But this is equivalent to the condition $a_0 \to \pm\infty$. By further increasing the strength of the potential, we obtain a curve as shown in Fig. 16b, resulting in a positive scattering length. Furthermore, $r\psi(r)$ has now an extremum for $r \leqslant r_0$ which is connected with the bound state of the potential. By further increasing the potential, a_0 will become negative again, etc. Whenever a_0 goes to $-\infty$, a new bound state with energy $E = 0$ is produced, and $\psi(r)$ will have as many extrema as there are bound states. On the other hand, for an repulsive potential the situation is quite different (Fig. 16c). Here a_0 is always positive and smaller than r_0 .

In the case of absorption, the potential $v(r)$ is complex and the scattering length for $E = 0$ also becomes complex: $a_0 = a_0' + ia_0''$. However, neutrons are very weakly absorbed and normally $|a_0''/a_0| \approx 10^{-5}$. Even for such a strong absorber as boron the ratio $|a_0''/a_0'|$ is only 0.04.

Multiple Scattering

For the scattering of neutrons the multiple-scattering equations can be simplified essentially. Because the range r_0 of the potential is very much smaller than both the wavelength and the lattice constant, in Eq. (8.9) $G_0(\mathbf{r} - \mathbf{r}')$ and $\varphi_n(\mathbf{r}'')$ can be replaced by $G_0(\mathbf{r} - \mathbf{R}^n)$ and $\varphi_n(\mathbf{R}^n)$. Then the single-scattering matrix $t(\mathbf{r}', \mathbf{r}'')$ only enters through the integral $\int d\mathbf{r}'\, d\mathbf{r}''\, t(\mathbf{r}', \mathbf{r}'')$. This approximation is equivalent to replacing the t-matrix by

$$t(\mathbf{r}, \mathbf{r}') = (\hbar^2/2m)\, 4\pi a\, \delta(\mathbf{r})\, \delta(\mathbf{r}'). \tag{10.3}$$

Here, a is the scattering length for the energy E. Together with (10.3), we obtain from (8.9)

$$\psi(\mathbf{r}) = e^{i\mathbf{K}\mathbf{r}} - \sum_n a \frac{e^{iK|\mathbf{r}-\mathbf{R}_n|}}{|\mathbf{r} - \mathbf{R}_n|} \varphi_n(\mathbf{R}_n). \tag{10.4}$$

Therefore, we only need the effective field φ_n at the position $\mathbf{R}_n$ of the nucleus number n. Then Eq. (8.10) gives

$$\varphi_n(\mathbf{R}_n) = e^{i\mathbf{K}\mathbf{R}_n} - \sum_{m(\neq n)} a \frac{e^{iK|\mathbf{R}_n-\mathbf{R}_m|}}{|\mathbf{R}_n - \mathbf{R}_m|} \varphi_n(\mathbf{R}_m). \tag{10.5}$$

These algebraic equations are quite analogous to Eqs. (9.11) for muffin-tin potentials. However due to the s scattering we have only one unknown constant per nucleus, namely $\varphi_n(\mathbf{R}_n)$.

The scattering length a for an energy E is connected in a simple way with the scattering length a_0 for $E = 0$. From (8.6) we have

$$\begin{aligned} t &= v + v \frac{1}{E + i\epsilon - H_0} t \\ &= v \frac{1}{i\epsilon - H_0} t + v + v \left(\frac{1}{E + i\epsilon - H_0} - \frac{1}{i\epsilon - H_0} \right) t. \end{aligned} \tag{10.6}$$

By taking the term $v(1/(i\epsilon - H_0))t$ to the left-hand side and dividing by $1 - v(1/(i\epsilon - H_0))$, we obtain

$$t = t_0 + t_0 \left(\frac{1}{E + i\epsilon - H_0} - \frac{1}{i\epsilon - H_0} \right) t \tag{10.7}$$

with

$$t_0 = \frac{1}{1 - v(1/(i\epsilon - H_0))} v.$$

Here t_0, being real, is the single-scattering matrix for zero energy. In the $\mathbf{r}$ representation, Eq. (10.7) is

$$t(\mathbf{r}, \mathbf{r}') = t_0(\mathbf{r}, \mathbf{r}') + \int d\mathbf{r}'' \, d\mathbf{r}''' \, t_0(\mathbf{r}, \mathbf{r}'). \\ \times \left(\frac{-2m}{\hbar^2 4\pi}\right) \frac{\exp[iK|\mathbf{r}'' - \mathbf{r}'''|] - 1}{|\mathbf{r}'' - \mathbf{r}'''|} t(\mathbf{r}''', \mathbf{r}'). \tag{10.8}$$

Now if both t and t_0 have the form (10.3) with scattering lengths a and a_0, we obtain from (10.8)

$$a = a_0 - iKa_0 a = a_0/(1 + iKa_0). \tag{10.9}$$

Because $|Ka_0| \ll 1$ normally, $a \cong a_0$. In principle, however, a_0 can be arbitrarily large. Then a is limited by $1/K$.

By setting $K = i\varkappa$, the scattering length diverges for $\varkappa = 1/a_0$ if $a_0 > 0$. This indicates that for $a_0 > 0$ the potential $v(r)$ has a bound state with the energy $E = -(\hbar^2/2m)(1/a_0^2)$, which, for instance, vanishes for $|a_0| \to \infty$.

Equation (10.3) represents the t-matrix for a potential of "zero range" i.e., in the limit $Kr_0 \to 0$. For a repulsive potential the scattering length a_0 vanishes in this limit because $0 \leqslant a_0 \leqslant r_0$ always. However, for an attractive potential, a finite value of a_0 can always be obtained by adjusting the potential depth. Furthermore, we have at most one bound state. All others have energies $E \sim -(\hbar^2/2m)(1/r_0^2)$ approaching $-\infty$ for $r_0 \to 0$.

To determine the eigenfunctions for potentials of zero range in an infinite crystal, we introduce the ansatz (10.3) for t in (8.19) and (8.20). Then the Bloch wave $\phi_\mathbf{k}$ is given by

$$\begin{aligned} \phi_\mathbf{k}(\mathbf{r}) &= G(\mathbf{r}, 0)(\hbar^2/2m)\, 4\pi a \varphi_0(0) \\ &= -\left(\sum_n (e^{iK|\mathbf{r}+\mathbf{R}_n|}/|\mathbf{r} + \mathbf{R}_n|)\, e^{-i\mathbf{k}\mathbf{R}_n}\right) a\varphi_0(0). \end{aligned} \tag{10.10}$$

Here, $\varphi_0(0)$ is a normalization constant only. Furthermore, Eq. (8.20) gives the dispersion condition connecting the allowed $\mathbf{k}$ values with the energy $\hbar^2 K^2/2m$.

$$1 = G'(0, 0) \frac{\hbar^2}{2m} 4\pi a = -a \sum_{n \neq 0} \frac{e^{iK|\mathbf{R}_n|}}{|\mathbf{R}_n|} e^{-i\mathbf{k}\mathbf{R}_n}. \tag{10.11}$$

Equations (10.10) and (10.11) can also be written in a different form with sums over the reciprocal lattice. For example, by Fourier transformation, we have from (10.10)

$$\phi_{\mathbf{k}}(\mathbf{r}) = 4\pi a\varphi_0(0) \int \frac{d\mathbf{K}'}{(2\pi)^3} \frac{e^{i\mathbf{K}'\mathbf{r}}}{K^2 - K'^2 + i\epsilon} \cdot \sum_n e^{i(\mathbf{K}'-\mathbf{k})\mathbf{R}_n}$$

$$= \frac{4\pi a}{V_c} \varphi_0(0) \sum_{\mathbf{h}} \frac{e^{i(\mathbf{k}+\mathbf{h})\mathbf{r}}}{K^2 - (\mathbf{k}+\mathbf{h})^2}. \tag{10.12}$$

The sum in (10.12) gives a δ function if $\mathbf{K}'$ is equal to $\mathbf{k}$ up to a reciprocal lattice vector $\mathbf{h}$. Similarly we have from (10.11) by adding and subtracting the term $n = 0$,

$$\frac{1}{a} = 4\pi \int \frac{d\mathbf{K}'}{(2\pi)^3} \frac{1}{K^2 + i\epsilon - K'^2} \left(\frac{(2\pi)^3}{V_c} \sum_{\mathbf{h}} \delta(\mathbf{K}' - \mathbf{k} - \mathbf{h}) - 1 \right). \tag{10.13}$$

The two terms in the parentheses cannot be integrated separately since both diverge. For the first term, which is a discrete sum, the $i\epsilon$ in the denominator is unnecessary, whereas the imaginary part of the second term is iK. With the expression (10.9) for a, the imaginary term iK cancels on both sides and the remaining quantities are real. Substituting $\mathbf{K}' = \mathbf{h} + \mathbf{k}'$, the second integral can also be written as a sum over $\mathbf{h}$ with integrals $d\mathbf{k}'$ over the first Brillouin zone. Thus we have

$$1 = \frac{4\pi a_0}{V_c} \sum_{\mathbf{h}} \left\{ \frac{1}{K^2 - (\mathbf{k}+\mathbf{h})^2} - \frac{V_c}{(2\pi)^3} \int_{\text{1.B.Z.}} d\mathbf{k}' \frac{1}{K^2 - (\mathbf{k}' + \mathbf{h})^2} \right\}. \tag{10.14}$$

The sum over $\mathbf{h}$ converges because the integral cancels the first term for large $\mathbf{h}$. However, the sum of the first term or second term alone diverges. This is directly connected with the difference between the effective field and the wave function. According to (10.10), $\phi(\mathbf{r})$ and $G(\mathbf{r}, 0)$ diverge as $1/r$ for $\mathbf{r} \to 0$ and similarly for $\mathbf{r} \to \mathbf{R}_n$. However, the effective field $\varphi_0(\mathbf{r})$ and $G'(\mathbf{r}, 0)$ do not diverge for $\mathbf{r} \to 0$. Due to the smallness of a, however, the difference between $\phi(\mathbf{r})$ and $\varphi_0(\mathbf{r})$ is of practical importance only in the immediate vicinity of the nucleus, but not anywhere else in the unit cell.

In the one-beam case, i.e., if the Bragg condition is not fulfilled, we get from (10.14)

$$1 = \frac{4\pi a_0}{V_c} \frac{1}{K^2 - \mathbf{k}^2} \qquad \text{or} \qquad K^2 - \frac{4\pi a_0}{V_c} - \mathbf{k}^2 = 0. \tag{10.15}$$

The other terms in (10.14) have the order of magnitude

$$(4\pi a_0/V_c)(1/K^2) \sim O(10^{-5})$$

and can be neglected. However, this would not be possible for extremely large $a_0 \approx O(1/K)$. For $a_0 > 0$ the refractive index $n = k/K$ is $n = 1 - 2\pi a_0/V_c K^2 < 1$ leading to total reflection for nearly glancing incidence. On the other hand, for $a_0 < 0$ and consequently $K^2 < k^2$ or $n > 1$, the neutron can be bounded by the crystal similarly to a band electron. However, the binding energy is only of the order of 10^{-7} eV. Nevertheless, such bounded states may have some physical significance in temporarily capturing neutrons.[38]

If the Bragg condition is fulfilled for a number of beams, say $\mathbf{k}$, $\mathbf{k} + \mathbf{h}_1, \ldots, \mathbf{k} + \mathbf{h}_n$, then we get from (10.14) by neglecting the integral as before,

$$1 = \frac{4\pi a_0}{V_c} \sum_{\mathbf{h}=0}^{\mathbf{h}_n} \frac{1}{K^2 - (\mathbf{k} + \mathbf{h})^2}. \tag{10.16}$$

This can also be written in a more familiar form. Namely, from (10.12) we obtain for $\mathbf{h} = 0, \ldots, \mathbf{h}_n$,

$$\{K^2 - (\mathbf{k} + \mathbf{h})^2\} C_{\mathbf{h}} = \frac{4\pi a_0}{V_c} \varphi_0(0) = \frac{4\pi a_0}{V_c} \sum_{\mathbf{h}'=0}^{\mathbf{h}_n} C_{\mathbf{h}'}, \tag{10.17}$$

where the last identity follows by using (10.16).

This equation is identical with the basic equation for electron diffraction, if all $v_{\mathbf{h}}$ are replaced by $4\pi a_0/V_c$. Therefore, one may derive (10.17) as well without any t-matrix formalism simply by using Fermi's pseudopotential

$$v_F(\mathbf{r}) = (\hbar^2/2m)\, 4\pi a_0\, \delta(\mathbf{r}). \tag{10.18}$$

By neglecting the integral in (10.14), the effective field $\varphi_0(\mathbf{r})$ and $\phi(\mathbf{r})$ become equal. Physically, this is due to the fact that with a small number of beams the difference between $\varphi_0(\mathbf{r})$ and $\phi(\mathbf{r})$, being only important at the position of the nucleus, cannot be resolved. Therefore, by restricting to a few strongly excited beams, the Fermi potential (10.18) can be used and the whole formalism of electron diffraction remains applicable, e.g., the two-beam case, boundary conditions, etc. However, in addition to the assumption of zero-range potentials ($r_0 \ll \lambda$, d_{nn}) used to derive the representation (10.3) for the t matrix or for the dispersion condition (10.14), we have used the condition $|a_0| \ll \lambda$, d_{nn} in order to derive (10.17) or Fermi's pseudopotential. Therefore deviations from (10.17) are expected for extremely large a_0, for instance, near resonances.

[38] Yu. Kagan, *JETP Lett.* **11**, 147 (1970).

11. Diffraction of X-Rays

The theory for X-ray diffraction is quite analogous to the theory for electron diffraction, except that we have to consider a vector field instead of a scalar field. Therefore, we have to start with Maxwell's equations replacing the Schrödinger equation. For simplicity, we will treat the electrons of the crystal classically. A more thorough quantum mechanical treatment is given in Part VI. The frequencies of the motion of atomic electrons are of the order $\omega_0 \approx v/a_B$ where v is the electron velocity and a_B is Bohr's radius. Because the X-ray wavelength is comparable to a_B, these frequencies are small compared to the X-ray frequency $\omega = 2\pi c/\lambda$ if $v \ll c$. In this case the electrons may be treated as free. Their motion due to an electric field $E \sim e^{-i\omega t}$ is described by

$$m\ddot{\mathbf{r}} = e\mathbf{E}(\mathbf{r}, t). \tag{11.1}$$

If we denote the space dependent density of electrons by $\rho(\mathbf{r})$, then the density of the charge current is

$$\mathbf{j}(\mathbf{r}) = e\rho(\mathbf{r})\,\dot{\mathbf{r}} = i(e^2/m\omega)\,\rho(\mathbf{r})\,\mathbf{E}(\mathbf{r}, t). \tag{11.2}$$

Now we introduce this current into Maxwell's equations, which for harmonic time dependence $\sim e^{-i\omega t}$ are

$$\partial_{\mathrm{r}} \times \mathbf{E} = i(\omega/c)\,\mathbf{H} \qquad \text{and} \qquad \partial_{\mathrm{r}} \cdot \mathbf{H} = 0, \tag{11.3}$$

$$\partial_{\mathrm{r}} \times \mathbf{H} = -i(\omega/c)\,\mathbf{E} + (4\pi/c)\,\mathbf{j} = -i(\omega/c)\,\epsilon\,\mathbf{E} = -i(\omega/c)\,\mathbf{D}; \qquad \partial_{\mathrm{r}} \cdot \mathbf{D} = 0. \tag{11.4}$$

Here, the dielectric constant ϵ is given by

$$\epsilon(\mathbf{r}, \omega) - 1 = \chi(\mathbf{r}) = -(4\pi e^2/m\omega^2)\,\rho(\mathbf{r}) = -(4\pi r_{\mathrm{e}}/K^2)\,\rho(\mathbf{r}), \tag{11.5}$$

with $\omega = cK = c2\pi/\lambda$ and the classical electron radius $r_{\mathrm{e}} = e^2/mc^2 = 2.82 \cdot 10^{-13}$ cm. Actually, for X-rays the deviation of ϵ from 1 is very small and for $\lambda \simeq 1$ Å we have $|\chi| \leqslant 10^{-4}$ for most elements.

By eliminating $\mathbf{H}$ from (11.3) and (11.4) we obtain

$$\partial_{\mathrm{r}} \times \partial_{\mathrm{r}} \times \mathbf{E} = (\omega/c)^2\,\mathbf{D}. \tag{11.6}$$

Furthermore, due to the smallness of $\chi(\mathbf{r})$ we have

$$\mathbf{E} = (1/\epsilon)\,\mathbf{D} \cong \mathbf{D} - \chi\mathbf{D}. \tag{11.7}$$

Substituting this result into Eq. (11.6) and using $\partial_{\mathbf{r}} \cdot \mathbf{D} = 0$ we get an equation for $\mathbf{D}$ alone:

$$(\partial_{\mathbf{r}}^2 + K^2)\,\mathbf{D}(\mathbf{r}) = -\partial_{\mathbf{r}} \times \partial_{\mathbf{r}} \times (\chi(\mathbf{r})\,\mathbf{D}(\mathbf{r})). \tag{11.8}$$

In an infinite crystal, the electron density $\rho(\mathbf{r})$ has the periodicity of the lattice. Therefore, $\chi(\mathbf{r})$ can be written

$$\chi(\mathbf{r}) = \sum_{\mathbf{h}} \chi_{\mathbf{h}} e^{i\mathbf{hr}}, \tag{11.9}$$

with

$$\chi_{\mathbf{h}} = -4\pi r_{\mathrm{e}}/V_{\mathrm{c}}K^2 \int_{V_{\mathrm{c}}} e^{-i\mathbf{hr}}\rho(\mathbf{r})\,d\mathbf{r} = -(4\pi r_{\mathrm{e}}/V_{\mathrm{c}}K^2) f_{\mathbf{h}}\,.$$

Here, $f_{\mathbf{h}}$ is the atomic scattering factor for X-rays (Eq. (1.3)). Furthermore, the eigenfunctions can be chosen as Bloch waves and can be expanded into plane waves analogously to Eq. (1.9):

$$\mathbf{D}_{\mathbf{k}}(\mathbf{r}) = \sum_{\mathbf{h}} \mathbf{D}_{\mathbf{h}} e^{i(\mathbf{k}+\mathbf{h})\mathbf{r}} \qquad \text{with} \qquad \mathbf{D}_{\mathbf{h}} \cdot (\mathbf{k} + \mathbf{h}) = 0. \tag{11.10}$$

Since $\partial_{\mathbf{r}} \cdot \mathbf{D} = 0$, the vectors $\mathbf{D}_{\mathbf{h}}$ are perpendicular to $\mathbf{k} + \mathbf{h}$. Introducing (11.9) and (11.10) into (11.8) and by comparing the coefficients of $e^{i(\mathbf{k}+\mathbf{h})\mathbf{r}}$ we obtain

$$(K^2 - (\mathbf{k} + \mathbf{h})^2)\,\mathbf{D}_{\mathbf{h}} = \sum_{\mathbf{h}'} \varkappa_{\mathbf{h}-\mathbf{h}'} \mathbf{D}_{\mathbf{h}'[\mathbf{h}]} \tag{11.11}$$

with

$$\varkappa_{\mathbf{h}-\mathbf{h}'} = \frac{(\mathbf{k} + \mathbf{h})^2}{K^2}\,\frac{4\pi r_{\mathrm{e}}}{V_{\mathrm{c}}} f_{\mathbf{h}-\mathbf{h}'} \cong \frac{4\pi r_{\mathrm{e}}}{V_{\mathrm{c}}} f\,.$$

Here, $\mathbf{D}_{\mathbf{h}'[\mathbf{h}]}$ is the component of $\mathbf{D}_{\mathbf{h}'}$ being perpendicular to $\mathbf{k} + \mathbf{h}$.

$$\mathbf{D}_{\mathbf{h}'[\mathbf{h}]} = \mathbf{D}_{\mathbf{h}'} - \frac{(\mathbf{k} + \mathbf{h})((\mathbf{k} + \mathbf{h}) \cdot \mathbf{D}_{\mathbf{h}'})}{(\mathbf{k} + \mathbf{h})^2}\,. \tag{11.12}$$

These equations determine the dynamical diffraction of X-rays and are very similar to the corresponding equations (1.10) and (5.1) for electrons or (10.17) for neutrons. The main difference is that the $\mathbf{D}_{\mathbf{h}'}$ are vectors. $\varkappa_{\mathbf{h}}$ is the analog to the potential coefficient $(2m/\hbar^2)\,V_{\mathbf{h}}$ for electrons or to $4\pi a_0/V_{\mathrm{c}}$ for neutrons.

For each plane wave $\mathbf{k} + \mathbf{h}$, we can introduce two polarization vectors $\mathbf{e}_{\mathbf{h}}^{s}$ $(s = 1, 2)$, which are perpendicular to $\mathbf{k} + \mathbf{h}$.

$$\mathbf{e}_{\mathbf{h}}^{s} \cdot (\mathbf{k} + \mathbf{h}) = 0 \qquad \text{and} \qquad \mathbf{e}_{\mathbf{h}}^{s} \cdot \mathbf{e}_{\mathbf{h}}^{s'} = \delta_{s,s'} \qquad \text{for } s \text{ and } s' = 1, 2. \tag{11.13}$$

For the scalar components $D_{\mathbf{h}}{}^s$ of

$$\mathbf{D}_{\mathbf{h}} = \sum_{s=1}^{2} D_{\mathbf{h}}{}^s \mathbf{e}_{\mathbf{h}}{}^s, \tag{11.14}$$

we obtain then the following equations:

$$(K^2 - (\mathbf{k} + \mathbf{h})^2)\, D_{\mathbf{h}}{}^s = \sum_{\mathbf{h}, s'} \varkappa_{\mathbf{h}-\mathbf{h}'}\, \mathbf{e}_{\mathbf{h}}{}^s \cdot \mathbf{e}_{\mathbf{h}'}^{s'} D_{\mathbf{h}'}^{s'} \,. \tag{11.15}$$

By setting the determinant of (11.15) equal to zero, we obtain for a given $\mathbf{k}$ vector the allowed frequencies $cK = \omega_\nu(\mathbf{k})$, which form bands. Analogously to electron diffraction, the Bloch waves for different $\mathbf{k}'$ and different bands ν are orthogonal, etc. For instance, instead of equation (1.12), we have now

$$\sum_{\mathbf{h}, s} D_{\mathbf{h}}{}^s(\mathbf{k}, \nu')\, D_{\mathbf{h}}{}^s(\mathbf{k}, \nu) = \delta_{\nu\nu'} \,. \tag{11.16}$$

Most of the other results of Part II are also valid for X-rays. For instance, we have the symmetries

$$\omega_\nu(\mathbf{k}) = \omega_\nu(\mathbf{k} + \mathbf{h}) = \omega_\nu(S\mathbf{k}) = \omega_\nu(-\mathbf{k}) = \omega_\nu{}^*(\mathbf{k}^*) \tag{11.17}$$

if $\rho(\mathbf{r})$ has the following properties (in the same sequence as (11.17)): periodic, symmetrical with respect to S, local, and real. Similarly, the theorems concerning the real lines and the behavior for complex $\mathbf{k}$ are valid without change.

There are two simple cases where the vector equation (11.10) reduces to two equal and decoupled equation for the components $D_{\mathbf{h}}{}^s$,thus leading to a scalar theory for each component. First, in the vacuum we have $\kappa_{\mathbf{h}-\mathbf{h}'} = 0$ and the two polarizations are degenerate. Second, for very small Bragg angles, all wave vectors $\mathbf{k} + \mathbf{h}$, $\mathbf{k} + \mathbf{h}'$ of the strongly excited waves are approximately equal. Then in (11.11), $\mathbf{D}_{\mathbf{h}'[\mathbf{h}]} \cong \mathbf{D}_{\mathbf{h}'}$ or in (11.15), $\mathbf{e}_{\mathbf{h}}{}^s \cdot \mathbf{e}_{\mathbf{h}'}^{s'} = \delta_{s,s'}$ and we obtain the same equation for the polarizations $s = 1$ and $s = 2$.

The density of the *energy current* is given by Poynting's vector

$$\mathbf{S} = (c/4\pi)\, \mathbf{E} \times \mathbf{H} \qquad \text{with real } \mathbf{E} \text{ and } \mathbf{H}. \tag{11.18}$$

By using complex quantities, the average of $\mathbf{S}$ over times $\tau \gg 1/\omega$ is[5,39]

$$\bar{\mathbf{S}}^t = (c/8\pi)\, \mathrm{Re}(\mathbf{E} \times \mathbf{H}^*). \tag{11.19}$$

[39] J. C. Slater and N. H. Frank, "Electromagnetism." McGraw-Hill, New York, 1947.

Since $|\chi| \ll 1$, we may as well as replace **E** by **D**. For a Bloch wave $\mathbf{D}_\mathbf{k}(\mathbf{r})$, the current contains contributions oscillating in space. However, the average over a unit cell is constant and analogously to (1.14) and (1.15) is given by

$$\bar{\mathbf{S}}^{t, V_c} = \frac{c}{8\pi K} \sum_{\mathbf{h}} (\mathbf{k} + \mathbf{h})|\mathbf{D}_\mathbf{h}|^2 = \frac{1}{8\pi} \frac{\partial \omega_\nu(\mathbf{k})}{\partial \mathbf{k}} . \tag{11.20}$$

Therefore, the current is perpendicular to the dispersion surface $\omega_\nu(\mathbf{k}) = \text{constant}$.

If no Bragg reflection is excited, we only have *one strong beam*. For both polarizations, the Bloch vector is determined by

$$(K^2 - \varkappa_0 - \mathbf{k}^2) = 0, \tag{11.21}$$

leading to a refractive index slightly smaller than 1.

$$n = k/K \cong 1 - (2\pi r_e/V_c K^2) f_0 . \tag{11.22}$$

If a Bragg reflection is excited, we have *two strong beams* **k** and **k** + **h**. In this case, a natural choice for the polarization vectors is (Fig. 17a):

$S = 1$ (σ polarization): $\mathbf{e}_0^1 = \mathbf{e}_\mathbf{h}^1$ perpendicular to both **k** and **k** + **h**,

$S = 2$ (π polarization): $\mathbf{e}_0^2$, $\mathbf{e}_\mathbf{h}^2$ in the plane of **k** and **k** + **h**.

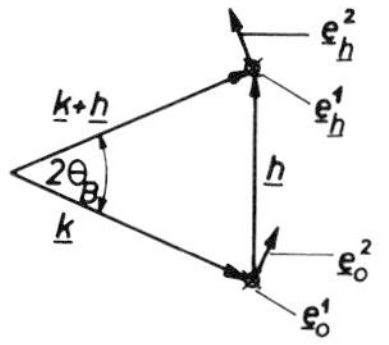

FIG. 17a. Direction of polarization vectors ($\mathbf{e}_0^1 = \mathbf{e}_\mathbf{h}^1$ are normal to the plane of drawing).

Then we have in addition to (11.13): $\mathbf{e}_0^1 \cdot \mathbf{e}_\mathbf{h}^2 = 0 = \mathbf{e}_\mathbf{h}^1 \cdot \mathbf{e}_0^2$ and $\mathbf{e}_0^2 \cdot \mathbf{e}_\mathbf{h}^2 = \cos(2\theta_\mathrm{B})$. Therefore the equations for the different polarizations σ and π are decoupled giving

$$\begin{aligned} (K^2 - \varkappa_0 - \mathbf{k}^2)\, D_0{}^s &= \varkappa_\mathbf{h}\, P_s D_\mathbf{h}{}^s, \\ (K^2 - \varkappa_0 - (\mathbf{k} + \mathbf{h})^2)\, D_\mathbf{h}{}^s &= \varkappa_\mathbf{h}\, P_s D_0{}^s, \end{aligned} \tag{11.23}$$

where the polarization factor $P_s = \mathbf{e}_0{}^s \cdot \mathbf{e}_\mathbf{h}{}^s$ is 1 for σ polarization and $\cos 2\theta_\mathrm{B}$ for π polarization.

By setting the determinant equal to zero, we obtain a different dispersion surface for each polarization. Far away from the Bragg condition both dispersion surfaces coincide and are identical with the spheres of radius $(K^2 - \varkappa_0)^{1/2}$ around $\mathbf{k} = 0$ and $\mathbf{k} = \mathbf{h}$. Near the Bragg condition the degeneracy is removed. For instance the smallest branch separation is

$$\Delta k_s = 2\pi/d_{\text{ext}} = P_s \varkappa_{\mathbf{h}}/K \cos \theta_{\mathrm{B}} \, . \tag{11.24}$$

The dispersion surfaces are qualitatively shown in Fig. 17b. The larger spheres with radius K represent the dispersion surface in the vacuum.

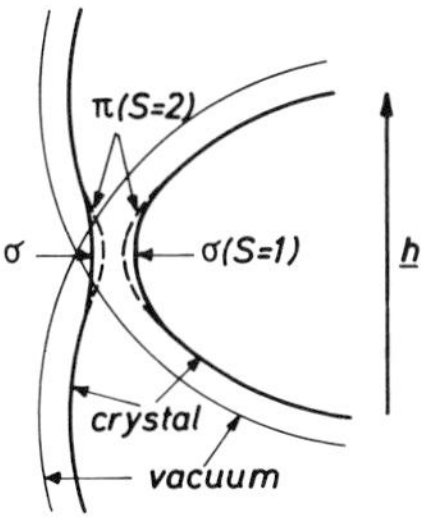

FIG. 17b. Dispersion surfaces for σ and π polarization.

The coefficients D_0^s and $D_{\mathbf{h}}^s$ can be calculated analogously to (2.10). For instance, for the Bragg condition we get for the σ polarization the fields

$$\mathbf{D}^\sigma(\mathbf{r}) = \mathbf{e}_0^1 (1/\sqrt{2})\, e^{i\mathbf{kr}} (1 \pm e^{i\mathbf{hr}}) \tag{11.25}$$

and for the π polarization

$$\mathbf{D}^\pi(\mathbf{r}) = (1/\sqrt{2})\, e^{i\mathbf{kr}} (\mathbf{e}_0^2 \pm e^{i\mathbf{hr}} \mathbf{e}_{\mathbf{h}}^2). \tag{11.26}$$

The upper sign refers to the inner branch and the lower to the outer one. The σ fields are identical with the Bloch waves ϕ^{I} and ϕ^{II} of Fig. 11. However, for the π polarization, we do not get pure sine or cosine waves, but always combinations of both, since $\mathbf{e}_0^2 \neq \mathbf{e}_{\mathbf{h}}^2$.

The absorption of X-rays can be described phenomenologically by a complex dielectric constant or by a complex density $\rho(\mathbf{r}) = \rho'(\mathbf{r}) + i\rho''(\mathbf{r})$, resulting in complex coefficients $\varkappa_{\mathbf{h}} = \varkappa_{\mathbf{h}}' + i\varkappa_{\mathbf{h}}''$.

Then the absorption of a Bloch wave $\mathbf{k}$ can be calculated according to (4.8) and (5.11). In particular, for the two-beam case we have

$$\mu^s = \mu_0 \pm \frac{P_s}{(1 + W_s^2)^{1/2}} \mu_{\mathbf{h}} \qquad \text{with} \qquad W_s = \frac{(\mathbf{k} + \frac{1}{2}\mathbf{h})\,\mathbf{h}}{P_s \varkappa_{\mathbf{h}}} \, . \tag{11.27}$$

The minimal absorption is therefore $\Delta\mu^{\sigma} = \mu_0 - \mu_h$ for σ polarization, but $\Delta\mu^{\pi} = \mu_0 - \cos 2\theta_B \cdot \mu_h$ for π polarization. Therefore, only the σ wave shows a strong anomalous transmission effect, but not the π wave. This is plausible from (11.26) because both π waves do not vanish at the atomic positions.

For multiple-beam cases, the different polarization will no longer be decoupled as in the two-beam case, thereby complicating the problem. Multiple-beam cases are interesting for X-ray diffraction, because even lower absorptions can be obtained than for the two-beam case, as has been demonstrated.[40] A number of symmetrical multiple-cases have been treated,[41] including the cases of Fig. 14. For instance, for the four-beam case of Fig. 14, the minimal absorption will be obtained for very small wave lengths, for which the theory for X-rays reduces to the scalar theory for electrons. Therefore, the smallest absorption is that of (6.10). The corresponding wave field vanishes quadratically at the atomic positions. Extensive treatments of the three-beam case have been given.[42]

IV. The Optical Potential

12. The Coherent Wave

In the dynamic scattering theory, as presented in the previous parts, the interaction with the crystal is described by a simple periodic potential $V(\mathbf{r})$ (1.1). Actually, however, the crystal is a very complex system of electrons and nuclei which can be excited by the scattered electrons or γ quanta. Therefore, the scattering process is a complicated many-body problem which cannot be solved exactly. In view of that the introduction of a stationary potential, $V(\mathbf{r})$ is a very naive approximation indeed. In order to improve this, one has tried to introduce an effective potential, called optical potential or pseudopotential, which should at least in principle include the complications of the many-body problem.

The oldest application of such an optical potential is the description of light propagation through a refracting medium. Here the use of a refractive index is equivalent to the introduction of an optical potential. This problem has been discussed and reviewed by Foldy[43] and Lax.[44]

[40] G. Bormann and W. Hartwig, *Z. Kristallogr.* **121**, 6 (1965). W. Übach and G. Hildebrandt, *Z. Kristallogr.* **129**, 1 (1968).

[41] T. Joko and A. Fukuhara, *J. Phys. Soc. Japan* **22**, 597 (1967).

[42] G. Hildebrandt, *Phys. Status Solidi* **24**, 245 (1967). Y. Héno and P. P. Ewald, *Acta Crystallogr.* **A24**, 5, 16 (1969).

[43] L. L. Foldy, *Phys. Rev.* **67**, 107 (1945).

[44] M. Lax, *Rev. Mod. Phys.* **23**, 287 (1951); *Phys. Rev.* **85**, 621 (1952).

In dynamical diffraction, an optical potential has first been used by Molière to describe the photoelectric absorption of X-rays[13] and later on by Yoshioka for the case of atomic excitations of electrons.[14] A formally exact derivation of the optical potential for elastic scattering has been given by Francis and Watson[45] and Feshbach.[46] These results are explicit but give rather complex expressions in terms of the many-body Hamiltonian of the interacting system. A detailed review of this theory as well as approximations expecially situated for high energy nuclear physics are given in the Goldberger–Watson book.[25]

Today the optical potential represents a very powerful method in nuclear physics, expecially for nucleon–nucleus scattering at intermediate energies ($\sim$100 MeV),[47] but also for pion–nucleon scattering at lower energies[48] as well as pion and nucleon scattering from nuclei at very high energies.[49] There are also applications to electron scattering by atoms.[50] Also, a number of review articles exists about the optical potential method.[51] Whereas in dynamical diffraction, an effective potential had been already introduced in 1936 by Molière,[13] the connection with the more general theory of Watson *et al.* has been realized only recently. Compared to nuclear physics, the theory is even simpler because the interaction is relatively weak, at least for high energy electrons and X-rays.

Now let us start by formulating the scattering problem. The total Hamiltonian is

$$H = h_0 + H_0 + V. \tag{12.1}$$

Here $h_0 = -(\hbar^2/2m)\,\partial_{\mathrm{r}}{}^2$ is the free Hamiltonian of the incident electron, whereas H_0 is the Hamiltonian for all the electrons and nuclei of the crystal containing all many-body interactions. V represents the inter-

[45] N. C. Francis and K. M. Watson, *Phys. Rev.* **92**, 291 (1953).

[46] H. Feshbach, *Ann. Phys.* **5**, 357 (1958).

[47] A. D. Kerman, H. Macmanns, and R. M. Thaler, *Ann. Phys.* **8**, 551 (1959). P. E. Hodson, *Advan. Phys.* **15**, 329 (1966).

[48] L. S. Kissinger, *Phys. Rev.* **98**, 761 (1955). M. Ericson and T. E. O. Ericson, *Ann. Phys.* **36**, 323 (1966).

[49] W. Czyz and L. Lesniak, *Phys. Lett.* **24B**, 227; **25B**, 319 (1967). V. Franco and R. J. Glauber, *Phys. Rev.* **142**, 1195 (1966).

[50] M. H. Mittleman and K. M. Watson, *Phys. Rev.* **113**, 198 (1959). M. H. Mittleman, *Advan. Theoret. Phys.* **1**, 283 (1965).

[51] A. L. Fetter and K. M. Watson, *Advan. Theoret. Phys.* **1**, 115 (1965). P. B. Jones, "The Optical Model in Nuclear and Particle Physics." Interscience, London, 1963. L. L. Foldy and J. D. Walecka, *Ann. Phys.* **54**, 447 (1969).

action of the incident electron with the nuclei $\mathbf{R}_n$ and the electrons $\mathbf{r}_j$ of the crystal (1.1).

$$V(\mathbf{r}; \cdots \mathbf{R}_n \cdots \mathbf{r}_j) = \sum_n (-Ze^2/|\mathbf{r} - \mathbf{R}_n|) + \sum_j (e^2/|\mathbf{r} - \mathbf{r}_j|). \quad (12.2)$$

By introducing the density operators $\rho_{\mathrm{N}}(\mathbf{r})$ of the nuclei and $\rho_{\mathrm{e}}(\mathbf{r})$ of the crystal electrons

$$\rho_{\mathrm{N}}(\mathbf{r}) = \sum_n \delta(\mathbf{r} - \mathbf{R}_n); \qquad \rho_{\mathrm{e}}(\mathbf{r}) = \sum_j \delta(\mathbf{r} - \mathbf{r}_j), \quad (12.3)$$

the interaction potential can be written as

$$V(\mathbf{r}) = \int d\mathbf{r}' \, (e^2/|\mathbf{r} - \mathbf{r}'|)(\rho_{\mathrm{e}}(\mathbf{r}') - Z\rho_{\mathrm{N}}(\mathbf{r}')). \quad (12.4)$$

Let us assume that the initial state of the crystal, before the scattering takes place, is $\phi_\alpha = \phi_\alpha(\cdots \mathbf{R}_n \cdots \mathbf{r}_j \cdots)$. Then we have

$$H_0\phi_\alpha = E_\alpha\phi_\alpha , \quad (12.5)$$

and similarly for the incident wave $\varphi_{\mathbf{K}} = e^{i\mathbf{K}\mathbf{r}}$,

$$h_0\varphi_{\mathbf{K}} = E_{\mathbf{K}}\varphi_{\mathbf{K}} \qquad \text{with} \qquad E_{\mathbf{K}} = \hbar^2\mathbf{K}^2/2m. \quad (12.6)$$

The total state vector $\Psi_{\mathbf{K},\alpha} = \Psi_{\mathbf{K},\alpha}(\mathbf{r}; \cdots \mathbf{R}_n \cdots \mathbf{r}_j)$ for an initial state $\varphi_{\mathbf{K}}\phi_\alpha$ satisfies the Lippmann–Schwinger equation:

$$\Psi_{\mathbf{K},\alpha} = \varphi_{\mathbf{K}}\phi_\alpha + \frac{1}{E_{\mathbf{K}} + E_\alpha + i\epsilon - h_0 - H_0} V\Psi_{\mathbf{K},\alpha} . \quad (12.7)$$

Of course, $\Psi_{\mathbf{K},\alpha}$ is extremely complicated and a solution is not feasible. Moreover, due to the finite temperature, a large number of states as well as initial state ϕ_α are excited. Therefore the intensities, if calculated by (2.7), have to be averaged over the thermal distribution of these states.

In order to make some progress, we restrict ourselves to the so-called "coherent wave." That is, we consider only those scattering processes in detail, for which the initial and final state of the crystal are identical. All such processes are necessarily elastic. For instance, the elastic component of $\Psi_{\mathbf{K},\alpha}$ is given by

$$\begin{aligned} \psi_{\mathbf{K},\alpha}(\mathbf{r}) &= (\phi_\alpha , \Psi_{\mathbf{K},\alpha}(\mathbf{r}))_{\text{crystal}} \\ &= \int \cdots d\mathbf{R}_n \cdots d\mathbf{r}_j \cdots \phi_\alpha{}^*(\cdots \mathbf{R}_n \cdots \mathbf{r}_j \cdots) \, \Psi_{\mathbf{K},\alpha}(r; \cdots \mathbf{R}_n \cdots \mathbf{r}_j). \quad (12.8) \end{aligned}$$

Since the scalar product only refers to the crystal subspace, $\psi_{\mathbf{K},\alpha}(\mathbf{r})$ is a wavefunction in the subspace of the scattering electron. By averaging (12.8) over the thermal distribution of the initial states ϕ_α, we obtain for the coherent wave

$$\psi_{\mathbf{K}}(\mathbf{r}) = \sum_\alpha p_\alpha(\phi_\alpha, \Psi_{\mathbf{K},\alpha}(\mathbf{r}))_{\text{crystal}}$$

$$\text{with} \quad p_\alpha = (1/Z)\, e^{-E_\alpha/KT} \quad \text{and} \quad \sum_\alpha p_\alpha = 1. \tag{12.9}$$

Whereas the thermal average is very important for phonons, the electrons are almost always in the ground state. In this case (12.9) reduces to a projection on the ground state.

The name "coherent" means that $\psi_{\mathbf{K}}(\mathbf{r})$ can interfere with the incident wave $\varphi_{\mathbf{K}}$. Namely by performing an interference experiment between the incident wave and the scattered one, the average intensity at a position $\mathbf{r}$ would be

$$I(\mathbf{r}) \sim \sum_\alpha p_\alpha(\varphi_{\mathbf{K}}(\mathbf{r})\,\phi_\alpha + \Psi_{\mathbf{K},\alpha}(\mathbf{r}), \varphi_{\mathbf{K}}(\mathbf{r})\,\phi_\alpha + \Psi_{\mathbf{K},\alpha}(\mathbf{r}))_{\text{crystal}} \tag{12.10}$$

By using (12.9) we obtain for $I(\mathbf{r})$

$$I(\mathbf{r}) \sim |\varphi_{\mathbf{K}}(\mathbf{r}) + \psi_{\mathbf{K}}(\mathbf{r})|^2 + \left\{\sum_\alpha p_\alpha(\Psi_{\mathbf{K},\alpha}(\mathbf{r}), \Psi_{\mathbf{K},\alpha}(\mathbf{r}))_{\text{crystal}} - |\psi_{\mathbf{K}}(\mathbf{r})|^2\right\}, \tag{12.11}$$

showing that only the coherent wave interferes with the incident wave. The second term in (12.11), while not interfering with $\varphi_{\mathbf{K}}$ represents the intensities due to the incoherent waves.[51a]

Now our goal is to calculate $\psi_{\mathbf{K}}(\mathbf{r})$ directly without calculating the much more complicated total waves $\Psi_{\mathbf{K},\alpha}$ for each α. Is there an equation for $\psi_{\mathbf{K}}(\mathbf{r})$ alone ? Indeed, this can be achieved by introducing an operator F so that

$$\Psi_{\mathbf{K},\alpha} = F\phi_\alpha\psi_{\mathbf{K}}\,. \tag{12.12}$$

Introducing (12.12) on the right-hand side of Eq. (12.7), we form the scalar product with ϕ_α on both sides. Due to (12.5), the operator H_0 in the Green's function can be replaced by E_α and cancels against the first E_α. After averaging over α we obtain the following equation for the coherent wave $\psi_{\mathbf{K}}(\mathbf{r})$

$$\psi_{\mathbf{K}} = \varphi_{\mathbf{K}} + \frac{1}{E_{\mathbf{K}} + i\epsilon - h_0}\, U\psi_{\mathbf{K}}\,, \tag{12.13}$$

[51a] For a more detailed discussion of the meaning of coherence, relative coherence, etc., we refer to Ref. 44.

where the optical potential U is given by

$$U = \sum_{\alpha} p_\alpha(\phi_\alpha, VF\phi_\alpha)_{\text{crystal}} \equiv \langle VF \rangle. \tag{12.14}$$

U is therefore the expectation value of the operator VF. Compared to the complicated equation (12.7) for $\Psi_{\mathbf{K},\alpha}$, which involves the whole dynamics of the crystal, Eq. (12.13) is a simple one-particle equation in the subspace of the scattering electron. So the scattering problem is simplified enormously so far as the optical potential U is known. Therefore, the determination of U represents the central problem. Exact equations for U and certain approximations are given in the next sections. Of course, we cannot expect U to be a simple potential, because it has to reflect all the crystal properties. The many-body complications have simply been transferred from Eq. (12.7) to the determination of U.

13. Properties of the Optical Potential

To obtain an equation for U or F, we introduce the ansatz (12.12) into Eq. (12.7). Then we multiply (12.13) by ϕ_α, introduce again the term $E_\alpha - H_0$ into the Green's function, and subtract this equation from (12.7). This gives the equation for F:

$$F = 1 + G_0(VF - \langle VF \rangle) \quad \text{with} \quad G_0 = \frac{1}{E_{\mathbf{K}} + E_\alpha + i\epsilon - h_0 - H_0}. \tag{13.1}$$

From (13.1), as well as from (12.12), it follows immediately that

$$\langle F \rangle = 1. \tag{13.2}$$

Equation (13.1) can also be written in a somewhat different form, which will be convenient later. Substituting

$$V = \langle V \rangle + \delta V \qquad \text{with} \qquad \delta V = V - \langle V \rangle \tag{13.3}$$

in (13.1), we obtain by using (13.2)

$$(1 - G_0\langle V \rangle)F = 1 - G_0\langle V \rangle + G_0(\delta V F - \langle \delta V F \rangle) \tag{13.4}$$

and by dividing this result by $(1 - G_0\langle V \rangle)$,

$$F = 1 + G_{\langle V \rangle}(\delta V F - \langle \delta V F \rangle)$$

with

$$G_{\langle V \rangle} = \frac{1}{E_{\mathbf{K}} + E_\alpha + i\epsilon - h_0 - H_0 - \langle V \rangle}. \tag{13.5}$$

A third form can be obtained by writing $V = U + (V - U)$ instead of (13.3). Then we have the expression

$$F = 1 + G_U(V - U)F; \qquad G_U = \frac{1}{E_{\mathbf{K}} + E_\alpha + i\epsilon - h_0 - H_0 - U} \tag{13.6}$$

since $\langle (V - U)F \rangle = 0$ according to (12.4) and (13.2). The formal solution of (13.6) is

$$\begin{aligned} F &= \frac{1}{1 - G_U(V - U)} \equiv 1 + \frac{1}{1 - G_U(V - U)} G_U(V - U) \\ &\equiv 1 + \frac{1}{E_K + E_\alpha + i\epsilon - H_0 - h_0 - V}(V - U). \end{aligned} \tag{13.7}$$

According to (12.14), the optical potential is given by the average $U = \langle VF \rangle$. Therefore, by knowing F we can evaluate U. But we can also directly derive an equation for U alone. Namely, by multiplying (13.1) with V and averaging we obtain

$$U = \langle V \rangle + \left\langle (V - U) \frac{1}{1 - G_U(V - U)} G_U(V - U) \right\rangle \tag{13.8}$$

or if we use the second line of (13.1)

$$U = \langle V \rangle + \left\langle (V - U) \frac{1}{E_{\mathbf{K}} + E_\alpha + i\epsilon - h_0 - H_0 - V} (V - U) \right\rangle. \tag{13.9}$$

These are implicit and rather complicated equations for the optical potential U. They are quite useful for the discussion of some general properties of U. First, however, we will derive these and equivalent equations for U more directly without introducing Watson's tricky F operator. For this we decompose $V = U + (V - U)$ in (12.7) and obtain

$$\Psi_{\mathbf{K},\alpha} = \frac{1}{1 - G_0 U} \varphi_{\mathbf{K}} \phi_\alpha + \frac{1}{1 - G_0 U} G_0(V - U) \Psi_{\mathbf{K},\alpha}. \tag{13.10}$$

Now, in the first term of the Green's function H_0 can be replaced by E_α again. Then the first term is equal to $\psi_{\mathbf{K}} \phi_\alpha$ if we define U by Eq. (12.13). Therefore, together with G_U of (13.6),

$$\begin{aligned} \Psi_{\mathbf{K},\alpha} &= \psi_{\mathbf{K}} \phi_\alpha + G_U(V - U) \Psi_{\mathbf{K},\alpha} \\ &= \left(1 + G_U(V - U) \frac{1}{1 - G_U(V - U)}\right) \psi_{\mathbf{K}} \phi_\alpha. \end{aligned} \tag{13.11}$$

Now, if we form $\psi_{\mathbf{K}}$ according to (12.9), the first term ($\psi_{\mathbf{K}}$) on the right-hand side cancels against the left-hand side, and we obtain the following condition determining U:

$$\left\langle (V-U)\frac{1}{1-G_U(V-U)}\right\rangle = 0. \tag{13.12}$$

This is the same condition as (13.8), as can be easily verified by writing the 1 in the nominator as $(1 - G_U(V-U)) + G_U(V-U)$.

Compared to a normal potential $V(\mathbf{r})$, the optical potential U has some strange properties. First we see from (13.9) that U is *energy dependent*, i.e., U depends explicitly on the incident energy $E_{\mathbf{K}}$. Furthermore, due to the occurrence of the Green's function, U is a *nonlocal* potential: $U(\mathbf{r}, \mathbf{r}')$. Moreover, the optical potential is *non-Hermitean* and, for example, from (13.9) it follows that $U(E_{\mathbf{K}} + i\epsilon) = U^{\dagger}(E_{\mathbf{K}} - i\epsilon)$. This is a very characteristic property of the optical potential, meaning that the coherent wave is absorbed in the crystal. Note however that there is no real absorption since the total interaction potential (12.2) is Hermitean. The apparent absorption of the coherent wave is merely due to production of incoherent and inelastic waves in the crystal diminishing the intensity of the coherent wave. This can be more clearly realized by the optical theorem, as will be shown below. Furthermore, the optical potential has to reflect all the properties of the crystal. For example, due to the atomic motion, U is *temperature dependent*, as will be discussed in Part V. In an infinite crystal, U is a *periodic* potential (Eq. (2.8)) since by the average in Eq. (12.14), no point in the crystal is especially preferred. With respect to the dynamical theory, this is the most important property of the optical potential. Therefore, all methods of the preceding sections can be applied simplifying the theory for the coherent wave considerably, provided the optical potential is known. Finally we note that the transpose $\hat{U}$ of U fulfills the same equation (2.8) as U does. Therefore U is *symmetrical*: $U = \hat{U}$, leading to the inversion symmetry $E_\nu(\mathbf{k}) = E_\nu(-\mathbf{k})$ (2.13a) for the band structure.

The *optical theorem* can be derived by using the equation (8.4) for the T matrix. For

$$T = V\frac{1}{1-G_0V} \qquad \text{with} \qquad G_0 = \frac{1}{E + i\epsilon - \mathscr{H}_0} \tag{13.13}$$

we obtain

$$T - T^{\dagger} = \frac{1}{(1-GV)^{\dagger}}\{V - V^{\dagger} + V^{\dagger}(G_0 - G_0{}^{\dagger})\,V\}\frac{1}{(1-G_0V)}\,. \tag{13.14}$$

Now let us decompose the operators T and V into their Hermitean and anti-Hermitean parts.

$$\mathcal{O} = \mathcal{O}' + i\mathcal{O}'' \quad \text{with} \quad \mathcal{O}' = (\mathcal{O} + \mathcal{O}^\dagger)/2 = \mathcal{O}'^\dagger, \\ \mathcal{O}'' = (\mathcal{O} - \mathcal{O}^\dagger)/2i = \mathcal{O}''^\dagger. \tag{13.15}$$

Then we get from (13.14) since $G_0'' = -\pi\,\delta(E - \mathscr{H}_0)$,

$$T'' = \frac{1}{(1 - G_0V)^\dagger}\,V''\,\frac{1}{(1 - G_0V)} - \pi T^\dagger\,\delta(E - \mathscr{H}_0)\,T. \tag{13.16}$$

This is the optical theorem in its most general form. For a Hermitean potential, $V'' = 0$ and the first term vanishes. With an incident wave $\varphi_a(\mathscr{H}_0\varphi_a = E_a\varphi_a)$ the Lippmann–Schwinger equation for the state vector ψ_a is

$$\psi_a = \varphi_a + G_0V\psi_a = (1 + G_0T)\,\varphi_a\,. \tag{13.17}$$

Therefore, we obtain from (13.16) for the matrix elements $T_{aa} = (\varphi_a\,,\,T\varphi_a)$

$$\operatorname{Im} T_{aa} = \operatorname{Im}(\psi_a\,,\,V\psi_a) - \pi\sum_b |T_{ab}|^2\,\delta(E - E_b). \tag{13.18}$$

We may now use this form of the optical theorem for the Lippmann–Schwinger equation (12.7) of the total state vector $\Psi_{\mathbf{K},\alpha}$. With $\varphi_a \to \varphi_{\mathbf{K}}\phi_\alpha$ and $V = V^\dagger$ we obtain for the average of T

$$\begin{aligned} \operatorname{Im}\langle T\rangle_{\mathbf{K},\mathbf{K}} &= \sum_\alpha p_\alpha \operatorname{Im} T_{\mathbf{K}\alpha,\mathbf{K}\alpha} \\ &= -\sum_{\alpha\mathbf{K}'\beta} \pi p_\alpha |T_{\mathbf{K}'\beta,\mathbf{K}\alpha}|^2\,\delta(E_{\mathbf{K}} + E_\alpha - E_{\mathbf{K}'} - E_\beta). \end{aligned} \tag{13.19}$$

On the other hand, we may also write down the optical theorem for the Lippmann–Schwinger equation (12.13) of the coherent wave $\psi_{\mathbf{K}}$. For this purpose we set $\varphi_a \to \varphi_{\mathbf{K}}$ and $V \to U$. Furthermore, we have

$$\Psi_{\mathbf{K}} = \varphi_{\mathbf{K}} + g_0U\psi_{\mathbf{K}} = \varphi_{\mathbf{K}} + g_0\langle T\rangle\,\varphi_{\mathbf{K}} \quad \text{with} \quad g_0 = \frac{1}{E_{\mathbf{K}} + i\epsilon - h_0}\,. \tag{13.20}$$

Therefore U and $\langle T\rangle$ are related by

$$\langle T\rangle = U + Ug_0\langle T\rangle, \tag{13.21}$$

so that we obtain for the optical theorem

$$\operatorname{Im}\langle T\rangle_{\mathbf{K},\mathbf{K}} = \operatorname{Im}(\psi_{\mathbf{K}}\,,\,U\psi_{\mathbf{K}}) - \pi\sum_{\mathbf{K}'} |\langle T\rangle_{\mathbf{K}',\mathbf{K}}|^2\,\delta(E_{\mathbf{K}} - E_{\mathbf{K}'}). \tag{13.22}$$

It is interesting to compare the two different versions (13.19) and (13.22) of the optical theorem. According to (13.19) the imaginary part of $\langle T\rangle_{\mathbf{K},\mathbf{K}}$ is proportional to the averaged total transition probability into all energetically allowed final states $(\mathbf{K}', \beta)$, which is essentially the total scattering cross section.

$$\sigma_{\text{total}} = -(2m/\hbar^2 K)\,\text{Im}\langle T\rangle_{\mathbf{K},\mathbf{K}}\,. \tag{13.23}$$

On the other hand, according to (13.22), $\text{Im}\langle T\rangle_{\mathbf{K},\mathbf{K}}$ is proportional to the total transition probability for scattering into coherent final states plus an absorption term. Therefore, we have

$$\sigma_{\text{total}} = \sigma_{\text{absorption}} + \sigma_{\text{coherent}}\,, \tag{13.24}$$

where the cross section for coherent scattering is given by

$$\sigma_{\text{coherent}} = (2m/\hbar^2 K)\sum_{\mathbf{K}'} |\langle T\rangle_{\mathbf{K}',\mathbf{K}}|^2\,\delta(E_{\mathbf{K}} - E_{\mathbf{K}'}). \tag{13.25}$$

The absorption cross section is directly related to U'' and given by

$$\sigma_{\text{absorption}} = -(2m/\hbar^2 K)\,\text{Im}(\psi_{\mathbf{K}}\,,\, U\psi_{\mathbf{K}}) = -(2m/\hbar^2 K)(\psi_{\mathbf{K}}\,,\, U''\psi_{\mathbf{K}}). \tag{13.26}$$

From (13.19) it is clear that this is nothing else but the cross section for incoherent scattering. Because of $\sigma_{\text{abs}} \geqslant 0$, we should expect that the operator U'' is negative definite. Indeed, from (13.21) we have

$$U = \langle T\rangle \frac{1}{1 + g_0\langle T\rangle} \tag{13.27}$$

and, therefore, analogously to (13.14) we obtain for U'',

$$U'' = \frac{U - U^\dagger}{2i} = \frac{1}{(1 + g_0\langle T\rangle)^\dagger}\{T'' - \pi\langle T^\dagger\rangle\,\delta(E_K - h_0)\langle T\rangle\}\frac{1}{1 + g_0\langle T\rangle}\,. \tag{13.28}$$

Further by using (13.14) for T'', this result can be written as

$$U'' = -\pi\langle \tau^\dagger\,\delta(E_{\mathbf{K}} + E - h_0 - H_0)\,\tau\rangle$$

with

$$\tau = (T - \langle T\rangle)\frac{1}{1 + g_0\langle T\rangle}\,. \tag{13.29}$$

Therefore, all expectation values of U'' are negative.

14. Approximations for the Optical Potential

Since it seems hopeless to solve the equations (13.8) and (13.9) exactly for the optical potential, we have to find reliable approximations. For this purpose, one interesting aspect is the energy dependence of U. For instance, the optical potential should be especially simple if the incident energy $E_{\mathbf{K}}$ is either much higher or much lower than the typical excitation energies $(E_\beta - E_\alpha)$ of the crystal.

By introducing a complete set of crystal state vectors ϕ_β, the Green's function G_0 (13.1) is

$$G_0 = \frac{1}{E_{\mathbf{K}} + E_\alpha + i\epsilon - h_0 - H_0} = \sum_\beta \frac{1}{E_{\mathbf{K}} + i\epsilon - h_0 + E_\alpha - E_\beta} |\phi_\beta\rangle\langle\phi_\beta|. \tag{14.1}$$

For high energies $E_{\mathbf{K}} \gg |E_\alpha - E_\beta|$ the difference $E_\alpha - E_\beta$ can be neglected in the denominator. Therefore, since $\sum_\beta |\phi_\beta\rangle \langle\phi_\beta| = 1$, we have

$$G_0 \cong 1/(E_{\mathbf{K}} + i\epsilon - h_0) \qquad \text{for} \quad E_{\mathbf{K}} \gg |E_\alpha - E_\beta|. \tag{14.2}$$

The physical meaning of this approximation is that during the scattering process the electrons and nuclei of the crystal practically do not move since the atomic and electronic velocities are considerable smaller than the velocity of the incident electron.

By introducing (14.2) in the Lippmann–Schwinger equation (12.7), it is convenient to write the wave function $\psi_{\mathbf{K},\alpha}$ in the form

$$\Psi_{\mathbf{K},\alpha}(\mathbf{r}; \cdots \mathbf{R}_n \cdots \mathbf{r}_j \cdots) = \tilde{\Psi}_{\mathbf{K}}(\mathbf{r}; \cdots \mathbf{R}_n \cdots \mathbf{r}_j \cdots)\,\phi_\alpha(\cdots \mathbf{R}_n \cdots \mathbf{r}_j \cdots). \tag{14.3}$$

Then the wave function ϕ_α cancels on both sides giving for $\tilde{\Psi}_{\mathbf{K}}$

$$\tilde{\Psi}_{\mathbf{K}} = \varphi_{\mathbf{K}} + \{1/(E_{\mathbf{K}} + i\epsilon - h_0)\}\, V\tilde{\Psi}_{\mathbf{K}}\,. \tag{14.4}$$

Due to the approximation (14.2) this is a one-particle equation describing the pure elastic scattering at nuclei and electrons fixed at the momentary positions $\mathbf{R}_n$, $\mathbf{r}_j$ respectively. Therefore the positions $\mathbf{R}_n$ and $\mathbf{r}_j$ enter only as parameters, but not as dynamic variables. Due to (12.9) the average of $\tilde{\Psi}_{\mathbf{K}}$ over the distribution of scatterer positions $\mathbf{R}_n$, $\mathbf{r}_j$ gives the coherent wave directly: $\psi_{\mathbf{K}} = \langle\tilde{\Psi}_{\mathbf{K}}\rangle$. Furthermore, the optical potential or the F operator is determined by the simplified equation

$$F = 1 + \{1/(E_{\mathbf{K}} + i\epsilon - h_0)\}(VF - \langle VF\rangle). \tag{14.5}$$

This approximation is known as *static approximation*. Lax[44] calls it "closure approximation," because a closed set of intermediate crystal states (14.2) is allowed instead of only those satisfying the energy conservation (14.1). Since the phonon energies are typically of the order of kT, it is a very good approximation for the thermal motion of the atoms (Section V,17), both for electron and X-ray diffraction. However, it cannot be applied to electronic excitations, especially not to inner-shell excitations.

In the opposite case, for very low energies $E_{\mathbf{K}} \ll |E_\alpha - E_\beta|$, the Green's function G_0 can be replaced by

$$G_0 = 1/(E_\alpha + i\epsilon - H_0) \tag{14.6}$$

and the equation for the F operator becomes

$$F = 1 + \{1/(E_\alpha + i\epsilon - H_0)\}(VF - \langle VF \rangle). \tag{14.7}$$

Now $F = F(\mathbf{r})$ is evidently a local operator because the electron coordinate $\mathbf{r}$ only enters in the potential V. Therefore, the optical potential U is local also, since $U = \langle V(\mathbf{r}) F(\mathbf{r}) \rangle$. In this limit, the target particles, moving much faster than the incident electron, adjust adiabatically to each position of the electron. Mittleman and Watson[50] have applied this so-called *adiabatic approximation* to study polarization effects in electron–atom scattering. It is not very useful in dynamical theory, except for very cold neutrons.

Particularly simple approximations for U can be derived in the weak-interaction limit, i.e., if the potential V is small. This suggests expanding U and F into powers of V. With

$$F = F^{(1)} + F^{(2)} + \cdots, \qquad U = U^{(1)} + U^{(2)} + \cdots, \tag{14.8}$$

we obtain from (13.1)

$$F^{(1)} = 1, \qquad F^{(2)} = G_0(V - \langle V \rangle). \tag{14.9}$$

Therefore, the first and second approximations for U are

$$U^{(1)} = \langle V \rangle, \tag{14.10}$$

$$U^{(2)} = \langle VG_0(V - \langle V \rangle)\rangle = \langle (V - \langle V \rangle)\, G_0(V - \langle V \rangle)\rangle. \tag{14.11}$$

In the first approximation, U is identical with the expectation value of the potential V (12.2). Therefore $U^{(1)}$ is a simple local and Hermitean potential. Since $\langle V \rangle$ contains only diagonal elements $(\phi_\alpha, V\phi_\alpha)$, replacing U by $\langle V \rangle$ means that all incoherent and inelastic intermediate states are neglected. The second approximation $U^{(2)}$, being a correction to the first one, is nonlocal, non-Hermitean, and energy dependent due to the occurrence of the Green's function G_0.

We can obtain somewhat better approximations for U by assuming that $\delta V = V - \langle V \rangle$ is small, thus exactly taking $\langle V \rangle$ into account. This means that we consider the fluctuations, rather than the potential, to be small. The starting point for these approximations is Eq. (13.5) for F. Analogously to (14.9), we obtain for F

$$F^{(1)} = 1, \qquad F^{(2)} = G_{\langle V \rangle}\, \delta V, \tag{14.12}$$

and subsequently for U

$$U^{(1)} = \langle V \rangle, \tag{14.13}$$

$$U^{(2)} = \langle (V - \langle V \rangle)\, G_{\langle V \rangle} (V - \langle V \rangle) \rangle. \tag{14.14}$$

The first approximation is identical to (14.9), whereas the second approximation contains the Green's function $G_{\langle V \rangle}$ (Eq. (13.5)) instead of G_0, reducing to Eq. (14.11) for weak potentials V. By decomposing $U^{(2)}$ according to (13.15) into the Hermitean part $U'^{(2)}$ and the anti-Hermitean part $iU''^{(2)}$ we obtain from (14.11) and analogously from (14.14),

$$U'^{(2)} = \left\langle (V - \langle V \rangle)\, P\left(\frac{1}{E_{\mathbf{K}} + E_\alpha - h_0 - H_0}\right) (V - \langle V \rangle) \right\rangle \tag{14.15}$$

and

$$U''^{(2)} = -\pi \left\langle (V - \langle V \rangle)\, \delta(E_{\mathbf{K}} + E_\alpha - h_0 - H_0)(V - \langle V \rangle) \right\rangle. \tag{14.16}$$

As a function of the energy $E_{\mathbf{K}}$, $U'^{(2)}$ and $U''^{(2)}$ are not independent, but connected by a dispersion relation

$$U'^{(2)}(E_{\mathbf{K}}) = -\frac{1}{\pi} \int dE_{\mathbf{K}}'\, P\left(\frac{1}{E_{\mathbf{K}} - E_{\mathbf{K}}'}\right) U''^{(2)}(E_{\mathbf{K}}'). \tag{14.17}$$

In order to evaluate $U^{(1)}$ and $U^{(2)}$ according to (14.10) and (14.11) or (14.13) and (14.14) we have to know the wavefunctions $\phi_\alpha(\cdots \mathbf{R}_n \cdots \mathbf{r}_j \cdots)$ of the crystals, which are quite complicated and rarely known. Fortunately, however, the approximations $U^{(1)}$ and $U^{(2)}$ only depend on two relatively simple properties of the crystal, namely, on the one- and two-particle density of the nuclei and electrons. For example, the local potential $U^{(1)}$ can be written

$$U^{(1)}(\mathbf{r}) = \langle V(\mathbf{r}) \rangle = \int d\mathbf{R}\, \frac{e}{|\mathbf{r} - \mathbf{R}|} \langle \eta(\mathbf{R}) \rangle, \tag{14.18}$$

where according to (12.2), $\eta(\mathbf{R})$ is the operator of the charge density and $\langle\eta(\mathbf{R})\rangle$ the charge density itself.

$$\eta(\mathbf{R}) = e\sum_j \delta(\mathbf{R} - \mathbf{r}_j) - Ze\sum_n \delta(\mathbf{R} - \mathbf{R}_n) = e\rho_e(\mathbf{R}) - Ze\rho_N(\mathbf{R}). \quad (14.19)$$

Therefore, $U^{(1)}$ only depends on the one-particle density of the nuclei $\langle\rho_N(\mathbf{R})\rangle$ and the electrons $\langle\rho_e(\mathbf{R})\rangle$.

Similarly, the nonlocal correction $U^{(2)}$ (Eq. (14.10)) becomes

$$U^{(2)}(\mathbf{r}, \mathbf{r}') = \int d\mathbf{R}\, d\mathbf{R}' \frac{e}{|\mathbf{r} - \mathbf{R}|} \frac{e}{|\mathbf{r}' - \mathbf{R}'|}$$

$$\times \Big\langle (\eta(\mathbf{R}) - \langle\eta(\mathbf{R})\rangle)\langle\mathbf{r}|G_0|\mathbf{r}'\rangle(\eta(\mathbf{R}') - \langle\eta(\mathbf{R}')\rangle)\Big\rangle, \quad (14.20)$$

showing that $U^{(2)}$ depends on the two-particle charge density. It is interesting to note that we may also introduce time-dependent correlation functions. With the identity

$$G_0 = \int_0^\infty (dt/i\hbar) \exp[(i/\hbar)(E_\mathbf{K} + E_\alpha + i\epsilon - h_0 - H_0)\, t], \quad (14.21)$$

we can introduce time-dependent operators $\mathbf{R}_n(t)$ and $\mathbf{r}_j(t)$ in the Heisenberg picture by

$$\mathbf{R}_n(t) = \exp[(i/\hbar)\, H_0 t]\, \mathbf{R}_n \exp[-(i/\hbar)\, H_0 t]. \quad (14.22)$$

For this we note that in (14.20) the scalar $\exp[(1/\hbar)\, E_\alpha t]$ commutes with $\eta(\mathbf{R})$ and can be replaced by $\exp[(i/\hbar)\, H_0 t]$ if standing in front of η. Then we get

$$U^{(2)}(\mathbf{r}, \mathbf{r}') = \int_0^\infty \frac{dt}{i\hbar} \int d\mathbf{R}\, d\mathbf{R}' \frac{e}{|\mathbf{r} - \mathbf{R}|} \frac{e}{|\mathbf{r}' - \mathbf{R}'|}$$

$$\times \langle\mathbf{r}| \exp\left[\frac{i}{\hbar}(E_\mathbf{K} + i\epsilon - h_0)\, t\right] |r'\rangle$$

$$\times \{\langle\eta(\mathbf{R}, t)\, \eta(\mathbf{R}', 0)\rangle - \langle\eta(\mathbf{R}, t)\rangle\langle\eta(\mathbf{R}', 0)\rangle\}. \quad (14.23)$$

Therefore $U^{(2)}$ depends on the time-dependent charge correlations. (Note that $\langle\eta(\mathbf{R}, t)\rangle$ is independent of t.)

In the static approximation (14.5) the H_0 dependence in the Green's function is neglected. Then in (14.23) we end up with the static correlation function

$$\{\langle\eta(\mathbf{R}, 0)\, \eta(\mathbf{R}', 0)\rangle - \langle\eta(\mathbf{R}, 0)\rangle\langle\eta(\mathbf{R}', 0)\rangle\} \quad (14.24)$$

instead of the time-dependent function. On the other hand, in the adiabatic approximation (14.1) only the total time integral of the correlation function enters.

The introduction of time in (14.21) seems to be highly artificial, since t is only an integration variable. However, one can also derive an optical potential for the time-dependent Schrödinger equation. Then, time-dependent correlation functions occur quite naturally since the analogous optical potential is nonlocal in time. For example, we get for the coherent wave an equation of the form

$$-(\hbar/i)\,\dot{\psi}(t) = h_0\psi(t) + \int^t dt'\; U(t-t')\,\psi(t'). \tag{14.25}$$

The time-dependent optical potential $U(t)$ is connected with the stationary optical potential U by

$$U = \int_0^\infty (dt/i\hbar)\,\exp[(i/\hbar)(E_{\mathbf{K}} + i\epsilon)\,t]\;U(t). \tag{14.26}$$

Whereas the preceding approximations are restricted to weak potentials or fluctuations, it is very difficult to get consistent approximations in more general cases. One way could be trying to solve Eq. (13.8) for U self-consistently, for instance in the approximation

$$U = \langle V\rangle + \langle (V-U)\,G_U(V-U)\rangle. \tag{14.27}$$

For strong single-particle interaction, one has to expand U in terms of single-scattering matrices. Such approximations, as discussed in the Goldberger–Watson book,[25] are especially situated for neutron diffraction.

V. Application to Electron Diffraction

15. The Averaged Potential

In this section, we apply the results of Part IV to the dynamic diffraction of electrons. Since the optical potential for an infinite crystal is periodical, we are especially interested in the Fourier coefficients $U_{\mathbf{k}+\mathbf{h},\mathbf{k}+\mathbf{h}'}$ of equation (2.10). Furthermore, the interaction potential is weak, $|V(\mathbf{r})| \ll E_{\mathbf{K}}$, and we can rely on the approximations (14.10) and (14.11) for U. In this subsection we discuss the first approximation (14.10) for U, whereas the correction $U^{(2)}$ (14.11) is evaluated subsequently for electron and phonon excitations.

Assuming the atomic electrons to follow adiabatically the motion of the nucleus (rigid ion model) we can make the ansatz for ϕ_α :

$$\phi_\alpha(\cdots \mathbf{R}_n \cdots \mathbf{r}_j \cdots) = \xi_\mu(\cdots \mathbf{R}_n \cdots)\, \varphi_n(\cdots \mathbf{r}_j \cdots ; \cdots \mathbf{R}_n \cdots). \qquad (15.1)$$

Here $\xi_\mu(\cdots \mathbf{R}_n \cdots)$ is the wavefunction of the nuclei, whereas $\varphi_n(\cdots \mathbf{r}_j \cdots; \cdots \mathbf{R}_n \cdots)$ is the electron wavefunction for a set of fixed positions $\cdots \mathbf{R}_n \cdots$ of the nuclei. Before the scattering, the electrons are in the ground state φ_0 .

The first approximation $U^{(1)}$, given by (14.18) is a simple, local, and real potential. Due to the rigid ion model the electron density can be written as

$$\langle \rho_e(\mathbf{R}) \rangle = \sum_n \int d\mathbf{R}'\, \rho_0(\mathbf{R} - \mathbf{R}') \langle \delta(\mathbf{R}' - \mathbf{R}_n) \rangle\, d\mathbf{R}'. \qquad (15.2)$$

Here $\rho_0(\mathbf{R} - \mathbf{R}')$ is the electron density of an atom in the ground state. Therefore, we obtain for the optical potential in the first approximation:

$$U^{(1)}(\mathbf{r}) = \langle V(\mathbf{r}) \rangle = \sum_n \int d\mathbf{R}\, v(\mathbf{r} - \mathbf{R}) \langle \delta(\mathbf{R} - \mathbf{R}_n) \rangle, \qquad (15.3)$$

where $v(\mathbf{r})$ is the potential of a single atom at $\mathbf{R}$.

$$v(\mathbf{r} - \mathbf{R}) = -\left\{ \frac{Ze^2}{|\mathbf{r} - \mathbf{R}|} - \int \frac{e^2}{|\mathbf{r} - \mathbf{R} - \mathbf{R}'|} \rho_0(\mathbf{R}')\, d\mathbf{R}' \right\}. \qquad (15.4)$$

The potential $v(\mathbf{r} - \mathbf{R})$ is eliminated due to the thermal motion, which becomes somewhat smoother. For instance, the $1/|\mathbf{r} - \mathbf{R}|$ singularity disappears in (15.3).

The Fourier coefficients $U^{(1)}_{\mathbf{k}+\mathbf{h},\mathbf{k}+\mathbf{h}'}$ depend only on $\mathbf{h} - \mathbf{h}'$ since $U^{(1)}$ is local. Furthermore, (15.3) is a convolution of $v(\mathbf{r} - \mathbf{R})$ and $\langle \delta(\mathbf{R} - \mathbf{R}_n) \rangle$. Therefore, its Fourier transform is a product of the single transforms and is given by

$$U^{(1)}_{\mathbf{k}+\mathbf{h},\mathbf{k}+\mathbf{h}'} = \langle V \rangle_{\mathbf{h}-\mathbf{h}'} = V_{\mathbf{h}-\mathbf{h}'} \langle \exp[-i(\mathbf{h} - \mathbf{h}')\, \mathbf{R}_n] \rangle = V_{\mathbf{h}-\mathbf{h}'} \exp[-M_{\mathbf{h}-\mathbf{h}'}] \qquad (15.5)$$

using $V_\mathbf{h}$ of (1.3). The factor $M_\mathbf{h}$ is the Debye–Waller factor depending on the thermal motion of the atom. It can be evaluated by the method of cumulants, i.e., by expanding $M_\mathbf{h}$ into powers of $\mathbf{h}$. For example, we obtain in general for a fluctuating quantity x

$$\langle e^{ix} \rangle = \exp\left\{ \sum_{n=0}^{\infty} (i^n/n!) \langle x^n \rangle_c \right\}. \qquad (15.6)$$

Here $\langle x^n \rangle_c$ are the so-called connected or cumulant-averages,[52] the lowest orders of which are

$$\begin{aligned} \langle x \rangle_c &= \langle x \rangle; \qquad \langle x^2 \rangle_c = \langle x^2 \rangle - \langle x \rangle^2, \\ \langle x^3 \rangle_c &= \langle x^3 \rangle - 3\langle x^2 \rangle \langle x \rangle + 2\langle x \rangle^3, \\ \langle x^4 \rangle_c &= \langle x^4 \rangle - 4\langle x^3 \rangle \langle x \rangle - 3\langle x^2 \rangle^2 + 12\langle x^2 \rangle \langle x \rangle^2 - 6\langle x \rangle^4. \end{aligned} \tag{15.7}$$

By introducing the equilibrium position $\overline{\mathbf{R}^n}$ in the averaged lattice we have

$$\mathbf{R}_n = \overline{\mathbf{R}}_n + \mathbf{s}_n \qquad \text{with} \qquad \langle \mathbf{R}_n \rangle = \overline{\mathbf{R}}_n\,, \quad \langle \mathbf{s}_n \rangle = 0, \tag{15.8}$$

where $\mathbf{s}_n$ is the thermal displacement of atom n. Because $\exp[i\mathbf{h}\overline{\mathbf{R}_n}] = 1$, we can replace $\mathbf{R}_n$ by $\mathbf{s}_n$ in (15.5). Therefore, the first-order average is zero. Further for centrosymmetrical crystal all uneven averages vanish, leading to a real and positive Debye–Waller factor $M_\mathbf{h}$.

$$M_\mathbf{h} = \tfrac{1}{2}\langle (\mathbf{hs}_n)^2 \rangle - \tfrac{1}{24}\{\langle (\mathbf{hs}_n)^4 \rangle - 3\langle (\mathbf{hs}_n)^2 \rangle^2\} + \cdots . \tag{15.9}$$

In the harmonic approximation the distribution of the displacements $\mathbf{s}_n$ is Gaussian and all averages higher than the second order vanish[53]

$$M_\mathbf{h} = \tfrac{1}{2}\langle (\mathbf{hs}_n)^2 \rangle = \tfrac{1}{6}h^2 \langle \mathbf{s}_n{}^2 \rangle \qquad \text{for cubic crystals.} \tag{15.10}$$

In this form $M_\mathbf{h}$ can be expressed by the spectral frequency distribution $z(\omega)$ of the crystal[53]

$$M_\mathbf{h} = \tfrac{1}{6}h^2 \int \{\epsilon(\omega, T)/M\omega^2\}\, z(\omega)\, d\omega \tag{15.11}$$

with

$$\epsilon(\omega, T) = \frac{\hbar\omega}{2} + \frac{\hbar\omega}{\hbar\omega/e^{kT} - 1} = \begin{cases} \hbar\omega/2 & \text{for } kT \ll \hbar\omega, \\ kT & \text{for } kT \gg \hbar\omega. \end{cases} \tag{15.12}$$

For a quick estimate we may use the Debye spectrum $z(\omega) = 3\omega^2/\omega_D{}^2$ for $\omega \leqslant \omega_D$. With the Debye temperature θ_D, defined by $\hbar\omega_D = k\theta_D$, the limiting values of $M_\mathbf{h}$ for high and low temperatures are[53]

$$M_\mathbf{h} = \begin{cases} \tfrac{1}{4}R/k\theta_D & \text{for } T \to 0 \\ RkT/(k\theta_D)^2 & \text{for } T \geqslant \theta_D \end{cases} \qquad \text{with} \qquad R = \frac{\hbar^2\mathbf{h}^2}{2M}. \tag{15.13}$$

R is the recoil energy of an atom when the momentum $\hbar\mathbf{h}$ is transferred

[52] See, e.g., R. Kubo, *J. Phys. Soc. Japan* **17**, 1100 (1962).

[53] W. Ludwig, "Recent Developments in Lattice Theory, Springer-Tracts in Modern Physics," Vol. 43. Springer, Berlin, Heidelberg, New York, 1967.

to it. From $T \simeq 0$ to $T \simeq \theta_D$ the Debye–Waller factor increases roughly by a factor 4, and is proportional to T for higher temperatures.

For anharmonic crystals the Debye–Waller factor is rather complicated.[53] Here are two different corrections to the harmonic expressions (15.11) and (15.13). The first and most important is an anharmonic contribution to the quadratic average $\langle \mathbf{s}_n{}^2 \rangle$ proportional to $(kT)^2$ for high temperatures. The second correction comes from the fourth-order average in (15.9). It is proportional to h^4, but also depends on the direction of $\mathbf{h}$, contrary to (15.10). For high temperatures, it is proportional to $(kT)^3$. Estimates of Maradudin and Flinn[54] show that this term should be very small. Therefore, the quadratic expression (15.10) can also be used in anharmonic crystals.

16. Electronic Excitations

The second approximation $U^{(2)}$ [Eq. (14.12)] is somewhat more complicated than $U^{(1)}$. According to (15.1) the total wavefunction ϕ_α of the crystal is a product of the wavefunction ξ_μ of the nuclei and the electronic wavefunction φ_n. Using the completeness relation for the φ_n,

$$1 = |\varphi_0\rangle\langle\varphi_0| + \sum_{n \neq 0} |\varphi_n\rangle\langle\varphi_n| \tag{16.1}$$

the correction $U^{(2)}$ can be split into two terms

$$U^{(2)} = U^{(2)}(\mathrm{el}) + U^{(2)}(\mathrm{ph}), \tag{16.2}$$

$$U^{(2)}(\mathrm{el}) = \Big\langle (V - \langle V \rangle)\, G_0 \sum_{n \neq 0} |\varphi_n\rangle\langle\varphi_n| (V - \langle V \rangle) \Big\rangle, \tag{16.3}$$

$$U^{(2)}(\mathrm{ph}) = \Big\langle (V - \langle V \rangle)\, G_0 |\varphi_0\rangle\langle\varphi_0| (V - \langle V \rangle) \Big\rangle. \tag{16.4}$$

$U^{(2)}(\mathrm{el})$ represents the effect of electron excitations on the optical potential, since in (16.3) only excited intermediate states for the electrons are allowed. On the other hand, $U^{(2)}(\mathrm{ph})$, depending only on the electronic ground state, describes the effect of phonon excitations, discussed in detail in Section V,17.

In (16.3), the averaged potential $\langle V \rangle$ does not depend explicitly on the electron coordinates $\mathbf{r}_j$ and drops out since $\langle \varphi_0 | \varphi_n \rangle = \delta_{0,n}$. The same argument applies to the interaction with the nuclei, so that only the unscreened Coulomb interaction with the crystal electrons remains. Further, we can use the static approximation (14.5) for the motion of the

[54] A. A. Maradudin and P. A. Flinn, *Phys. Rev.* **129**, 2529 (1963).

nuclei since the phonon frequencies $\hbar\omega \ll E_{\mathbf{K}}$ or $E_n - E_0$. Therefore, we obtain from (16.3)

$$U^{(2)}(\text{el}) = \sum_{n\neq 0} \left\langle V_{0n}^{\text{el}} \frac{1}{E_{\mathbf{K}} - (E_n - E_0) + i\epsilon - h_0} V_{n0}^{\text{el}} \right\rangle_{\text{th}} \tag{16.5}$$

or in $\mathbf{r}$ representation

$$U^{(2)(\text{el})}(\mathbf{r}, \mathbf{r}') = \sum_{n\neq 0} -\frac{2m}{\hbar^2 4\pi} \frac{e^{iK_n|\mathbf{r}-\mathbf{r}'|}}{|\mathbf{r}-\mathbf{r}'|} \langle V_{0n}^{\text{el}}(\mathbf{r})\, V_{n0}^{\text{el}}(\mathbf{r}')\rangle_{\text{th}} \tag{16.5a}$$

with

$$V^{\text{el}}(\mathbf{r}) = e^2 \sum_j \frac{1}{|\mathbf{r}-\mathbf{r}_j|}, \qquad K_n = \left(K^2 - \frac{2m}{\hbar^2}(E_n - E_0)\right)^{1/2}.$$

Here $\langle\ \rangle_{\text{th}}$ means the thermal average over the phonon states ξ_μ only. Thus we obtain for the Fourier coefficients of $U^{(2)}$:

$$U^{(2)}_{\mathbf{k}+\mathbf{h},\mathbf{k}+\mathbf{h}'}(\text{el}) = \frac{1}{NV_{\text{c}}} \int \frac{d\mathbf{K}'}{(2\pi)^3} \sum_{n\neq 0} \frac{\langle \tilde{V}_{0n}(-\mathbf{k}-\mathbf{h}+\mathbf{k}')\, \tilde{V}_{n0}(\mathbf{k}+\mathbf{h}'-\mathbf{k}')\rangle_{\text{th}}}{E_{\mathbf{K}} - (E_n - E_0) + i\epsilon - \hbar^2\mathbf{K}'^2/2m}, \tag{16.6}$$

where N is the number of unit cells of the crystal and V_{n0} is given by

$$\begin{aligned}\tilde{V}_{n0}(q) &= \int d\mathbf{r}\, e^{i\mathbf{q}\mathbf{r}} V_{n0}^{\text{el}}(\mathbf{r}) \\ &= -\{4\pi e^2/q^2\} \sum_j e^{i\mathbf{q}\mathbf{R}_{n_j}} \langle \varphi_n | e^{i\mathbf{q}\tilde{\mathbf{r}}_j} | \varphi_0\rangle.\end{aligned} \tag{16.7}$$

R_{n_j} is the position of the atom to which electron j is attached ($\mathbf{r}_j = \mathbf{R}_{n_j} + \tilde{\mathbf{r}}_j$). By using the Hartree–Fock approximation for the φ_n, we take only those intermediate states n into account which differ from the ground state φ_0 by the excitation of electron j. Otherwise $V_{n0} = 0$. Then we obtain from (16.6)

$$\begin{aligned}&U^{(2)}_{\mathbf{k}+\mathbf{h},\mathbf{k}+\mathbf{h}'}(\text{el}) \\ &= \langle \exp[i(\mathbf{h}'-\mathbf{h})\,\mathbf{R}_n]\rangle_{\text{th}} \frac{1}{V_{\text{c}}} \int \frac{d\mathbf{K}'}{(2\pi)^3} \frac{(4\pi e^2)^2}{|\mathbf{k}+\mathbf{h}-\mathbf{K}'|^2 |\mathbf{k}+\mathbf{h}'-\mathbf{K}'|^2} \\ &\quad\times \sum_{\alpha,\mu} f_{\alpha\mu}(-\mathbf{k}-\mathbf{h}+\mathbf{K}') f_{\mu\alpha}(\mathbf{k}+\mathbf{h}'-\mathbf{K}') \frac{1}{E_{\mathbf{K}} - (E_\mu - E_\alpha) + i\epsilon - E_{\mathbf{K}'}},\end{aligned} \tag{16.8}$$

with

$$f_{\mu\alpha}(q) = \int d\mathbf{r}\, \chi_\mu^*(\mathbf{r})\, e^{i\mathbf{q}\mathbf{r}} \chi_\alpha(\mathbf{r}). \tag{16.9}$$

The χ_ν are the single particle states of an atom. The summation over α in (16.8) ranges over the occupied levels only, whereas the sum over μ contains the nonoccupied levels only. It can be seen that the temperature dependence of $U^{(2)}(\mathrm{el})$, given by the Debye–Waller factor $e^{-M_{\mathbf{h}-\mathbf{h}'}}$, is the same as that of $U^{(1)}$ (15.5). This results from neglecting the correlations between electrons of different atoms.

Equation (16.8) is still relatively complicated since it involves a sum over an infinite number of nonoccupied states μ. Therefore we simplify (16.8) by replacing $E_\mu - E_\alpha$ by a suitably chosen mean energy loss ΔE. Furthermore, we use the completeness relation in the form

$$\sum_\mu |\mu\rangle\langle\mu| = 1 - \sum_{\alpha'} |\alpha'\rangle\langle\alpha'|, \tag{16.10}$$

where α' refers to occupied levels only. Thus we obtain finally

$$\begin{aligned} U^{(2)}_{\mathbf{k}+\mathbf{h},\mathbf{k}+\mathbf{h}'}(\mathrm{el}) &= e^{-M_{\mathbf{h}-\mathbf{h}'}} \frac{1}{V_c} \int \frac{d\mathbf{K}'}{(2\pi)^3} \frac{(4\pi e^2)^2}{|\mathbf{k}+\mathbf{h}-\mathbf{K}'|^2 |\mathbf{k}+\mathbf{h}'-\mathbf{K}'|^2} \\ &\quad \times \frac{F_{\mathbf{hh}'}(\mathbf{K}')}{E_\mathbf{K} - \Delta E + i\epsilon - E_{\mathbf{K}'}}, \end{aligned} \tag{16.11}$$

with

$$\begin{aligned} F_{\mathbf{hh}'}(\mathbf{K}') &= \sum_\alpha \left\{ f_{\alpha\alpha}(\mathbf{h}'-\mathbf{h}) - f_{\alpha\alpha}(-\mathbf{k}-\mathbf{h}+\mathbf{K}') f_{\alpha\alpha}(\mathbf{k}+\mathbf{h}'-\mathbf{K}') \right\} \\ &\quad - \sum_{\alpha,\alpha';(\alpha\neq\alpha')} f_{\alpha\alpha'}(-\mathbf{k}-\mathbf{h}+\mathbf{K}') f_{\alpha'\alpha}(\mathbf{k}+\mathbf{h}'-\mathbf{K}'). \end{aligned} \tag{16.12}$$

We are particularly interested in the Fourier coefficients of $U''^{(2)}$ [Eq. (14.16)] representing the absorption of the coherent wave. Due to the δ function, only the angle integration in (16.11) remains.

$$U''^{(2)}_{\mathbf{k}+\mathbf{h},\mathbf{k}+\mathbf{h}'}(\mathrm{el}) = -e^{-M_{\mathbf{h}-\mathbf{h}'}} \cdot \frac{2mK'e^4}{\hbar^2 V_c} \int d\Omega' \frac{F_{\mathbf{hh}'}(\mathbf{K}')}{|\mathbf{k}+\mathbf{h}-\mathbf{K}'|^2 |\mathbf{k}+\mathbf{h}'-\mathbf{K}'|^2}$$

where

$$K' = |\mathbf{K}'| = (K - (2m/\hbar^2)\,\Delta E)^{1/2}. \tag{16.13}$$

The results (16.11) and (16.13) have first been given by Yoshioka[14] (besides the temperature dependence). Neglecting the exchange term $\sum_{\alpha'\neq\alpha}$ in (16.12), Yoshioka calculated the Fourier coefficients explicitly for MgO, using the Thomas–Fermi model. Whelan[55] improved these calculations using Freeman's Hartree–Fock wave functions,[56] again

[55] M. J. Whelan, *J. Appl. Phys.* **36**, 2099 (1965).

[56] A. J. Freeman, *Phys. Rev.* **113**, 176 (1959); *Acta Crystallogr.* **12**, 274, 929 (1959); **13**, 190, 618 (1960).

neglecting the exchange terms. Recently, Radi[57] evaluated the effect of electronic excitations using the improved approximation $U^{(2)}$ of (14.14) obtaining good agreement with Whelan. However, we have to point out that due to the approximations involved in (16.11) and in these calculations (rigid ion model, mean energy loss ΔE, exchange terms, etc.), these results should not be regarded as completely reliable.

The rigid ion model cannot be applied to band electrons, as they behave more or less like free electrons. However, for the absorption, band electrons are quite important due to the possibility of *plasmon excitations*, i.e., excitations of collective, longitudinal vibrations of the electron plasma. In the usual theory[58] the electrons are treated as free, moving relative to a uniform and positively charged background. Therefore, only the coefficients $U_{00}^{(2)}$ has to be calculated. The mixed ones $U_{0\mathbf{h}}^{(2)}$ vanish due to translation invariance. From (16.6) we obstain for $U_{00}''^{(2)}$

$$U_{00}''^{(2)}(\mathrm{el}) = \frac{1}{NV_c} \int \frac{d\mathbf{K}'}{(2\pi)^3} \frac{(4\pi e^2)^2}{|\mathbf{k} - \mathbf{K}'|^4} S\left(\mathbf{k} - \mathbf{K}'; \frac{E_\mathbf{K} - E_{\mathbf{K}'}}{h}\right). \quad (16.14)$$

Here

$$S(\mathbf{q}, \omega) = -\pi \sum_{n \neq 0} \left|\left\langle \varphi_n \right| \sum_j e^{i\mathbf{q}\mathbf{r}_j} \left| \varphi_0 \right\rangle\right|^2 \delta(\hbar\omega - (E_n - E_0)) \quad (16.15)$$

is van Hove's correlation function,[58] which is connected with the dielectric constant $\epsilon(\mathbf{q}, \omega)$ by

$$S(\mathbf{q}, \omega) = \begin{cases} -(q^2/4\pi^2 e^2)\,\mathrm{Im}\{1/\epsilon(\mathbf{q}, \omega)\} & \text{for} \quad \omega > 0, \\ 0 & \text{for} \quad \omega < 0. \end{cases} \quad (16.16)$$

Now for small momentum transfers q, the imaginary part of $1/\epsilon$ can be approximated by[58a]

$$\mathrm{Im}\{1/\epsilon(\mathbf{q}, \omega)\} = -\pi\,\delta(1 - \omega_\mathrm{p}^2/\omega^2) \qquad \text{where} \quad \omega_\mathrm{p}^2 = 4\pi\rho_\mathrm{b} e^2/m \quad (16.17)$$

ω_p is the plasma frequency, which is of the order of 20 eV for most materials. By substituting $\mathbf{k} - \mathbf{K}' = \mathbf{q}$, we have for small q

$$E_\mathbf{K} - E_{\mathbf{K}'} \cong (\hbar^2 \mathbf{k}/m) \cdot \mathbf{q}. \quad (16.18)$$

Therefore, we obtain from (16.17)

$$U_{00}''^{(2)} = -\frac{\omega_\mathrm{p}}{2} \int \frac{d\mathbf{q}}{(2\pi)^3} \frac{4\pi e^2}{q^2} \delta\left(\frac{\hbar\mathbf{K}}{m} \cdot \mathbf{q} - \omega_\mathrm{p}\right) \quad (16.19)$$

[57] G. Radi, *Z. Phys.* **212**, 456 (1968).

[58] D. Pines, "Elementary Excitations in Solids." Benjamin, New York, 1963.

[58a] ρ_b is the density of band electrons.

The integral, diverging logarithmically for large q, is cut off at the maximal momentum which can be transferred in a head-on collision: $\hbar q_{\max} = \hbar K$. We obtain finally

$$U''^{(2)}_{00}(\text{el}) = -\frac{\hbar\omega_{\text{p}}}{2Ka_{\text{B}}}\ln\frac{2E_{\mathbf{K}}}{\hbar\omega_{\text{p}}}, \qquad \text{where} \quad a_{\text{B}} = \frac{\hbar^2}{me^2}. \tag{16.20}$$

Therefore, the excitation of plasmons leads to a uniform and temperature-independent contribution to the absorption. By applying this result to electron diffraction, special care is necessary since the transferred momentum q is extremely small. For example, the scattering angles are of the order of $\theta_E = \hbar\omega_{\text{p}}/2E$ and comparable to the width of the Bragg reflection. Therefore the inelastic scattered waves satisfy the Bragg condition partially and show essentially the same interferences as the coherent wave, as has been pointed out by Howie.[59]

Therefore, Eq. (16.20) can only be used rigorously if by energy selection the coherent wave is measured alone.

An anomaly in the mean free path for plasmon excitation has been observed by Ishida *et al.*[60] and Tonomura and Watanabe.[61] Whereas the present theory only gives an homogeneous absorption due to the free electron model, the experiments show a somewhat increased absorption for the sin wave field avoiding the atoms and a lower absorption for the cos field passing through the atomic planes. This effect is contrary to the usual anomalous transmission effect. An explanation could be that the conduction electrons forming the plasma have a somewhat larger density between the atomic planes than on these planes. However there are also two other possible explanations[62,63] so that the correct interpretation is not clear.

17. Phonon Excitations

To calculate the effect of the phonons on the optical potential we have to evaluate $U^{(2)}(\text{ph})$ of Eq. (16.4).

$$U^{(2)}(\text{ph}) = \left\langle (V_{00} - \langle V_{00}\rangle_{\text{th}}) \frac{1}{E_{\mathbf{K}} + E_{\mu} + i\epsilon - H_0^{\text{ph}} - h_0} (V_{00} - \langle V_{00}\rangle_{\text{th}}) \right\rangle_{\text{th}}. \tag{17.1}$$

[59] A. Howie, *Proc. Phys. Soc.* **A271**, 268 (1963).
[60] K. Ishida, M. Mannami, and K. Tanaka, *J. Phys. Soc. Japan* **23**, 1362 (1967).
[61] A. Tonomura and H. Watanabe, *Jap. J. Appl. Phys.* **6**, 1163 (1967).
[62] Y. H. Othsuki, *Phys. Lett.* **27A**, 65 (1968).
[63] G. Radi, *Z. Phys.* **213**, 244 (1968).

Here H_0^{ph} is the Hamiltonian of the phonon system and $V_{00}(\mathbf{r})$ the expectation value of V in the electronic ground state

$$V_{00}(\mathbf{r}) = \sum_n v(\mathbf{r} - \mathbf{R}_n) \qquad \text{with} \qquad v(\mathbf{r} - \mathbf{R}_n) \text{ of Eq. (15.4).} \tag{17.2}$$

Analogously to (14.23), $U^{(2)}(\text{ph})$ can be transformed to

$$U^{(2)} = \int d\mathbf{R}\, d\mathbf{R}' \int_0^\infty (dt/i\hbar)\, v_{\mathbf{R}} \exp[(i/h)(E_{\mathbf{K}} + i\epsilon - h_0)\, t]\, v_{\mathbf{R}'}$$
$$\times \{\langle \rho_{\text{N}}(\mathbf{R}, t)\, \rho_{\text{N}}(\mathbf{R}', 0)\rangle - \langle \rho_{\text{N}}(\mathbf{R}, t)\rangle\langle \rho_{\text{N}}(\mathbf{R}', 0)\rangle\}, \tag{17.3}$$

where $\rho_{\text{N}}(\mathbf{R}, t) = \sum_n \delta(\mathbf{R} - \mathbf{R}_n(t))$ is the time-dependent density operator of the nuclei. In this representation the properties of the incident electron and those of the crystal are clearly separated. Since the energies of the phonons are much smaller than the energy of the incident electron, we may use the static approximation by replacing the correlation function by the static function for $t = 0$ (Eq. (14.24)). Then we obtain for the matrix elements of $U^{(2)}(\text{ph})$

$$U^{(2)}_{\mathbf{k}+\mathbf{h},\mathbf{k}+\mathbf{h}'}(\text{ph}) = \{V_c/(2\pi)^3\} \int d\mathbf{K}'\, V_{\mathbf{k}+\mathbf{h}-\mathbf{K}'} V_{\mathbf{K}'-\mathbf{k}-\mathbf{h}'}$$
$$\times 1/(E_{\mathbf{K}} + i\epsilon - E_{\mathbf{K}'}) \cdot S_{\mathbf{hh}'}(\mathbf{k} - \mathbf{K}'). \tag{17.4}$$

The correlation function $S_{\mathbf{h},\mathbf{h}'}$, depending only on crystal properties, is given by

$$S_{\mathbf{hh}'}(\mathbf{k} - \mathbf{K}') = \sum_{m-n} \{\langle \exp -i\, \boldsymbol{\varkappa}\mathbf{R}_m + i\, \boldsymbol{\varkappa}'\mathbf{R}_n\rangle - \langle \exp -i\, \boldsymbol{\varkappa}\mathbf{R}_m\rangle\langle \exp i\, \boldsymbol{\varkappa}'\mathbf{R}_n\rangle\}$$

$$\text{where} \qquad \boldsymbol{\varkappa} = \mathbf{k} + \mathbf{h} - \mathbf{K}' \qquad \text{and} \qquad \boldsymbol{\varkappa}' = \mathbf{k} + \mathbf{h}' - \mathbf{K}'. \tag{17.5}$$

Due to the crystal periodicity the correlation depends on the difference $\mathbf{m} - \mathbf{n}$ only, to which the summation is restricted. From (17.4) we can easily evaluate the matrix elements of the Hermitean and anti-Hermitean terms $U'^{(2)}$ and $iU''^{(2)}$.

$$U'^{(2)}_{\mathbf{k}+\mathbf{h},\mathbf{k}+\mathbf{h}'}(\text{ph}) = \{V_c/(2\pi)^3\} \int d\mathbf{K}'\, V_{\mathbf{k}+\mathbf{h}-\mathbf{K}'} V_{\mathbf{K}'-\mathbf{k}-\mathbf{h}'}$$
$$\times P(1/\{E_{\mathbf{K}} - E_{\mathbf{K}'}\})\, S_{\mathbf{hh}'}(\mathbf{k} - \mathbf{K}') \tag{17.6}$$

$$U''^{(2)}_{\mathbf{k}+\mathbf{h},\mathbf{k}+\mathbf{h}'}(\text{ph}) = -(2mK/\hbar^2)(V_c/(4\pi)^2) \int d\Omega_{\mathbf{K}'}\, V_{\mathbf{k}+\mathbf{h}-\mathbf{K}'} V_{\mathbf{K}'-\mathbf{k}-\mathbf{h}'}$$
$$\times S_{\mathbf{hh}'}(\mathbf{k} - \mathbf{K}'). \tag{17.7}$$

Now, the main problem consists of obtaining simple analytical approximations for $S_{\mathbf{hh}'}$. First, we set $\mathbf{R}_n = \bar{\mathbf{R}}_n + \mathbf{s}_n$ according to (15.8). In the harmonic approximation the displacements $\mathbf{s}_n$ and $\mathbf{s}_m$ have a Gaussian distribution, and, therefore, any linear combination of $\mathbf{s}_n$ and $\mathbf{s}_m$ also. Consequently, we obtain in the harmonic approximation with $M_\varkappa$ of (15.10)

$$S_{\mathbf{h},\mathbf{h}'}(k - \mathbf{K}') = \exp[-M_\varkappa - M_{\varkappa'}] \sum_{m-n} \exp[-i(\mathbf{k} - \mathbf{K}')(\bar{\mathbf{R}}_m - \bar{\mathbf{R}}_n)]$$
$$\times \{\exp\langle(\varkappa \cdot \mathbf{s}_m)(\varkappa' \cdot \mathbf{s}_n)\rangle - 1\} \tag{17.8}$$

In Einstein's model, the different atoms are treated as independent oscillators. In this approximation the correlations for $m \neq n$ vanish in (17.5) and (17.8). Only the self-correlation for $m = n$ remains, which can be expressed by Debye–Waller factors:

$$S^{\mathrm{ES}}_{\mathbf{hh}'} = \exp -M_{\mathbf{h}-\mathbf{h}'} - \exp -M_{\mathbf{k}+\mathbf{h}-\mathbf{K}'} \cdot \exp -M_{\mathbf{k}+\mathbf{h}'-\mathbf{K}'} . \tag{17.9}$$

In Einstein's model, the dynamics of the crystal is only unrealistically reflected. It is true that the correlations between different atoms are normally smaller than the self-correlation. For instance, the average $\langle(s_x{}^n)^2\rangle$ is typically a factor 2 or 3 larger than the average $\langle s_x{}^n s_x{}^m\rangle$ for neighboring atoms.[64] However, the correlations have an extremely long range decreasing for large distances $\bar{\mathbf{R}}_m - \bar{\mathbf{R}}_n$ as $1/|\bar{\mathbf{R}}_m - \bar{\mathbf{R}}_n|$. This behavior is taken into account in the "one-phonon approximation" by expanding (17.8) linearly in $\langle(\varkappa \cdot \mathbf{s}_m)(\varkappa' \cdot \mathbf{s}_n)\rangle$.

$$S^{\mathrm{1ph}}_{\mathbf{hh}'} = \exp[-M_\varkappa - M_{\varkappa'}] \sum_{m-n} \exp[-i(\mathbf{k} - \mathbf{K}')(\bar{\mathbf{R}}_m - \bar{\mathbf{R}}_n)]\langle(\varkappa \cdot \mathbf{s}_m)(\varkappa' \cdot \mathbf{s}_n)\rangle$$
$$= \exp[-M_\varkappa - M_{\varkappa'}] \sum_s (\varkappa \cdot \mathbf{e}_\mathbf{q}{}^s)(\varkappa' \cdot \mathbf{e}_\mathbf{q}{}^s)\, \epsilon(\omega_s(\mathbf{q}), T)/M\omega_s{}^2(\mathbf{q}). \tag{17.10}$$

The last result follows by expanding $\mathbf{s}_n$ in phonon creation and annihilation operators and by taking the thermal average.[53] $\mathbf{e}_\mathbf{q}{}^s$ is the normalized polarization vector for the polarization s. The phonon vector $\mathbf{q}$ lying in the first Brillouin zone is chosen such that $\mathbf{K}' - \mathbf{k} = \mathbf{g} + \mathbf{q}$, where $\mathbf{g}$ is a reciprocal lattice vector. Characteristically, $S_{\mathbf{hh}'}$ diverges for small $\mathbf{q}$ as $1/q^2$, since $\omega_s(q) \approx c_s q$ for small q. This is due to the slow decrease $\sim 1/R$ of the correlations for large distances R.

[64] G. Leibfried, "Gittertheorie der mechanischen und thermischen Eigenschaften der Kristalle," *in* "Handbuch der Physik," Vol. VII, p. 1, Springer Verlag, Berlin and New York, 1955.

By combining the Einstein and one-phonon approximations we can obtain one having the advantages of both. We take into account the somewhat larger self-correlation exactly and expand the terms $m \neq n$ linearly.

$$S_{\mathbf{hh}'} = S_{\mathbf{hh}'}^{\mathrm{ES}} + S_{\mathbf{hh}'}^{\mathrm{C}}. \tag{17.11}$$

The correlation term $S_{\mathbf{hh}'}^{\mathrm{C}}$ is evaluated by adding and subsequently subtracting the term $m = n$,

$$S_{\mathbf{hh}'}^{\mathrm{C}} = \exp(-M_{\varkappa} - M_{\varkappa'}] \sum_s \Big\{ (\varkappa \cdot \mathbf{e}_{\mathbf{q}}{}^s)(\varkappa' \cdot \mathbf{e}_{\mathbf{q}}{}^s) \frac{\epsilon(\omega_s(\mathbf{q}'), T)}{M\omega_s{}^2(\mathbf{q})}$$

$$- \frac{V_{\mathrm{C}}}{(2\pi)^3} \int_{\mathrm{I.B.Z.}} d\mathbf{q}' \, (\varkappa \cdot \mathbf{e}_{\mathbf{q}'}^s)(\varkappa' \cdot \mathbf{e}_{\mathbf{q}'}^s) \frac{\epsilon(\omega_s(\mathbf{q}'), T)}{M\omega_s{}^2(\mathbf{q}')} \Big\}. \tag{17.12}$$

The second term, arising from the subtraction of the self-correlation, is the q-average of the first one-phonon term over the first Brillouin zone.

By introducing the one-phonon approximation or (17.12) into (17.7), a problem arises, because the integration for $U''^{(2)}_{\mathbf{h},\mathbf{h}'}$ diverges due to the $1/q^2$ behavior of the integrand for small q if the Bragg condition $\mathbf{k}^2 = (\mathbf{k} + \mathbf{h})^2$ is fulfilled. However, as will be shown later on, the integration has to be cut off at a minimal q value of $q_c = |v_{\mathbf{h}}|/h$. The integrals have then to be evaluated numerically.

The influence of thermal motion on dynamical electron diffraction has first been evaluated by Yoshioka and Kainuma[65] by perturbation theory. More general treatments equivalent to our derivation have been given by a number of authors,[66–73] especially Othsuki and Yanagawa,[66] Othsuki,[67] Dederichs,[68] Kainuma and Yoshioka.[69] Detailed numerical calculations of the phonon absorption have been performed by Hall and Hirsch[70,71] and Whelan.[72] These calculations show that normally the phonon excitation gives a larger contribution to the absorption than does the electron excitation. Furthermore, the phonon excitation gives an important anomalous transmission effect since the displacement distri-

[65] H. Yoshioka and Y. Kainuma, *J. Phys. Soc. Japan* **17**, Suppl. BII, 134 (1962).
[66] Y. H. Othsuki and S. Yanagawa, *J. Phys. Soc. Japan* **21**, 326 (1966).
[67] Y. H. Othsuki, *J. Phys. Soc. Japan* **21**, 2300 (1966); **24**, 555 (1968).
[68] P. H. Dederichs, *Phys. Kondens. Mater.* **5**, 347 (1966).
[69] Y. Kainuma and H. Yoshioka, *J. Phys. Soc. Japan* **21**, 1352 (1966).
[70] C. R. Hall and P. B. Hirsch, *Proc. Roy. Soc.* **A286**, 158 (1965).
[71] C. R. Hall, *Phil. Mag.* **12**, 815 (1965).
[72] M. J. Whelan, *J. Appl. Phys.* **36**, 2103 (1965).
[73] L. B. Krashnina, V. B. Molodkin, and E. A. Tikhonova, *Sov. Phys.—Solid State* **9**, 1306 (1967).

bution is very much localized near the lattice positions. For more details we refer to these publications. Figure 18a, taken from Hall and Hirsch,[70] shows the temperature dependence of the normal absorption μ_0 and $\mu_{\mathbf{h}}$ and the anomalous absorption $\Delta\mu_{\mathbf{h}} = \mu_0 - \mu_{\mathbf{h}}$ for a 111-reflection of Pb. Figure 18b–f shows the variation of $\Delta\mu_{\mathbf{h}}$ with temperature for three reflections and for Al, Cu, Ag, Au, and Pb.

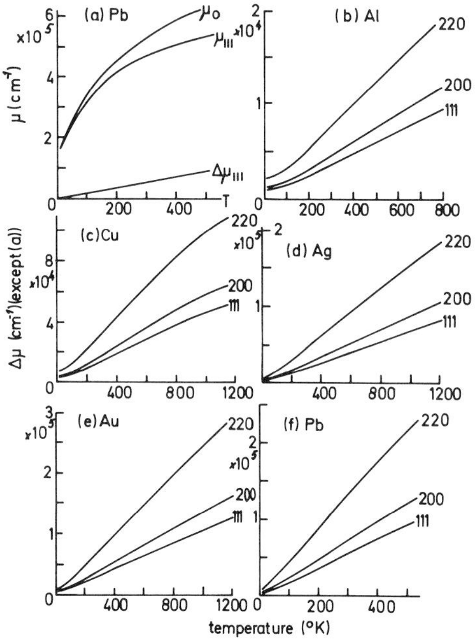

FIG. 18. (a) Variation of μ_0, μ_{111}, and $\Delta\mu_{111}$, with temperature for Pb. (From Hall and Hirsch[70]). (b)–(f) Variation of $\Delta\mu_{\mathbf{h}}$ with temperature for three reflections for Al, Cu, Ag, Au, and Pb.

There is also a recent review on the influence of thermal vibrations on electron diffraction, being mostly concerned with experiments using polycrystalline samples.[74]

Divergence Difficulties

The divergence of $U''^{(2)}_{\mathbf{hh}'}$ for small q is not a failure of the one-phonon approximation: the approximation $U^{(2)}$ of (14.11) fails for long-range correlations $\sim 1/|\mathbf{R}_m - \mathbf{R}_n|$. Since $U^{(2)}$ contains the free Green's function

[74] G. Horstmann, *Fortschr. Phys.* **16**, No. 3, 1968.

G_0 (or g_0 in the static approximation), the electron is regarded as free, while intermediately scattered from atom m to atom n. However, for large distances $|\mathbf{R}_m - \mathbf{R}_n|$, being comparable to or larger than the extinction length, extinction effects, etc., become very important and the effect of the periodic potential on the motion can no longer be neglected. Therefore, the intermediately scattered electron has to be treated as a Bloch wave rather than a plane wave. Precisely this effect is described by the improved approximation (14.14) for $U^{(2)}$. It contains the Green's function $G_{\langle V\rangle}$ instead of G_0, i.e., in the static approximation the Green's function

$$G_{\langle V\rangle} = 1/(E_{\mathbf{K}} + i - h_0 - \langle V\rangle)$$

describing the motion in the averaged potential $\langle V\rangle$. Therefore, we expect no divergence difficulties with the approximation (14.14). This can be shown by expanding the Bloch wave $\phi_{\mathbf{k},\nu}$ according to (4.5) in terms of Bloch waves $\phi^{(0)}_{\mathbf{k},\nu'}$, which are eigenfunctions to $h_0 + \langle V\rangle$. By treating $U^{(2)}$ as a perturbation, we need the expectation value $U^{(2)}_{\nu\nu}(\mathbf{k})$ (4.6) to determine the Bloch vector $\mathbf{k}$. For this we obtain from (14.14) after some straightforward calculations

$$\begin{aligned} U''^{(2)}_{\nu\nu}(\mathbf{k}) &= -\pi\{V_c/(2\pi)^3\} \int \{dS_{\mathbf{k}'}/|\mathrm{grad}\, E_\nu(\mathbf{k}')|\} \\ &\quad\times \Big\{ \sum_{\mathbf{hh'gg'}} C_{\mathbf{h}}^{0*}(\mathbf{k})\, C_{\mathbf{h'}}^{0}(k)\, C_{\mathbf{g}}^{0}(\mathbf{k}')\, C_{\mathbf{g'}}^{0*}(\mathbf{k}') \\ &\quad\times V_{\mathbf{k}+\mathbf{h}-\mathbf{k}'-\mathbf{g}} V_{-\mathbf{k}-\mathbf{h}'+\mathbf{k}'+\mathbf{g}'} S_{\mathbf{h}-\mathbf{g},\mathbf{h}'-\mathbf{g}'}(\mathbf{k}') \Big\}. \end{aligned} \tag{17.13}$$

This is a surface integral over the dispersion surfaces $E_\nu(\mathbf{k}') = \hbar^2K^2/2m$. No divergence occurs since for $\mathbf{k}' \to \mathbf{k}$ the integral contains a factor

$$|C_{\mathbf{h}}^{0*}(\mathbf{k})\, C_{\mathbf{g}}^{0}(\mathbf{k})\, V_{\mathbf{h}-\mathbf{g}} e^{-M_{\mathbf{h}-\mathbf{g}}}(\mathbf{h} - \mathbf{g}) \cdot \mathbf{e}_{\mathbf{q}}{}^{s}|^2 = 0, \tag{17.14}$$

which compensates the diverging $1/\omega^2$ factor in $S_{\mathbf{h}-\mathbf{g},\mathbf{h},-\mathbf{g}}(\mathbf{k}')$. In (17.13) the Bloch waves $\mathbf{k}'$ degenerate into plane waves and similarly the dispersion surface into spheres if the Bragg reflection is not excited. For such $\mathbf{k}'$ values the integrand is identical with the analogous expression obtained from the approximation (14.9) with the free Green's function. Therefore, we may simply use the approximation (14.9) for $U^{(2)}$, but at the same time, cut off the divergencies in the integrands for such q values for which the Bloch wave properties of $\phi^{(0)}_{\mathbf{k}'\nu}$ are important. Using as condition $W = 1$ (5.6), we get a cutoff value $q_c = |v_{\mathbf{h}}|/h$. Smaller q values are then not allowed. Thus we may say that the effect of the Green's function $G_{\langle V\rangle}$ is to cutoff the long-range correlations at the extinction length.

VI. Optical Potential Method for X-Ray Diffraction

18. The Photoelectric Effect

In this section we shall give a semiclassical description of X-ray diffraction following Molière's approach.[13] The X-ray field is treated classically and the crystal electrons quantum mechanically. The influence of thermal motion is discussed separately in the next section. Here the nuclei are considered as fixed.

In order to solve Maxwell's equations, one normally tries to find a solution in terms of a vector potential $\mathbf{A}(\mathbf{r}, t)$ and a scalar potential $\phi(\mathbf{r}, t)$, by setting

$$\mathbf{H}(\mathbf{r}, t) = \partial_{\mathbf{r}} \times \mathbf{A}(\mathbf{r}, t), \tag{18.1}$$

$$\mathbf{E}(\mathbf{r}, t) = -(1/c)\, \partial_t \mathbf{A}(\mathbf{r}, t) - \partial_{\mathbf{r}} \cdot \phi(\mathbf{r}, t). \tag{18.2}$$

By Maxwell's equations alone, $\mathbf{A}$ and ϕ are not determined uniquely since an additional gauge condition can be chosen. Our choice is $\phi(\mathbf{r}, t) = 0$, which guarantees that $\mathbf{A}(\mathbf{r}, t)$ vanishes in the field free case $\mathbf{E} = 0$ and $\mathbf{H} = 0$. Then we obtain from Maxwell's equations

$$-\partial_{\mathbf{r}} \times \partial_{\mathbf{r}} \times \mathbf{A} - (1/c^2)\, \partial_t^2 \mathbf{A} = -(4\pi e/c)\, \mathbf{J}_0(\mathbf{r}, t), \tag{18.3}$$

where $\mathbf{J}_0(\mathbf{r}, t)$ is the current density. In principle we could understand (18.3) as an equation for microscopic quantities $\mathbf{A}(\mathbf{r}; \cdots \mathbf{r}_j \cdots; t)$ and $\mathbf{J}(\mathbf{r}; \cdots \mathbf{r}_j \cdots; t)$ depending explicitly on the electron coordinates $\mathbf{r}_j$ of the crystal. However, we will assume $\mathbf{A}(\mathbf{r}, t)$ and $\mathbf{J}_0(\mathbf{r}, t)$ to be macroscopic quantities, namely the expectation values of microscopic ones in the electronic state of the crystal at time t. Then $\mathbf{A}(\mathbf{r}, t)$ is the coherent field in the sense of Section 4.12 and $\mathbf{J}_0(\mathbf{r}, t)$ is the expectation value of the current density operator $\mathbf{J}(\mathbf{r})$ in the state $\phi(t)$ of the electrons.

$$\mathbf{J}_0(\mathbf{r}, t) = (\phi(t), \mathbf{J}(\mathbf{r})\, \phi(t)). \tag{18.4}$$

In the presence of a magnetic field, the current density operator consists of two parts, one due to the momenta p_j of the electrons (**j**) and another one due to the field $(\mathbf{j}^A)$[74a]:

$$\mathbf{J}(\mathbf{r}) = \sum_j (1/2m)(\mathbf{p}_j\, \delta(\mathbf{r} - \mathbf{r}_j) + \delta(\mathbf{r} - \mathbf{r}_j)\, \mathbf{p}_j) - (e/mc)\, \mathbf{A}(\mathbf{r}, t) \sum_j \delta(\mathbf{r} - \mathbf{r}_j)$$

$$= \mathbf{j}(\mathbf{r}) + \mathbf{j}^A(\mathbf{r}). \tag{18.5}$$

Now we want to evaluate $\mathbf{J}_0(\mathbf{r}, t)$ linearly in the field and neglect the quadratic terms in **A**. The Hamiltonian of the crystal electrons is up to linear terms in **A**,

$$H = H_0 + H_1(t). \tag{18.6}$$

H_0 is the unperturbed Hamiltonian and $H_1(t)$ the interaction with the **A** field

$$H_1(t) = -(e/c) \sum_j (1/2m)(\mathbf{p}_j \mathbf{A}(\mathbf{r}_j\,, t) + \mathbf{A}(\mathbf{r}_j\,, t)\, p_j)$$

$$= -(e/c) \int d\mathbf{r}'\, \mathbf{A}(\mathbf{r}', t)\, \mathbf{j}(\mathbf{r}'). \tag{18.7}$$

In principle, the vector potential **A**, occurring in $H_1(t)$ as well as in $\mathbf{J}(\mathbf{r})$ [Eq. (18.5)], is the microscopic **A** field, depending explicitly on the electron coordinates $\mathbf{r}_j$ via the microscopic Maxwell equations. However, we identify **A** approximately with the coherent field $\mathbf{A}(\mathbf{r}, t)$.

By expanding the state vector $\phi(t)$, given by

$$-(\hbar/i)\, \partial_t \phi(t) = H(t)\, \phi(t) \tag{18.8}$$

in powers of **A**, we get

$$\phi(t) = e^{-(i/\hbar) E_0 t} \phi_0 + \phi_1(t). \tag{18.9}$$

Here ϕ_0 is the initial (ground) state of the crystal $(H_0 \phi_0 = E_0 \phi_0)$. $\phi_1(t)$ is obtained by perturbation theory

$$\phi_1(t) = -(i/\hbar) \int^t dt'\, e^{-(i/\hbar) H_0 (t-t')} H_1(t')\, e^{-(i/\hbar) E_0 t'} \phi_0\,. \tag{18.10}$$

[74a] This formula for **J** can be derived by defining $\mathbf{J}(\mathbf{r}, t)$ by the continuity equation

$$\partial_t \rho(\mathbf{r}, t) + \partial_{\mathbf{r}} \cdot \mathbf{J}(\mathbf{r}, t) = 0.$$

Here $\rho(\mathbf{r}, t) = \sum_j \partial(\mathbf{r} - \mathbf{r}_j(t))$ is the density operator in the Heisenberg picture. Considering the equation of motion $\partial_t \rho(\mathbf{r}, t) = (i/\hbar)[H, \rho]$, the commutator can be written as the divergence of a vector, namely **J** of Eq. (18.5).

By introducing (18.7) in (18.10), and (18.9) and (18.5) in (18.4), we can evaluate $\mathbf{J}_0(\mathbf{r}, t)$ linearly in $\mathbf{A}$. (The zeroth-order term vanishes.)

$$\mathbf{J}_0(\mathbf{r}, t) = -(e/mc)\langle\rho_{\text{el}}(\mathbf{r})\rangle\, \mathbf{A}(\mathbf{r}, t) \\ -(e/c)\int^t dt' \int d\mathbf{r}'(\eta(\mathbf{r}, \mathbf{r}'; t-t') + \eta^*(\mathbf{r}, \mathbf{r}'; t-t'))\, \mathbf{A}(\mathbf{r}', t'). \tag{18.11}$$

Here the tensor η is given by

$$\eta_{\mu\nu}(\mathbf{r}, \mathbf{r}'; t-t') = -(i/\hbar)\langle j_\mu(\mathbf{r})\, e^{-(i/\hbar)H_0(t-t')} j_\nu(\mathbf{r}')\rangle. \tag{18.12}$$

$\langle\ \rangle$ means the expectation value in the ground state ϕ_0. By introducing the time-dependent operator $\mathbf{j}(\mathbf{r}, t)$ in the Heisenberg picture (14.22), η can be expressed by the current–current correlations of the crystal.

$$\eta_{\mu\nu}(\mathbf{r}, \mathbf{r}'; t-t') = -(i/\hbar)\langle j_\mu(\mathbf{r}, t)\, j_\nu(\mathbf{r}', t')\rangle. \tag{18.13}$$

Therefore, we get from (18.3) and (18.4) the basic equation

$$-\partial_{\mathbf{r}} \times \partial_{\mathbf{r}} \times \mathbf{A}(\mathbf{r}, t) - (1/c^2)\,\partial_t^2\mathbf{A}(\mathbf{r}, t) \\ = \int^t dt' \int d\mathbf{r}'\, \xi(\mathbf{r}, \mathbf{r}'; t-t')\, \mathbf{A}(\mathbf{r}', t'). \tag{18.14}$$

where the elements $\xi_{\mu\nu}$ of the tensor ξ are[74b]

$$\xi_{\mu\nu}(\mathbf{r}, \mathbf{r}'; t-t') = 4\pi(e^2/mc^2)\langle\rho_{\text{el}}(\mathbf{r})\rangle\, \delta(\mathbf{r}-\mathbf{r}')\, \delta(t-t') \\ + (4\pi e^2/c^2)(2/\hbar)\, \text{Im}\langle j_\mu(\mathbf{r}, t)\, j_\nu(\mathbf{r}', t')\rangle\, s(t-t'). \tag{18.15}$$

Whereas the first term, being local in space and time, describes the scattering by the electron density in the ground state, the second term contains the dynamics of the crystal. It only depends on the imaginary part of the correlation function, which according to van Hove[75] describes the local disturbance of the current of the crystal induced by the electromagnetic field.

In the stationary case we have

$$\mathbf{A}(\mathbf{r}, t) = \mathbf{A}(\mathbf{r}, \omega)\, e^{-i(\omega+i\epsilon/\hbar)t} \tag{18.16}$$

[74b] $s(t)$ is the step function: $s(t) = 1$ for $t > 0$, $s(t) = 0$ for $t < 0$.

[75] L. van Hove, *Physica* **24**, 404 (1958).

and obtain from (18.14) for $\mathbf{A}(\mathbf{r}, \omega)$,

$$-\partial_{\mathbf{r}} \times \partial_{\mathbf{r}} \times \mathbf{A}(\mathbf{r}, \omega) + (\omega/c)^2 \mathbf{A}(\mathbf{r}, \omega) = \int d\mathbf{r}' \, \xi(\mathbf{r}, \mathbf{r}'; \omega) \, \mathbf{A}(\mathbf{r}', \omega), \tag{18.17}$$

with

$$\xi(\mathbf{r}, \mathbf{r}'; \omega) = \int_0^\infty dt \, e^{i(\omega + i\epsilon/\hbar)t} \xi(\mathbf{r}, \mathbf{r}'; t) = \xi^{(1)} + \xi^{(2)}, \tag{18.18}$$

$$\xi_{\mu\nu}^{(1)} = 4\pi(e^2/mc^2)\langle \rho_{\mathrm{el}}(\mathbf{r})\rangle \, \delta(\mathbf{r} - \mathbf{r}'), \tag{18.19}$$

$$\begin{aligned} \xi_{\mu\nu}^{(2)} &= \frac{4\pi e^2}{c^2} \left\{ \left\langle j_\mu(\mathbf{r}) \frac{1}{E_0 + \hbar\omega + i\epsilon - H_0} j_\nu(\mathbf{r}') \right\rangle \right. \\ &\quad \left. + \left\langle j_\mu(\mathbf{r}) \frac{1}{E_0 - \hbar\omega + i\epsilon - H_0} j_\nu(\mathbf{r}') \right\rangle^* \right\}. \end{aligned} \tag{18.20}$$

Evidently

$$\xi(\mathbf{r}, \mathbf{r}'; \omega) = \xi^*(\mathbf{r}, \mathbf{r}'; -\omega) \tag{18.21}$$

since $\xi(\mathbf{r}, \mathbf{r}'; t)$ is real. Therefore we may split up $\xi(\mathbf{r}, \mathbf{r}'; \omega)$ into a real tensor ξ' and into an imaginary tensor ξ''. Of course, the first approximation $\xi^{(1)}$ (18.19) is real.

The real part of $\xi^{(2)}$, obtained by replacing the Green's function by the principal value, describes the influence of virtual excitations and is normally known as Hönl's corrections to $\xi^{(1)}$. The imaginary part of (18.20) is

$$\begin{aligned} \xi''_{\mu\nu} &= -\pi(4\pi e^2/c^2) \sum_n \{ j_\mu^{0n}(\mathbf{r}) \, j_\nu^{n0}(\mathbf{r}') \, \delta(\hbar\omega - (E_n - E_0)) \\ &\quad - j_\mu^{n0}(\mathbf{r}) \, j_\nu^{0n}(\mathbf{r}') \, \delta(\hbar\omega - (E_0 - E_n))\}, \end{aligned} \tag{18.22}$$

where

$$j_\mu^{0n}(\mathbf{r}) = (\phi_0 \, , j_\mu(\mathbf{r}) \, \phi_n).$$

For $\omega > 0$, the first term in the parentheses describes the absorption of the X-rays due to the photoelectric effect. The second term, giving the induced emission, vanishes since the crystal electrons are in the ground state before the scattering ($E_n > E_0$). For high frequencies ω, $\xi^{(2)}$ decreases with increasing ω, e.g., $\xi'^{(2)}$ decreases as $1/\omega$. Therefore, only $\xi^{(1)}$ remains in this limit.

Let us approximate the intermediate states ϕ_n in (18.22) and (18.20) by the excited states of the individual atoms. Then evidently $\xi(\mathbf{r}, \mathbf{r}')$ can be written as a sum of single-atom contributions ξ^0, namely,

$$\xi(\mathbf{r}, \mathbf{r}') = \sum_m \xi^0(\mathbf{r} - \mathbf{R}_m \, , \mathbf{r}' - \mathbf{R}_m). \tag{18.23}$$

For $\xi^0(\mathbf{r} - \mathbf{R}_m, \mathbf{r}' - \mathbf{R}_m)$, we obtain the same equations (18.18)–(18.20) as for ξ, except that the averages are formed with the ground state of atom m.

Since the X-ray energies (≈ 10 keV) are comparable with the binding energies of the inner shells, the electrons of these shells are most important for the photoelectric effect. Therefore, the ranges, over which ξ^0 is nonlocal, are very small and restricted to the inner shells.

To compare Eq. (18.17) with the classical theory of Section III,11, we introduce the dielectric field $\mathbf{D}(\mathbf{r}, \omega)$.

$$\mathbf{D}(\mathbf{r}, \omega) = \mathbf{E}(\mathbf{r}, \omega) + i(4\pi e/\omega)\,\mathbf{J}(\mathbf{r}, \omega)$$
$$= \mathbf{E}(\mathbf{r}, \omega) + \int d\mathbf{r}'\, \chi(\mathbf{r}, \mathbf{r}'; \omega)\,\mathbf{E}(\mathbf{r}', \omega), \tag{18.24}$$

where $\chi = -(c/\omega)^2\xi$. Since $\chi \approx 10^{-4}$, $\mathbf{E}(\mathbf{r}', \omega)$ may as well be replaced by $\mathbf{D}(\mathbf{r}', \omega)$. By multiplying Eq. (18.17) with $i\omega/c$, we obtain the following equation for $\mathbf{D}$, since $\mathbf{E} = i\omega/c\mathbf{A}$ and $\omega = cK$:

$$\partial_{\mathbf{r}}^2\mathbf{D}(\mathbf{r}, \omega) + K^2\mathbf{D}(\mathbf{r}, \omega) = -\partial_{\mathbf{r}} \times \partial_{\mathbf{r}} \times \int d\mathbf{r}'\, \chi(\mathbf{r}, \mathbf{r}'; \omega)\,\mathbf{D}(\mathbf{r}', \omega). \tag{18.25}$$

This is a generalization of Eq. (11.8). Instead of the scalar $\chi(\mathbf{r})$ in (11.8), we have now a nonlocal tensor $\chi(\mathbf{r}, \mathbf{r}'; \omega)$.

In an infinite crystal χ is periodic: $\chi(\mathbf{r}, \mathbf{r}') = \chi(\mathbf{r} + \mathbf{R}_n, \mathbf{r}' + \mathbf{R}_n)$. Therefore, the eigenfunctions, chosen as Bloch waves, can be expanded in plane waves. With

$$\mathbf{D}(\mathbf{r}) = \sum_{\mathbf{h}, s} e^{i(\mathbf{k}+\mathbf{h})\mathbf{r}}\mathbf{e}_{\mathbf{h}}{}^{s} D_{\mathbf{h}}{}^{s}, \tag{18.26}$$

we obtain from (18.25)

$$(K^2 - (\mathbf{k} + \mathbf{h})^2)\, D_{\mathbf{h}}{}^{s} = \sum_{\mathbf{h}' s'} \chi_{\mathbf{h}\mathbf{h}'}^{ss'} D_{\mathbf{h}'}^{s'}, \tag{18.27}$$

where the elements $\chi_{\mathbf{h}\mathbf{h}'}^{ss'}$ are given by

$$\chi_{\mathbf{h}\mathbf{h}'}^{ss'} = (1/V_c) \int_{-\infty}^{+\infty} d\mathbf{r}\, d\mathbf{r}'\, e^{-i(\mathbf{k}+\mathbf{h})\mathbf{r}}\mathbf{e}_{\mathbf{h}}{}^{s} \cdot \xi^0(\mathbf{r} - \mathbf{R}_m, \mathbf{r}' - \mathbf{R}_m) \cdot \mathbf{e}_{\mathbf{h}'}^{s'}$$
$$\times\, e^{i(\mathbf{k}+\mathbf{h})\mathbf{r}'}. \tag{18.28}$$

With the first approximation $\xi^{(1)}$ [Eq. (18.19)] we obtain the same result as in Part III, 11.

$$\chi_{\mathbf{h}\mathbf{h}'}^{(1)ss'} = \frac{4\pi r_e}{V_c} f_{\mathbf{h}-\mathbf{h}'}\mathbf{e}_{\mathbf{h}}{}^{s} \cdot \mathbf{e}_{\mathbf{h}'}^{s'}. \tag{18.29}$$

For the second approximation we get from (18.20),

$$\chi_{\mathbf{hh}'}^{(2)ss'} = \frac{4\pi e^2}{c^2 V_c} \sum_\alpha \Big\{ (\mathbf{e_h}^s \cdot \tilde{\mathbf{j}}^{0\alpha}(\mathbf{k}+\mathbf{h}))(\mathbf{e}_{\mathbf{h}'}^{s'} \cdot \tilde{\mathbf{j}}^{\alpha 0}(-\mathbf{k}-\mathbf{h}')) \frac{1}{\hbar\omega + i\epsilon - (E_n - E_0)}$$

$$+ (\mathbf{e_h}^s \cdot \tilde{\mathbf{j}}^{\alpha 0}(\mathbf{k}+\mathbf{h}))(\mathbf{e}_{\mathbf{h}'}^{s'} \cdot \tilde{\mathbf{j}}^{0\alpha}(-\mathbf{k}-\mathbf{h}')) \frac{1}{-\hbar\omega - i\epsilon - (E_n - E_0)} \Big\}, \tag{18.30}$$

with

$$\tilde{\mathbf{j}}^{0\alpha}(\mathbf{K}) = \int d\mathbf{r}\, e^{-i\mathbf{K}\mathbf{r}} (\varphi_0 \,, \mathbf{j}(\mathbf{r})\, \varphi_\alpha). \tag{18.31}$$

Here φ_0 and φ_α are the ground and excited states of an invidual atom.

The coefficients of $\chi^{(2)}$ simplify to a great extent if we consider that the spatial extent of the inner shells is much smaller than the wave length and the lattice constant. Therefore, the factors $e^{-i(\mathbf{k}+\mathbf{h})\mathbf{r}}$ and $e^{i(\mathbf{k}+\mathbf{h}')\mathbf{r}}$ in (18.28) can be replaced by 1 for the calculation of $\chi^{(2)}$ (but not for $\chi^{(1)}$!). This is equivalent to replacing $\tilde{\mathbf{j}}^{0\alpha}(\mathbf{k}+\mathbf{h})$ by $\tilde{\mathbf{j}}^{0\alpha}(0)$, which is known as the dipole approximation.

$$\chi_{\mathbf{hh}'}^{(2)ss'} = \frac{4\pi e^2}{m^2 c^2} \frac{\hbar^2}{V_c} \mathbf{e_h}^s \cdot e_{\mathbf{h}'}^{s'}$$

$$\times \sum_{j,\alpha_j} |\langle 0_j | \partial_{x_j} | \alpha_j \rangle|^2 \left(\frac{1}{\hbar\omega + i\epsilon - E_n + E_0} - \frac{1}{\hbar\omega + i\epsilon + E_n - E_0} \right) \tag{18.32}$$

0_j and α_j refer to the ground and excited states of electron j, x_j is the x-component of $\mathbf{r}_j$. Similar to $\chi^{(1)}$ [Eq. (18.29)], $\chi^{(2)}$ contains a factor $\mathbf{e_h}^s \cdot \mathbf{e}_{\mathbf{h}}^{s'}$ also since due to the rotational invariance of the atom no other direction is preferred. Furthermore, as a function of $\mathbf{h}$ and $\mathbf{h}'$, the coefficients are all equal besides the trivial factor $\mathbf{e_h}^s \cdot \mathbf{e}_{\mathbf{h}'}^{s'}$.

In the dipole approximation, we can therefore justify the phenomenological ansatz of Section III,11 with a complex density $\rho(\mathbf{r}) = \rho'(\mathbf{r}) + i\rho''(\mathbf{r})$. Since $\chi_{\mathbf{hh}'}^{(2)ss'}$ has essentially the same form as $\chi_{\mathbf{hh}}^{(1)ss'}$, we can take $\chi^{(2)}$ in $\chi^{(1)}$ into account by choosing an appropriate complex structure factor $f_\mathbf{h} = f_\mathbf{h}' + if_\mathbf{h}''$.

Photoelectric-absorption cross sections, i.e., the imaginary part of (18.32), have been calculated by several authors in the dipole approximation using hydrogenlike eigenfunctions. Extensive calculations of dipole as well as quadrupole terms have been performed by Wagenfeld[76]

[76] H. Wagenfeld, *Phys. Rev.* **144**, 216 (1966).

and Gutmann and Wagenfeld.[76a] The Hönl corrections to the atomic scattering factor, i.e., the real parts of (18.32), have been calculated by Hönl[77] and Eisenlohr and Müller.[78]

19. Influence of Thermal Motion

Whereas in the preceding section, the nuclei of the crystal have been considered as fixed, we will now discuss the effects of thermal motion. We assume that the atoms are not deformed during the motion (rigid ion model). Further, we can apply the static approximation (Section IV,14) since the X-rays energies (≈ 10 keV) are much larger than the typical phonon energies ($\approx kT$). Therefore we can calculate $\mathbf{A}(\mathbf{r}, \omega)$ or $\mathbf{D}(\mathbf{r}, \omega)$ for a given, momentary displacement configuration of the atoms and then average the result over the thermal distribution of the displacements to obtain the coherent wave $\langle \mathbf{D}(\mathbf{r}) \rangle$.

$$\langle \mathbf{D}(\mathbf{r}) \rangle = \langle \mathbf{D}(\mathbf{r}; \cdots \mathbf{R}_n \cdots) \rangle. \tag{19.1}$$

Starting with Eq. (18.25) for specific displacements $\mathbf{R}_n$, the tensor χ is according to (18.23) a sum of single atom contributions χ^0.

$$\chi(\mathbf{r}, \mathbf{r}') = \sum_n \chi^0(\mathbf{r} - \mathbf{R}_n, \mathbf{r}' - \mathbf{R}_n). \tag{19.2}$$

In the dipole approximation, we may replace $\chi(\mathbf{r}, \mathbf{r}')$ by the scalar $\chi(\mathbf{r})$ of Section III,11 if the electron density is chosen as complex, which has been shown in the preceding section. Therefore, we have from(18.25) for the scattering at the momentary positions $\mathbf{R}_n = \bar{\mathbf{R}}_n + \mathbf{s}_n$ of the nuclei

$$(\partial_\mathbf{r}^2 + K^2)\,\mathbf{D}(\mathbf{r}; \cdots \mathbf{R}_n \cdots) = -\partial_\mathbf{r} \times \partial_\mathbf{r} \times \{\chi(\mathbf{r})\,\mathbf{D}(\mathbf{r}; \cdots \mathbf{R}_n \cdots)\}, \tag{19.3}$$

with

$$\chi(\mathbf{r}) = -(4\pi r_e/K^2) \sum_n \rho_0(\mathbf{r} - \mathbf{R}_n), \tag{19.4}$$

where $\rho_0 = \rho_0' + i\rho_0''$ is the electron density of an individual atom. Since Eq. (19.1) for $\mathbf{D}$ has the same form as the Schrödinger equation for ψ, the theory for the coherent field $\langle \mathbf{D}(\mathbf{r}) \rangle$ and the analogous optical potential is the same as in Section IV. Therefore, we obtain for the coherent field (19.1) an equation with an optical potential U, which is a tensor in this case,

$$(\partial_\mathbf{r}^2 + K^2)\langle \mathbf{D}(\mathbf{r}) \rangle = \int d\mathbf{r}'\, \mathsf{U}(\mathbf{r}, \mathbf{r}')\langle \mathbf{D}(\mathbf{r}') \rangle. \tag{19.5}$$

[76a] A. Gutmann and H. Wagenfeld, *Acta Crystallogr.* **22**, 334 (1967).
[77] H. Hönl, *Ann. Phys.* **18**, 625 (1933).
[78] H. Eisenlohr and G. L. Müller, *Z. Phys.* **136**, 491, 511 (1954).

Approximations for U can be obtained analogously to those in Section IV,14. Since $|\chi| \leqslant 10^{-4}$, the approximations (14.10) and (14.11) are appropriate.

$$\mathsf{U} = \mathsf{U}^{(1)} + \mathsf{U}^{(2)}. \tag{19.6}$$

The first approximation $\mathsf{U}^{(1)}$ is local and essentially given by the averaged electron density $\langle \rho_e(\mathbf{r}) \rangle$ of (15.2).

$$\begin{aligned} \mathsf{U}^{(1)}(\mathbf{r}) &= -\partial_{\mathbf{r}} \times \partial_{\mathbf{r}} \times \langle \chi(\mathbf{r}) \rangle \\ &= \partial_{\mathbf{r}} \times \partial_{\mathbf{r}} \times (4\pi e^2/K^2) \sum_n \langle \rho_0(\mathbf{r} - \mathbf{R}_n) \rangle. \end{aligned} \tag{19.7}$$

The second approximation $\mathsf{U}^{(2)}$, describing the effects of phonon excitations, is obtained to

$$\mathsf{U}^{(2)}(\mathbf{r}, \mathbf{r}') = \partial_{\mathbf{r}} \times \partial_{\mathbf{r}} \times \left\langle \delta\chi(\mathbf{r}) \frac{(-1)}{4\pi} \frac{e^{i\mathbf{K}|\mathbf{r}-\mathbf{r}'|}}{|\mathbf{r} - \mathbf{r}'|} \partial_{\mathbf{r}'} \times \partial_{\mathbf{r}'} \times \delta\chi(\mathbf{r}') \right\rangle, \tag{19.8}$$

where

$$\delta\chi(\mathbf{r}) = \chi(\mathbf{r}) - \langle \chi(\mathbf{r}) \rangle.$$

Since $\mathsf{U}^{(2)}$ is a correction to $\mathsf{U}^{(1)}$, we can replace $\chi(\mathbf{r})$ in (19.8) by its real value. In an infinite crystal we can expand $\langle \mathbf{D}(\mathbf{r}) \rangle$ into plane waves (18.26) because of the periodicity of U. Therefore, analogously to (18.27), we have for the coefficient $D_{\mathbf{h}}{}^s$,

$$(K^2 - (\mathbf{k} + \mathbf{h})^2)\, D_{\mathbf{h}}{}^s = \sum_{s'\mathbf{h}'} U_{\mathbf{hh}'}^{ss'} D_{\mathbf{h}'}^{s'} . \tag{19.9}$$

For the first approximation $\mathsf{U}^{(1)}$, the coefficients $U_{\mathbf{hh}'}^{ss'}$ are obtained to

$$\begin{aligned} U_{\mathbf{hh}'}^{(1)ss'} &= \frac{(\mathbf{k} + \mathbf{h})^2}{K^2} \frac{4\pi r_e}{V_c} f_{\mathbf{h}-\mathbf{h}'} \mathbf{e}_{\mathbf{h}}{}^s \cdot \mathbf{e}_{\mathbf{h}'}^{s'} \langle e^{-i(\mathbf{h}-\mathbf{h}')\mathbf{R}_n} \rangle \\ &= \frac{4\pi r_e}{V_c} f_{\mathbf{h}-\mathbf{h}'} \mathbf{e}_{\mathbf{h}}{}^s \cdot \mathbf{e}_{\mathbf{h}'}^{s'} e^{-M_{\mathbf{h}-\mathbf{h}'}}. \end{aligned} \tag{19.10}$$

Therefore, in the first approximation the coefficients of the ideal crystal are simply multiplied with e^{-M}, where M is the Debye–Waller factor (Section V,15). Since $f_{\mathbf{h}} = f_{\mathbf{h}}' + if_{\mathbf{h}}''$, the coefficients $\chi_{\mathbf{hh}'}^{(2)ss'}$ [Eq. (18.30)] of the photoelectric absorption get a factor e^{-M} too. The results for the Fourier coefficients of $\mathsf{U}^{(2)}$ are

$$U_{\mathbf{hh}'}^{(2)ss'} = \frac{(4\pi r_e)^2}{V_c}\int\frac{d\mathbf{K}'}{(2\pi)^3}\frac{1}{K^2+i\epsilon-K'^2}\frac{(\mathbf{k}+\mathbf{h})^2}{K^2}f_{\mathbf{k}+\mathbf{h}-\mathbf{k}'}\frac{K'^2}{K^2}f_{\mathbf{K}'-\mathbf{k}-\mathbf{h}'}$$

$$\times\left(\mathbf{e_h}^s\cdot\mathbf{e_{h'}}^{s'}-\left(\mathbf{e_h}^s\cdot\frac{\mathbf{K}'}{K'}\right)\left(\mathbf{e_{h'}}^{s'}\cdot\frac{\mathbf{K}'}{K'}\right)\right)S_{\mathbf{hh}'}(\mathbf{k}-\mathbf{K}'). \tag{19.11}$$

Here the correlation function $S_{\mathbf{hh}'}(\mathbf{k}-\mathbf{K}')$ is given by (17.5). Besides the polarization factors, (19.11) is essentially the same expression as for electron diffraction (17.4). For the coefficients of $U''^{(2)}$, describing the absorption due to thermal diffuse scattering, we obtain an angle integral over all directions of $\mathbf{K}'$, with $|\mathbf{K}'| = K$:

$$U_{\mathbf{hh}'}''^{(2)ss'} = -\frac{Kr_e^2}{V_c}\int d\Omega_{\mathbf{K}'}f_{\mathbf{k}+\mathbf{h}-\mathbf{K}'}f_{\mathbf{K}'-\mathbf{k}-\mathbf{h}'}$$

$$\times\left(\mathbf{e_h}^s\cdot\mathbf{e_{h'}}^{s'}-\left(\mathbf{e_h}^s\cdot\frac{\mathbf{K}'}{K}\right)\left(\mathbf{e_{h'}}^{s'}\cdot\frac{\mathbf{K}'}{K}\right)\right)S\ '(\mathbf{k}-\mathbf{K}'). \tag{19.12}$$

Let us now discuss the anomalous absorption of the σ polarized wave in the two-beam case (11.25). Since the two polarization vectors $\mathbf{e}_0{}^1$ and $\mathbf{e_h}{}^1$ are equal, we have with Einstein's approximation (17.1),

$$U_{\mathbf{hh}'}''^{\mathrm{ES}} = -\frac{Kr_e^2}{V_c}\int d\Omega_{\mathbf{K}'}f_{\mathbf{k}+\mathbf{h}-\mathbf{K}'}f_{\mathbf{K}'-\mathbf{k}-\mathbf{h}'}\left(1-\left(\mathbf{e}^1\cdot\frac{\mathbf{K}'}{K}\right)^2\right)$$

$$\times\left(\exp[-M_{\mathbf{h}-\mathbf{h}'}]-\exp[-M_{\mathbf{k}+\mathbf{h}-\mathbf{K}'}-M_{\mathbf{K}'-\mathbf{k}-\mathbf{h}'}]\right). \tag{19.13}$$

Since in the Bragg condition $K^2 = \mathbf{k}^2 = (\mathbf{k}+\mathbf{h})^2$, we obtain $U_{00}''^{\mathrm{ES}} = U_{\mathbf{hh}}''^{\mathrm{ES}}$ and furthermore, $U_{0\mathbf{h}}''^{\mathrm{ES}} = U_{\mathbf{h}0}''^{\mathrm{ES}}$. The integration has to be performed numerically or by expanding $f_{\mathbf{K}}$ into Gauss functions.[79]

From the correlation term (17.12), we do not obtain, in general, a large contribution since the second term cancels the first one on the average. For an isotropic Debye model and for high temperatures $S_{hh'}$ is given by

$$S_{\mathbf{hh}'}^{\mathrm{c}} = \frac{kT}{M\bar{c}^2}\left(\frac{1}{q^2}-\frac{1}{q^2}\right)(\mathbf{k}+\mathbf{h}-\mathbf{K}')\cdot(\mathbf{k}+\mathbf{h}'-\mathbf{K}'), \tag{19.14}$$

where $1/q^2$ is the average of $1/q^2$ over the first Brillouin zone and $\bar{c}$ is the average velocity of sound. From (19.14) we obtain a large contribution, only if the $1/q^2$ singularity lies on or very near to the integration surface

[79] J. B. Forsyth and M. Wells, *Acta Crystallogr.* **12**, 412 (1959).

$\mathbf{K}'^2 = \mathbf{K}^2$ and if the $1/q^2$ behavior is not suppressed by the factor $(\mathbf{k} + \mathbf{h} - \mathbf{K}') \cdot (\mathbf{k} + \mathbf{h}' - \mathbf{K}')$. For example, in the Bragg condition we have $S_{00} \sim 1/q^2$ for $\mathbf{K}' \cong \mathbf{k} + \mathbf{h}$ and $S_{\mathbf{hh}} \sim 1/q^2$ for $\mathbf{K}' \cong \mathbf{k}$, whereas, $S_{0\mathbf{h}}$ and $S_{\mathbf{h}0}$ behave as $1/q$ only in these regions. Therefore, in the Bragg condition, we have

$$U''^{\mathrm{C}}_{00} = U''^{\mathrm{C}}_{\mathbf{hh}} \cong -\frac{r_{\mathrm{e}}^2}{V_{\mathrm{c}}} f^2 \frac{kT}{M\bar{c}^2} \frac{h^2}{K} 2\pi \ln \frac{q_{\mathrm{D}}}{q_{\mathrm{c}}} \tag{19.15}$$

whereas $U''^{\mathrm{C}}_{0\mathbf{h}}$ and $U''^{\mathrm{C}}_{\mathbf{h}0}$ can be neglected. q_{c} is the minimal q value ($q_{\mathrm{c}} = (4\pi r_{\mathrm{e}}/V_{\mathrm{c}}) f_h/h$) and q_{D} the maximal one of the Debye model.

In practice, the photoelectric absorption is by far the dominant absorption process. The diffraction as well as the absorption of X-rays is therefore essentially governed by the coefficients $U^{(1)}_{\mathbf{hh}'}$ [Eq. (19.10)], being temperature dependent via the Debye–Waller factor e^{-M}. For instance, the anomalous absorption for the σ wave field is in the first approximation (photoelectric absorption)

$$\Delta\mu_{\mathbf{h}}^{\mathrm{PE}} = (\mu_0 - \mu_{\mathbf{h}} e^{-M_{\mathbf{h}}}) \cong \mu_0(1 - e^{-M_{\mathbf{h}}}) \tag{19.16}$$

since $\mu_0 = \mu_{\mathbf{h}}$ in the dipole approximation. Therefore, the absorption is strongly temperature-dependent. Since generally $M \ll 1$, $\Delta\mu_{\mathbf{h}}$ gives a direct measure of the Debye–Waller factor. Thus $\Delta\mu_{\mathbf{h}}$ is proportional to kT for $T \geqslant \theta_{\mathrm{D}}$ and limited by the zero-point motion for low temperature.

Calculations of the absorption due to thermal diffuse scattering (phonon excitations) give corrections of the order of a few percent of the photoelectric absorption. For the anomalous absorption $\Delta\mu$, we have with (19.13) and (19.15) an additional contribution

$$\Delta\mu_{\mathbf{h}}^{\mathrm{TDS}} = \Delta\mu_{\mathbf{h}}^{\mathrm{ES}} + \Delta\mu_{\mathbf{h}}^{\mathrm{c}} = -\frac{1}{K}(U''^{\mathrm{ES}}_{00} - U''^{\mathrm{ES}}_{0\mathbf{h}}) - \frac{1}{K} U''^{\mathrm{C}}_{00} \tag{19.17}$$

giving, for instance, for Cu with MoK_α radiation a 3 % correction to (19.16). Also this contribution is proportional to kT for high temperature and behaves roughly as h^2 by varying the reflections.

The theory for X-ray diffraction has been treated by several authors.[68,80–85]

[80] R. Parthasarathy, *Acta Crystallogr.* **13**, 802 (1960).
[81] Y. H. Othsuki, *J. Phys. Soc. Japan* **19**, 2285 (1964); **20**, 314 (1964).
[82] Y. H. Othsuki and S. Yanagawa, *J. Phys. Soc. Japan* **21**, 506 (1966).
[83] A. M. Afanasev and Yu. Kagan, *Acta Crystallogr.* **A24**, 163 (1967).
[84] E. A. Tikhonova, *Sov. Phys.—Solid State* **9**, 394 (1967).
[85] H. Sano, K. Ohtaka, and Ohtsuki, *J. Phys. Soc. Japan* **27**, 1254 (1969).

A number of experiments[86–94a] have been performed, showing reasonable agreement with theory.

For illustration, Fig. 19, taken from Baldwin,[93] shows the integrated intensity of the anomalous transmitted beam versus temperature for several Cu crystals of different thickness t. Figures 20(a) and (b) show some recent low temperature measurements of Ludewig[94] in Ge.

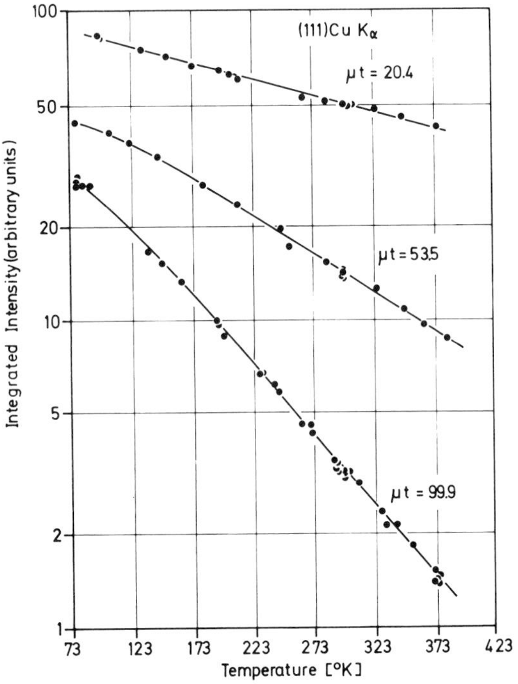

FIG. 19. A plot of the relative integrated intensities of the Bormann reflected beam versus temperature for three crystals examined. (From Baldwin.[93])

[86] B. Okkerse, *Philips Res. Rep.* **17**, 464 (1962).
[87] B. W. Batterman, *Phys. Rev.* **134**, A1354 (1964).
[88] A. Merlini and S. Pace, *Nuovo Cimento* **35**, 377 (1965).
[89] C. Ghezzi, S. Pace, and A. Merlini, *Nuovo Cimento* **49B**, 58 (1967).
[90] D. Ling and H. Wagenfeld, *Phys. Lett.* **15**, 8 (1965).
[91] O. N. Efimov, *Phys. Status Solidi* **22**, 297 (1967).
[92] T. O. Baldwin, A. Merlini, and F. W. Young, *Phys. Rev.* **163**, 591 (1967).
[93] T. O. Baldwin, *Phys. Status Solidi* **25**, 71 (1968).
[94] J. Ludewig, *Acta Crystallogr.* **A25**, 116 (1969).
[94a] C. Ghezzi, A. Merlini, and S. Pace, *Phys. Rev.* **B4**, 1833 (1971).

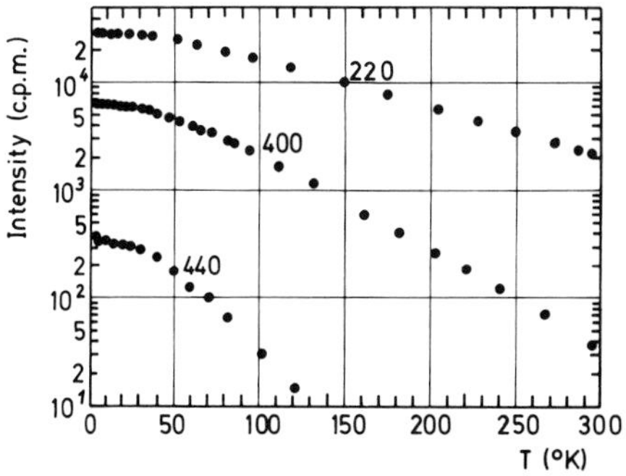

FIG. 20a. Relative intensity (log I) versus temperature T curves of three Ge reflections; MoK_α, $\mu_0 t = 96$.

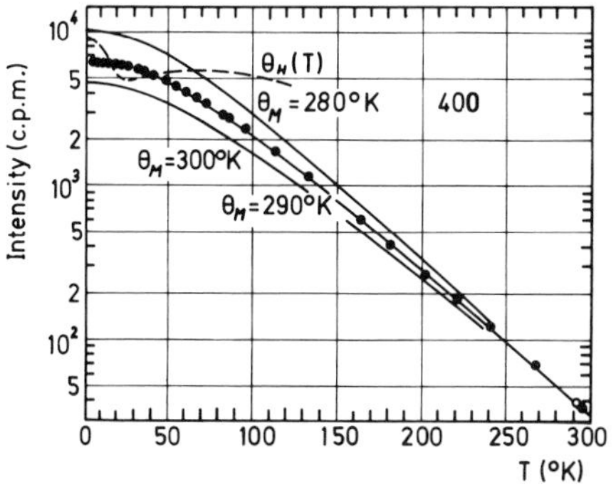

FIG. 20b. Ge 400 intensity calculated with 3 constant values of the characteristic temperature θ_M. (●): experiment. Dashed line: calculated with θ_H (specific heat data of the characteristic temperature).

VII. Influence of Statistically Distributed Defects

Whereas the theory presented in the last sections refers to ideal crystals, we will show in this section that the coherent-wave theory can also be applied to real crystals, namely to crystals with a statistical distribution of defects, such as point defects, small defect clusters or substitutional disorder. Here, one has to assume that the measured intensity does not depend on the accidental microscopic defect configurations, but only on macroscopic quantities such as the average defect densities, their correlations, etc. Therefore, one has to average over all the possible microscopic defect configurations. Significant deviations from such an average intensity are not expected for crystals with point defects or many small clusters.

Further, we restrict ourselves to the coherent waves as in the preceding sections. The diffuse, incoherent waves are scattered by an oblique angle and normally do not satisfy the Bragg condition, so that the Bragg intensities are totally due to the coherent wave. However special attention is necessary for large defect clusters of the order of an extinction length or, e.g., for dislocations. In this case the diffuse waves are scattered by a very small angle and partially satisfy the Bragg condition, leading to "line-broadening" in the Bragg case. Since no theory taking this effect into account is available, we restrict ourselves to such defects being small compared to an extinction length. Moreover, we consider the effect of statistical defects alone, by neglecting the thermal motion, since both effects are essentially additive, as has been discussed.[95] Eventually we will describe the effects of inelastic excitation on electron diffraction by a complex atomic potential $v(\mathbf{r} - \mathbf{R}_n) = v' + iv''$ and similarly the photoelectric absorption of X-rays by a complex atomic density $\rho_0(\mathbf{r} - \mathbf{R}_n)$, as in the preceding section.

The basic equations for the coherent wave remain the same as for the thermal motion. For instance, for X-rays we get the coherent field $\langle \mathbf{D} \rangle$ by averaging the microscopic field over all defect configurations [Eq. (19.1)]. Furthermore, the optical potential is in the first and second approximation given by (19.7) and (19.8), with the only difference that the thermal average has to be replaced by the configuration average. Similarly for electron diffraction, the optical potential is approximately given by (14.10) and (14.11). Since in the crystal $U(\mathbf{r}, \mathbf{r}')$ is periodic, we only have to evaluate the coefficients $U_{\mathbf{hh}'}$ or $U_{\mathbf{hh}'}^{\mathrm{ss}'}$ for X-rays.

20. Isolated Point Defects

We shall now evaluate the optical potential for the case of point defects being statistically independent distributed or for substitutional disorder. We consider a nonprimitive lattice, the lattice points (n, j) of which can be occupied by different kinds of atoms (index α, say A or B). Since the lattice points are given by all possible equilibrium positions of the various atoms, interstitial positions enter as lattice points, too. Let us define random number

$$p_\alpha^{nj} = \begin{cases} 1 & \text{if } (n, j) \text{ is occupied by an } \alpha \text{ atom,} \\ 0 & \text{if } (n, j) \text{ is not occupied by an } \alpha \text{ atom.} \end{cases} \tag{20.1}$$

Then evidently

$$p_\alpha^{nj} p_{\alpha'}^{nj} = \delta_{\alpha\alpha'} p_\alpha^{nj} \qquad \text{and} \qquad \langle p_\alpha^{nj} \rangle = c_\alpha^{\,j}, \tag{20.2}$$

[95] P. H. Dederichs, *Phys. Status Solidi* **23**, 377 (1967).

where $c_\alpha{}^j$ is the mean concentration of α atoms on the position j in the unit cell. In the case of electrons, we have then

$$V(\mathbf{r}) = \sum_{nj\alpha} p_\alpha^{nj} v_\alpha(\mathbf{r} - \mathbf{R}^{nj}) \tag{20.3}$$

and obtain for the first approximation (14.10)

$$U_{\mathbf{hh}'}^{(1)} = \sum_{j,\alpha} c_\alpha{}^j v_{\mathbf{h}-\mathbf{h}'}^\alpha \langle \exp[-i(\mathbf{h} - \mathbf{h}')\,\mathbf{R}_\alpha^{nj}]\rangle. \tag{20.4}$$

The result for X-rays is quite analogous.

$$U_{\mathbf{hh}'}^{(1)ss'} = \frac{4\pi r_e}{V_c}\,\mathbf{e_h}^s \cdot \mathbf{e}_{\mathbf{h}'}^{s'} \sum_{\alpha j} c_\alpha{}^j f_{\mathbf{h}-\mathbf{h}'}^\alpha \langle \exp[-i(\mathbf{h} - \mathbf{h}')\,\mathbf{R}_\alpha^{nj}]\rangle, \tag{20.5}$$

where $f_\mathbf{h}{}^\alpha$ is the atomic scattering-factor of the α atoms. Due to the defects, the original lattice is uniformly expanded. Denoting the lattice points in this averaged, expanded lattice by $\bar{\mathbf{R}}^{nj}$, we introduce static displacement $\mathbf{s}_\alpha^{nj}$ to describe the real position $\mathbf{R}_\alpha^{nj}$.

$$\mathbf{R}_\alpha^{nj} = \bar{\mathbf{R}}^{nj} + \mathbf{s}_\alpha^{nj} \qquad \text{with} \qquad \langle \mathbf{R}_\alpha^{nj}\rangle = \overline{\mathbf{R}^{nj}}, \quad \langle \mathbf{s}_\alpha^{nj}\rangle = 0. \tag{20.6}$$

Of course, the reciprocal lattice vectors $\mathbf{h}$ and $\mathbf{h}'$ refer to the expanded lattice, too. Thus we get, e.g., from (20.4) the same result as for an ideal lattice:

$$U_{\mathbf{hh}'}^{(1)} = \sum_j \bar{v}_{\mathbf{h}-\mathbf{h}'}^j \exp[-i(\mathbf{h} - \mathbf{h}')\,\bar{\mathbf{R}}^j], \tag{20.7}$$

but with $\bar{v}_{\mathbf{h}-\mathbf{h}'}^j$ given by

$$\bar{v}_{\mathbf{h}-\mathbf{h}'}^j = \sum_\alpha c_\alpha{}^j v_{\mathbf{h}-\mathbf{h}'}^\alpha \langle \exp[-i(\mathbf{h} - \mathbf{h}')\mathbf{s}_\alpha^{nj}]\rangle. \tag{20.8}$$

The problem remains to determine the Debye–Waller factor due to static displacements being defined by

$$\exp[-L_\mathbf{h}^{\alpha j}] = \langle \exp[-i\mathbf{h}\mathbf{s}_\alpha^{nj}]\rangle \tag{20.9}$$

analogously to the thermal one $M_\mathbf{h}$ [Eq. (15.5)]. Due to the central limit theorem, the displacement $\mathbf{s}^{nj}$ has a Gaussian distribution, if a large number of defects gives a more or less equal contribution to $\mathbf{s}^{nj}$. Then we have by dropping the indices α, j

$$\exp[-L\,] = \exp[-\tfrac{1}{2}\langle(\mathbf{h} \cdot \mathbf{s})^2\rangle\}. \tag{20.10}$$

However, neighboring atoms or defects normally give a larger contribution than do distant ones. Then Eq. (20.10) no longer holds. For this case we write (by considering a primitive lattice with only one kind of defect)

$$\mathbf{s} = \sum_m (p^m \mathbf{t}^m - c\mathbf{t}^m), \tag{20.11}$$

where $\mathbf{t}^m$ is the displacement of the atom at position 0 due to a defect at the position m. The first term is the total displacement from the original lattice position for a specific defect configuration, the second part is the average displacement, so that the difference is the displacement counted from the expanded lattice position. For a statistical distribution, the random numbers p^m are independently distributed for different m. Therefore, we have

$$\begin{aligned}\langle \exp[-i\mathbf{hs}]\rangle &= \prod_m \exp[ic\mathbf{ht}^m]\langle \exp[-ip^m\mathbf{ht}^m]\rangle \\ &= \prod_m \exp[ic\mathbf{ht}^m]\{1 + c(\exp[-i\mathbf{ht}^m] - 1)\} \\ &\cong \exp\left[-c\sum_m (1 - \exp[-i\mathbf{ht}^m] - i\mathbf{ht}^m)\right] \quad \text{for } c \ll 1. \end{aligned} \tag{20.12}$$

It is important to note that the displacements $\mathbf{t}^m$ decrease as $1/\mathbf{R}_m{}^2$ for large distances. By writing $\exp[-i\mathbf{ht}^m] = \cos \mathbf{ht}^m - i \sin \mathbf{ht}^m$, the sine function cancels the term $i\mathbf{ht}^m$ for large m. Moreover, the imaginary term in the exponent vanishes completely if the displacement field has inversion symmetry ($\mathbf{t}^{\mathbf{m}} = -\mathbf{t}^{-\mathbf{m}}$). Then we have finally,

$$L_{\mathbf{h}} = c\sum_m (1 - \cos \mathbf{ht}^m), \tag{20.13}$$

where we have used the condition $c \ll 1$ only (but not $L \ll 1$). In particular, if $\mathbf{ht}^m \ll 1$, we can expand the cosine function, obtaining an expression quadratically in $\mathbf{h}$ as in (20.10). However, in general the displacements of point defects are not that small, expecially the displacements of the nearest neighbors, which give the largest contribution to $L_{\mathbf{h}}$.

However, to get a rough estimate of $L_{\mathbf{h}}$, we expand $L_{\mathbf{h}}$ quadratically in $\mathbf{h}$. For the displacement field we use $\mathbf{t}(\mathbf{R}_m) = A\mathbf{R}_m/|\mathbf{R}_m|^3$. Furthermore, we replace the sum by an integral, cutting off the integration for small $\mathbf{R}_m$ at the ion radius ρ_c. Then we obtain

$$L = c\,\frac{2\pi}{3}\,\frac{A^2h^2}{V_c\rho_c} \tag{20.13a}$$

Let us discuss the anomalous absorption of the Bloch wave of type II exactly in the Bragg condition (Fig. 21). We have (for X-rays: for the

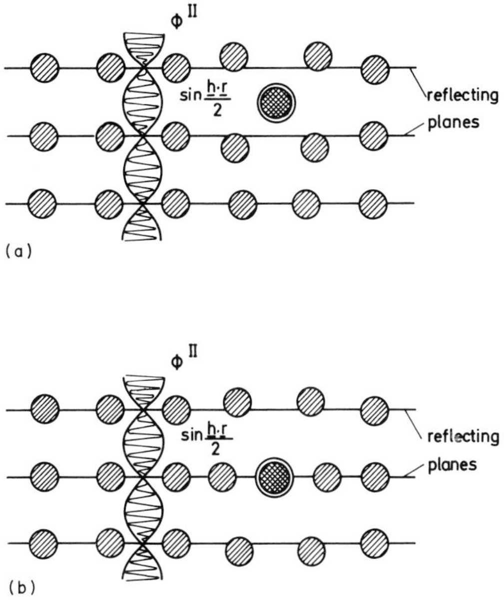

FIG. 21. Diagram of the sin-modulated Bloch wave ϕ^{II} in a defect lattice. (a) Impurity interstitial between the reflecting lattice planes. (b) Impurity interstitial lying in the reflecting planes.

σ polarization)

$$\Delta\mu_{\mathbf{h}} = -(1/K)\{\tfrac{1}{2}(U''_{00} + U''_{\mathbf{hh}}) - \tfrac{1}{2}(U''_{0\mathbf{h}} + U''_{\mathbf{h}0})\}. \tag{20.14}$$

First we consider a primitive lattice of A and B substitutional atoms distributed at random, for which we obtain

$$\Delta\mu_{\mathbf{h}} = (c_{\mathrm{A}}\mu_0{}^{\mathrm{A}} + c_{\mathrm{B}}\mu_0{}^{\mathrm{B}}) - (c_{\mathrm{A}}\mu_{\mathbf{h}}{}^{\mathrm{A}} \exp[-L_{\mathbf{h}}{}^{\mathrm{A}}] + c_{\mathrm{B}}\mu_{\mathbf{h}}{}^{\mathrm{B}} \exp[-L_{\mathbf{h}}{}^{\mathrm{B}}]), \tag{20.15}$$

where $L_{\mathbf{h}}{}^{\mathrm{A}}$, $L_{\mathbf{h}}{}^{\mathrm{B}}$ and $\mu_{\mathbf{h}}{}^{\mathrm{A}}$, $\mu_{\mathbf{h}}{}^{\mathrm{B}}$ are the Debye–Waller factors and absorption factors of atom A and B (photoelectric absorption for X-rays, electronic excitations for electrons). We can obtain the results for a lattice with vacancies by identifying the B atoms with vacancies and the A atoms with the atoms of the matrix

$$(\mu_{\mathbf{h}}{}^{\mathrm{B}} = \mu_{\mathbf{h}}{}^{\mathrm{V}} = 0,\ c_{\mathrm{A}} = 1 - c_{\mathrm{V}} \cong 1)$$

$$\Delta\mu_{\mathbf{h}} \cong (\mu_0 - \mu_{\mathbf{h}} \exp -L_{\mathbf{h}}). \tag{20.16}$$

Second, we consider a lattice with impurity interstitials at the position $\mathbf{R}^I$ in the basic call and obtain

$$\Delta\mu_{\mathbf{h}} = (\mu_0 - \mu_{\mathbf{h}} \exp -L_{\mathbf{h}}) + c_{\mathrm{I}}(\mu_0{}^{\mathrm{I}} - \cos(\mathbf{h} \cdot \mathbf{R}^{\mathrm{I}})\, \mu_{\mathbf{h}}{}^{\mathrm{I}} \exp -L_{\mathbf{h}}{}^{\mathrm{I}}). \quad (20.17)$$

This absorption consists of two parts. The first term is the absorption of the lattice atoms which are displaced into the space between the reflecting lattice planes and therefore absorb the wave II more strongly (Fig. 21a,b). The second term is the absorption of the impurity interstitials themselves. This contribution is largest, if the interstitials are located in the middle between the reflecting planes ($\cos \mathbf{hR}^{\mathrm{I}} = -1$) where they absorb wave II most strongly (Fig. 21a). For X-rays, (20.17) especially simplifies due to $\mu_0 \cong \mu_{\mathbf{h}}$. If further $L_{\mathbf{h}} \ll 1$, one has

$$\Delta\mu_{\mathbf{h}} = \mu_0 \cdot L_{\mathbf{h}} + c_{\mathrm{I}}\mu_0{}^{\mathrm{I}}(1 - \cos \mathbf{h} \cdot \mathbf{R}^{\mathrm{I}}). \quad (20.18)$$

Then the second term has a very characteristic $\mathbf{h}$ dependence and practically disappears for $\cos(\mathbf{h} \cdot \mathbf{R}^I) = 1$, when the interstitials are lying in the reflecting planes (Fig. 21b). The direct absorption of the impurities would be even more important if one could choose the X-ray energy to be near the absorption edge of the impurities. In this case the second term in (20.18) would be enlarged by a factor of 6 to 8.

Whereas (20.15)–(15.18) only represents the photoelectric absorption or absorption due to electronic excitations, respectively, we get an additional contribution to the absorption from the second approximation $U^{(2)}$ since due to the production of "diffuse scattering," the intensity of the coherent wave is diminished. For a small concentration of point defects, i.e., linear in c, we obtain again two contributions to $U^{(2)}$, which are, for instance, for electron diffraction

$$U^{(2)}_{\mathbf{hh}'} = U^{(2)\,\mathrm{dis}}_{\mathbf{hh}'} + U^{(2)\,\mathrm{pot}}_{\mathbf{hh}'}. \quad (20.19)$$

The first term $U^{(2)\,\mathrm{dis}}_{\mathbf{hh}'}$ arises from the diffuse scattering originating at the displaced lattice atoms. The coefficients for this absorption process are the same as (17.4), with the difference that the displacements in the correlation function $S_{\mathbf{hh}'}$ are static displacements. Approximations for $S_{\mathbf{hh}'}$ can be obtained similar to those in Section V,17, e.g., by Einstein's model (17.9). The second term is due to the fluctuation of the potential at the various lattice sides. For example, for interstitials we obtain

$$U''^{(2)\,\mathrm{pot}}_{\mathbf{hh}'} = -\pi(V_c/(2\pi)^3)(K/2)\, c_{\mathrm{I}} \exp[-L^{\mathrm{I}}_{\mathbf{h}-\mathbf{h}}] \exp[-i(\mathbf{h} - \mathbf{h}')\, \mathbf{R}^{\mathrm{I}}]$$
$$\times \int d\Omega_{\mathbf{K}'}\, V^{\mathrm{I}}_{\mathbf{k}+\mathbf{h}-\mathbf{K}'} V^{\mathrm{I}}_{\mathbf{K}'-\mathbf{k}-\mathbf{h}'}. \quad (20.20)$$

Here we have the same factor $\exp[-i(\mathbf{h} - \mathbf{h}')\,\mathbf{R}^{l}]$ leading to $\cos(\mathbf{h}\cdot\mathbf{R}^{l})$ in Eq. (20.18). Therefore, this term also gives a maximal absorption when the interstitials are lying between the reflecting planes.

The effect of point defects on absorption has first been discussed by Hall *et al.*[96] These authors calculate the absorption due to diffuse scattering, i.e., $U_{\mathbf{hh}'}^{(2)\mathrm{dis}}$ and $U_{\mathbf{hh}'}^{(2)\mathrm{pot}}$ and compare it with measurements of the absorption of Si doped with impurities. For X-rays, the absorption due to diffuse scattering is very small compared to the photoelectric absorption (20.18), quite in analogy to the case of thermal motion. It is hoped that the dependence of $\Delta\mu$ [Eq. (20.18)] on $\mathbf{R}^{l}$ can be used to obtain information about the interstitial position. A recent experiment of Edelheit *et al.*[97] with electron irradiated Cu crystal has demonstrated the effects of point defects on absorption. Figure 22 shows the decrease of the measured intensity as a function of the irradiation dose (left side)

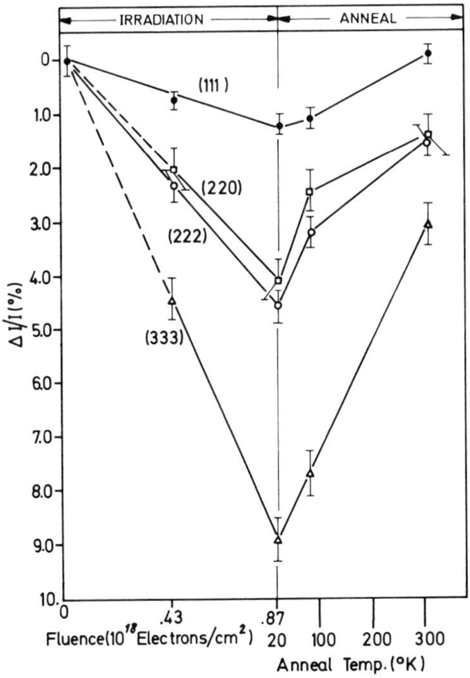

FIG. 22. The change of the anomalous transmitted X-ray intensity for four sets of reflecting planes as a function of the 2-MeV-electron fluence and the subsequent isochronal annealing temperature.

[96] C. R. Hall, P. B. Hirsch, and G. R. Booker, *Phil. Mag.* **14**, 976 (1966).

[97] L. S. Edelheit, J. C. North, J. G. Ring, J. S. Koehler, and F. W. Young, *Phys. Rev.* **B2**, 2907 (1970).

and as a function of the annealing temperature (right side). However a definite localization of the interstitial could not yet be obtained by this experiment.

The effect of point defects on anomalous transmission, as presented in this section follows the discussion of Dederichs[95] and Young *et al.*[98] The problem also has been considered by other authors.[99–102] The expression (20.13) for the static Debye–Waller factor L has been derived by Krivoglaz.[103]

21. Effects of Defect Clustering

In the preceding section we assumed that the point defects were randomly distributed. Now we will discuss the effects on absorption which can occur if point defects are clustered together. We will see that such effects can be quite large. In order to do this, we have to know something about the displacement fields of such defect aggregates. Because calculations from first principles are not available, we make two models for defect clustering.

a. Cluster Model. We assume that the point defects are clustered together such that their displacement fields simply superimpose on each other. This would be the case for a very loose cluster with a small defect concentration. For the displacement fields of the point defects we still use the continuum expression $\mathbf{t}(\mathbf{r}) = A\mathbf{r}/r^3$. Furthermore, we will assume that the cluster centers are randomly distributed and that the clusters have spherical symmetry.

b. Loop Model. For high defect densities the point defects of a cluster may collapse into a flat disk and form a dislocation loop. Then, from continuum theory, the displacement field in the asymptotic region $r \gtrsim 2R_0$ is

$$\mathbf{t}(\mathbf{r}) = \frac{bR_0^2}{8r^2}\left\{\frac{1-2\nu}{1-\nu}(2\mathbf{b}^0(\mathbf{b}^0\cdot\mathbf{e})-\mathbf{e}) + \frac{3}{1-\nu}\mathbf{e}(\mathbf{b}^0\cdot\mathbf{e})^2\right\} \tag{21.1}$$

with $\mathbf{b}^0 = \mathbf{b}/|\mathbf{b}|$, $\mathbf{e} = \mathbf{r}/r$, where ν is Poisson's number and $\pi R_0^2 = F$ is the surface area of the loop which is not necessarily circular. $\mathbf{b}$ is the

[98] F. W. Young, T. O. Baldwin, and P. H. Dederichs, *in* "Vacancies and Interstitials in Metals" (A. Seeger *et al.*, eds.), p. 619. North-Holland, Amsterdam.

[99] V. B. Molodkin and E. A. Tikhonova, *Fiz. Metal. Metalloved.* **24**, 385 (1967).

[100] V. B. Molodkin, *Fiz. Metal. Metalloved.* **25**, 410 (1968).

[101] V. I. Iveronova, A. A. Katsnel'son, and V. I. Kisin, *Sov. Phys.—Solid State* **11**, 2557 (1970).

[102] M. Kuriyama, *Phys. Status Solidi* **24**, 743 (1967).

[103] M. A. Krivoglaz, "Theory of X-Ray and Thermal-Neutron Scattering by Real Crystals." Plenum, New York, 1969.

Burgers vector, assumed to be perpendicular to the loop plane and defined as positive for interstitial loops and negative for vacancy loops. Also in this model, we assume that the loops are distributed at random.

The main difference between the two models is that in the cluster model we neglect any nonlinear relaxation process of the interacting point defects, whereas this is taken into account in the loop model. However in both models the asymptotic displacement field of an aggregate of n_{cl} defects is proportional to n_{cl}/r^2, which is most important for the following.

a. Photoelectric Absorption (Debye–Waller Factor)

In the following we only consider explicitly the anomalous transmission of X-rays. For electrons the results are the same with some slight modifications. However, the sizes of the clusters are much more restricted, since the extinction length is much smaller.

As in the case of randomly distributed point defects, the photoelectric absorption of impurity aggregates, calculated from the imaginary part of the first approximation $\mathsf{U}^{(1)}$, consists of two parts: the photoelectric absorption of the defects themselves and the absorption of the displaced lattice atoms [Eq. (20.18)]. In the cluster model, Eq. (20.18) still holds, and the direct defect absorption expressed by the last term does not change at all if the impurities are clustered. On the other hand, the indirect absorption via the displaced lattice atoms, given by $\mu_0 \cdot L_{\mathbf{h}}$, can increase remarkably due to the increase of the total lattice displacement.

Assuming the displacement $\mathbf{s}$ of a lattice atom to be Gaussian distributed, we get for the Debye–Waller factor

$$L_{\mathbf{h}} = \tfrac{1}{2}\langle(\mathbf{hs})^2\rangle = \tfrac{1}{6}h^2\langle s^2\rangle \tag{21.2}$$

for cubic crystals. In the cluster model $\mathbf{s}$ is the sum of all displacements $\mathbf{t}_i$ due to all defects $i = 1, \ldots, N_{\mathrm{d}}$. Assuming, moreover, that we have N_{cl} clusters, each containing n_{cl} defects ($N_{\mathrm{d}} = N_{\mathrm{cl}} n_{\mathrm{cl}}$) and that the cluster radii are very small, then

$$\langle \mathbf{s}^2\rangle = \sum_{i,j=1}^{N_{\mathrm{d}}} \langle \mathbf{t}_i \cdot \mathbf{t}_j\rangle = n_{\mathrm{cl}} \sum_{i=1}^{N_{\mathrm{d}}} \langle \mathbf{t}_i^{\,2}\rangle = n_{\mathrm{cl}} N_{\mathrm{d}} \langle \mathbf{t}_1^{\,2}\rangle = N_{\mathrm{cl}} n_{\mathrm{cl}}^2 \langle \mathbf{t}_1^{\,2}\rangle, \tag{21.3}$$

because the displacement $\mathbf{t}_j$ of all the defects in the cluster of defect i is the same and because each defect gives on the average the same contribution. We see that n_{cl} enters quadratically, which is due to the "coherent" displacement addition of the defects in the same cluster, whereas N_{cl} enters linearly, which is due to the "incoherent" addition of the displacements of the different clusters. If the defects are randomly distributed, we get $\langle \mathbf{s}^2 \rangle = N_d \langle \mathbf{t}_1{}^2 \rangle$, because now all displacements add incoherently. Therefore, we see that under the above assumption $\langle \mathbf{s}^2 \rangle$ increases due to clustering by a factor n_{cl}.

Actually, this formula is an overestimation of the clustering effect because we have assumed the cluster radius $R_{cl} \approx 0$. For clusters with finite radius, this coherent displacement addition is only correct for atoms far outside the cluster, because only the asymptotic displacement of the cluster is proportional to n_{cl}. Inside the cluster the displacements add more or less incoherently. A more realistic calculation gives, therefore,

$$L_{\mathbf{h}} \cong L_{\mathbf{h}}{}^0 + \tfrac{1}{6} C_d \, 4\pi A^2 h^2 n_{cl} / V_c R_{cl} \, ; \qquad C_d = C_{cl} n_{cl} \, , \tag{21.4}$$

where R_{cl} is an average cluster radius and $L_{\mathbf{h}}{}^0$ is the Debye–Waller factor for statistically distributed point defects [Eq. (20.13a)]. The increase in the Debye–Waller factor is essentially given by $n_{cl} \rho_c / R_{cl}$. In,[104] corrections to the quadratic expression (21.22) have been discussed, being small for not too large or too dense clusters.

For the Debye–Waller factor of randomly distributed dislocation loops, we get analogously to Eq. (20.13)

$$L_{\mathbf{h}} = C_L \sum_n \{1 - \cos(\mathbf{h} \cdot \mathbf{t}_n{}^L)\} \cong C_L \int (d\mathbf{r}/V_c)\{1 - \cos(\mathbf{h} \cdot \mathbf{t}^L(\mathbf{r}))\}, \tag{21.5}$$

where C_L is the loop density. Because the displacements $\mathbf{t}^L$ of the loops are quite large ($\approx b$), the cosines cannot be expanded. In general one can show that the function $\mathbf{t}(\mathbf{r})$ of a loop with an average radius R_0 is only a function of the reduced coordinate $\mathbf{r}/R_0 : \mathbf{t}(\mathbf{r}) = \boldsymbol{\tau}(\mathbf{r}/R_0)$, where now $\boldsymbol{\tau}$ depends only on the orientation and the shape of the loop and not on the size. Therefore $L_{\mathbf{h}}$ is proportional to $R_0{}^3$, i.e.,

$$L_{\mathbf{h}} = C_L (R_0{}^3 / V_c) \int d\tilde{\mathbf{r}} (1 - \cos(\mathbf{h} \cdot \boldsymbol{\tau}(\tilde{\mathbf{r}}))). \tag{21.6}$$

Taking into account that the number n_L of defects forming a loop is proportional to $(R_0/a_0)^2$, we have $L_{\mathbf{h}} \sim C_d R_0/a_0 \sim C_d n_L a_0/R_0$ with $C_d = C_L n_L$. Therefore, if point defects form a loop, the Debye–Waller factor of $R_0/a_0 \approx n_L a_0/R_0$, and so does the photoelectric absorption of the lattice atoms. The enhancement factor is essentially the same as in the cluster case.

The remaining integral in (21.6) cannot be evaluated in general due

[104] P. H. Dederichs, *Phys. Rev.* **B1**, 1306 (1970).

to the complicated displacement field $\boldsymbol{\tau}(\mathbf{r})$. Following Krivoglaz,[105] we replace $\boldsymbol{\tau}(\mathbf{r})$ by the asymptotic expression for $r \gtrsim 2R_0$, an approximation which is allowed for $hb \gg 1$ but which is a surprisingly good approximation also for moderate values of hb. Considering that we may have loops on {111} and/or {110}, we must average (21.6) over all the cubic equivalent planes. Thus, $L_{\mathbf{h}}$ depends only slightly on the direction of $\mathbf{h}$ and is approximately given by

$$L_{\mathbf{h}} \cong C_L R_0^3 (hb)^{3/2} / 2V_c \,. \tag{21.7}$$

Characteristically, the h dependence, resulting from the large displacements of the loop, is less than the "normal" h^2 behavior.

b. Diffuse Scattering Absorption

The absorption due to the diffuse waves also consists of two parts, one due to the diffuse scattering at the defects themselves and one due to the diffuse scattering at the displaced lattice atoms. As for the photoelectric absorption, the latter is the most important in the case of clustering and will be considered here exclusively. This is essentially determined by the correlation $\langle s_i^n s_j^{n'} \rangle$ of the displacements of different atoms n and n', in contrast to the Debye–Waller factor depending on $\langle (\mathbf{s}^n)^2 \rangle$. For $|\mathbf{R}^n - \mathbf{R}^{n'}| \ll R_{\mathrm{cl}}$, the increase in these correlations due to clustering is essentially a factor $n_{\mathrm{cl}} \cdot a_0/R_{\mathrm{cl}}$, as it is for the Debye-Waller factor. But the long-range correlations for $|\mathbf{R}^n - \mathbf{R}^{n'}| \gg R_{\mathrm{cl}}$ increase much faster and indirect proportion to n_{cl}, because they are determined by the asymptotic displacement fields of the clusters. They cause a very strong diffuse scattering near the Bragg reflections, which gives rise to an extra absorption. This absorption can be calculated straightforwardly from the coefficients $U''^{(2)ss'}_{\mathbf{hh}'}$ of Eq. (19.12), where the function $S_{\mathbf{hh}'}(\mathbf{k} - \mathbf{K}')$, given by (17.5), describes the correlations of the static displacements $\mathbf{s}_m$ and $\mathbf{s}_n$. Since the correlations for long distances $|\bar{\mathbf{R}}^m - \bar{\mathbf{R}}^n|$ are small, we expand $S_{\mathbf{hh}'}$ linearly in $\langle s_i^m s_j^n \rangle$,

$$S_{\mathbf{hh}'}(\mathbf{k} - \mathbf{K}') = \sum_{m-n} \exp[i\mathbf{q}(\bar{\mathbf{R}}_m - \bar{\mathbf{R}}_n)\}$$
$$\times \langle ((\mathbf{k} + \mathbf{h} - \mathbf{K}') \cdot \mathbf{s}_m)((\mathbf{k} + \mathbf{h}' - \mathbf{K}') \cdot \mathbf{s}_n) \rangle, \tag{21.8}$$

from which, for the cluster model, we obtain

$$S_{\mathbf{hh}'}(\mathbf{k} - \mathbf{K}') = C_{\mathrm{d}}((\mathbf{k} + \mathbf{h} - \mathbf{K}') \cdot \mathbf{t}(\mathbf{q}))((\mathbf{k} + \mathbf{h}' - \mathbf{K}') \cdot \mathbf{t}(\mathbf{q}))(1 + \tilde{g}(\mathbf{q})), \tag{21.9}$$

[105] M. A. Krivoglaz and K. P. Ryaboshapka, *Phys. Metals Metallogr. (USSR)* **17**, 1 (1953).

with

$$\mathbf{t}(\mathbf{q}) = \sum_n \exp[i\mathbf{q}\mathbf{R}^n]\,\mathbf{t}(\mathbf{R}^n) \cong i(4\pi A/V_c)(\mathbf{q}/q^2),$$

$$\tilde{g}(\mathbf{q}) = \sum_n \exp[i\mathbf{q}\mathbf{R}^n]\,g(\mathbf{R}^n),$$

and $\mathbf{K}' - \mathbf{k} = \mathbf{g} + \mathbf{q}$; $\mathbf{g}$ is a reciprocal lattice vector, defined such that $\mathbf{q}$ lies always in the first Brillouin zone. $g(\mathbf{R}^n - \mathbf{R}^{n'})$ is the conditional probability of finding a defect at position $\mathbf{R}^n$ if one knows already that there is a defect of the same cluster at point $\mathbf{R}^{n'}$. For $q \ll 1/R_{cl}$, we have therefore, $\tilde{g}(\mathbf{q}) \cong (n_{cl} - 1) \cong n_{cl}$, the number of particles in one cluster, whereas $\tilde{g}(\mathbf{q})$ is very small compared to this for $q \gg 1/R_{cl}$. For randomly distributed point defects, we have $\tilde{g}(\mathbf{q}) = 0$ and only the factor 1 remains. The enhancement factor $\tilde{g}(\mathbf{q}) \approx n_{cl}$ is most important for these regions of the integration, for which the remaining strain factors $(\mathbf{k} + \mathbf{h} - \mathbf{K}') \cdot \tilde{\mathbf{t}}(\mathbf{q})$ have a divergence. Otherwise the enlargement of the integrand in a very small area cannot lead to substantial enlargement of the integral $U''_{\mathbf{hh}'}$. In the two-beam case $(0, \mathbf{h})$ the coefficients U''_{00}, $U''_{\mathbf{hh}'}$, $U''_{0\mathbf{h}}$, and $U''_{\mathbf{h}0}$ are important. For instance, the U''_{00} integral has a $1/q^2$ diverge near $\mathbf{K}' \approx \mathbf{k} + \mathbf{h}$, which is due to the well known strong tails of the diffuse scattering near the Bragg reflections. For large clusters ($n_{cl} \gg 1$) the only important contribution to U''_{00} comes therefore from this region. The coefficient $U''_{\mathbf{hh}}$ turns out to be equal to U''_{00}, whereas the coefficients $U''_{0\mathbf{h}}$ and $U''_{\mathbf{h}0}$ are an order of magnitude smaller because they have only a $1/q$ divergence. They will be neglected in the following. Therefore, we get for the anomalous absorption of the σ polarized wave due to diffuse scattering

$$\Delta\mu_{\mathbf{h}}^{\mathrm{DS}} = \frac{\pi}{V_c} r_e^2 f_h^2 \left(\frac{h}{K}\right)^2 C_L n_{cl} \left(\frac{4\pi A}{V_c}\right)^2 \cos^2\theta_B \ln\frac{1}{R_{cl} q_c}. \qquad (21.10)$$

Due to dynamical effects the $1/q^2$ divergence has been cut off for $q \leqslant q_c$. Because of $U''_{0\mathbf{h}} \approx 0$, this absorption is the same for the two Bloch waves on the different branches of the dispersion surfaces. Moreover, the absorption changes only very slightly if the Bragg condition is not exactly fulfilled. In this case q_c has to be replaced by q_0, the smallest q value which can occur; $q_0 = ||\mathbf{k} + \mathbf{h}| - \mathbf{k}|$. This absorption decreases logarithmically with the excitation error as long as $q_0 \leqslant 1/R_{cl}$, and is essentially zero otherwise.

According to Eq. (21.10), $\Delta\mu^{\mathrm{DS}}$ is proportional to $C_d n_{cl} = C_{cl} n_{cl}^2$ and is therefore very sensitive to clustering. Whereas for isolated point defects $\Delta\mu^{\mathrm{DS}}$ is negligible compared with the photoelectric absorption, in the case of clustering it can become comparable or even much larger than the photoelectric absorption, depending on the cluster size. Moreover, it shows a different wavelength dependence, being proportional to λ^2 for small Bragg angles.

For dislocation loops we get a formula similar to (21.10). But C_d must be replaced by the loop concentration C_L, and $\tilde{\mathbf{t}}(\mathbf{q})$ by the Fourier transform $\tilde{\mathbf{t}}^L(\mathbf{q})$ of the loop displacement field, whereas $\tilde{g}(q) = 0$. For small q we can evaluate $\tilde{\mathbf{t}}^L(\mathbf{q})$ from the asymptotic expression in Eq. (21.1). As in the cluster case, we have very intense diffuse scattering near the Bragg reflection, which gives rise to an extra absorption. The calculations are similar to the cluster case, but more lengthy. If we assume that the loops can occupy all planes and that no directions are preferred, we get finally

$$\Delta\mu_{\mathbf{h}}^{\mathrm{DS}} = \frac{1}{V_c} r_e^2 f_{\mathbf{h}}^2 \left(\frac{h}{K}\right)^2 C_L \left(\frac{b\pi R_0^2}{V_c}\right)^2 \ln\frac{1}{R_0 q_c} \times \left\{\frac{8}{15} + \frac{\pi}{15}\frac{(-1 + 6\gamma + 3\nu^2)}{(1-\nu)^2}\cos^2\theta_B\right\}. \tag{21.11}$$

In analogy to the cluster expression (31), $\Delta\mu^{\mathrm{DS}}$ is proportional to $C_L R_0^4 \approx C_L n_L^2 = C_d n_L$. Therefore, it is also very sensitive to clustering and increases for increasing R_0 much faster than the photoelectric absorption (21.7), which is proportional to $C_L R_0^3$. Whereas for very small loops the diffuse scattering absorption is negligible compared with the photoelectric absorption, the diffuse absorption becomes much more important for larger loops.

To illustrate the importance of defect clustering, we have plotted in Fig. 23, on a double logarithmic scale, the absorption when a constant number of point defects build up dislocation loops of varying radius R_0. Therefore, $\Delta\mu(R_0)$ divided by the absorption of fictitious loops of radius 1 Å is essentially the absorption per point defect as a function of the loop radius R_0. The photoelectric absorption increases linearly with R_0, giving a straight line with slope 1 in Fig. 23, whereas the diffuse scattering absorption increases proportionally to R_0^2, giving a line with slope 2. As an example, we have taken the absorption of MoK_α and CuK_α radiation in Cu. The photoelectric absorption is coincidentally nearly the same for both wavelengths ($\mu_0^{\mathrm{Mo}K_\alpha} = 438$; $\mu_0^{\mathrm{Cu}K_\alpha} = 446$ cm^{-1}), whereas, according to (21.11), the diffuse scattering absorption is a factor of $(\lambda_{\mathrm{Cu}}/\lambda_{\mathrm{Mo}})^2 \simeq 4.5$ larger for CuK_α than for MoK_α. For small radii, only the photoelectric absorption is important, which corresponds to the case of randomly distributed point defects. For larger radii, the diffuse scattering absorption dominates and $\Delta\mu^{\mathrm{PE}}$ becomes more and more negligible. Our theory is valid for radii $R_0 \ll d_{\mathrm{extinction}}$, say

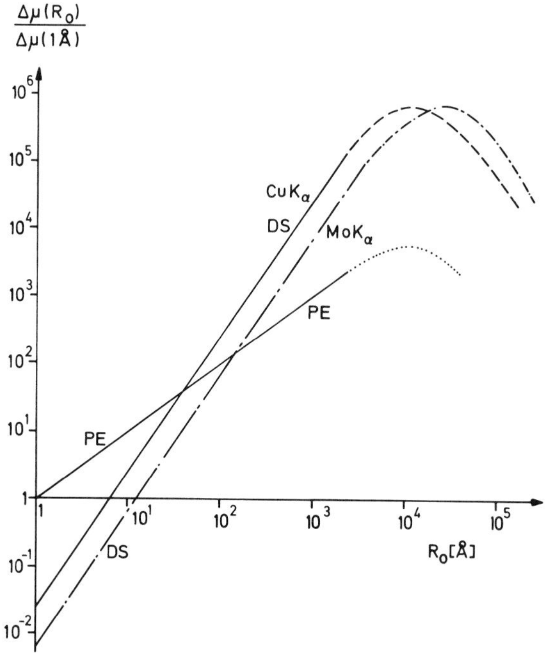

FIG. 23. Absorption "per point defect" as a function of the loop radius R_0.

$R_0 \lesssim 10^3$ Å. Therefore, the dashed lines above $R_0 \gtrsim 10^3$ Å have only a qualitative meaning. From the scale of the ordinate, one sees that the absorption per point defect can increase by a factor of 10^5. For more details of the theory, we refer to Dederichs.[104]

The effects of defect clustering were first observed by Patel and Batterman.[106] Efimov *et al.*[107] have investigated the effects of impurities, vacancies, and cluster. More detailed experiments have been done by a number of authors.[98,108–110] Wenzl[109] has studied the precipitation of Li interstitials in Ge crystals. Baldwin *et al.*[108] and Larson and Young[110] investigated the anomalous transmission through neutron irradiated Cu crystals, where small dislocation loops are present. A detailed comparison with the theory has been made.[110] Figure 24 shows the measured

[106] J. R. Patel and B. W. Batterman, *J. Appl. Phys.* **34**, 2716 (1963).

[107] O. N. Efimov and A. M. Elistratov, *Sov. Phys.—Solid State* **5**, 1364, 1543 (1963). O. N. Efimov, *Sov. Phys.—Solid State* **12**, 1235 (1970). O. N. Efimov, E. G. Sheikhet, and L. I. Datsenko, *Phys. Status Solidi* **38**, 489 (1970).

[108] T. O. Baldwin, F. A. Sherill, and F. W. Young, *J. Appl. Phys.* **39**, 1541 (1968).

[109] H. F. Wenzl, *Z. Naturforsch.* **26a**, 495 (1971).

[110] B. C. Larson and F. W. Young, Jr., *Phys. Rev.* **B4**, 1709 (1971).

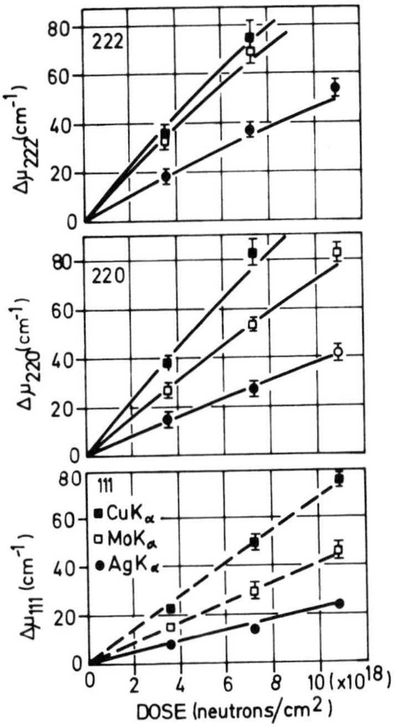

FIG. 24. Anomalous absorption $\Delta\mu_h$ in Cu for various radiations and reflections versus neutron dose. The solid lines were calculated from the dashed lines.

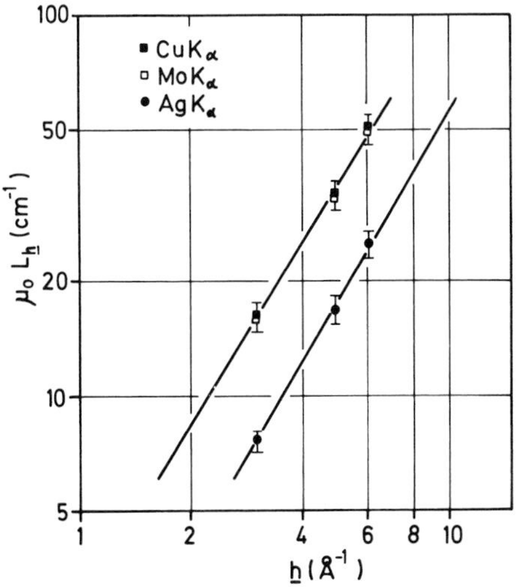

FIG. 25. **h** dependence of the static Debye–Waller factor.

absorption $\Delta\mu_{\mathbf{h}}$ versus the fast neutron dose for various reflections and wavelengths. The difference between the curves for CuK_α and MoK_α is entirely due to the diffuse-scattering absorption since in Cu the photoelectric absorption is practically the same for both radiations. The solid lines were calculated by using the dashed lines to determine the defect quantities $C_L R_0^3$ and $C_L R_0^4$ entering in Eqs. (21.7) and (21.11). Figure 25 shows in a semilogarithmic plot the **h** dependence of the Debye–Waller factor $L_{\mathbf{h}}$ determining the photoelectric absorption. The experimental value for the slope is 1.6 compared to $\frac{3}{2}$ exponent in Eq. (21.7).

Acknowledgments

The author would like to thank T. O. Baldwin, K. Dettmann, G. Leibfried, and F. W. Young for helpful and stimulating discussions.

Ferromagnetic Thin Films

A. Corciovei, G. Costache, and D. Vamanu

Institute for Atomic Physics, Bucharest, Romania

I. Introduction

From the practical point of view, a thin ferromagnetic film is a ferromagnetic body presenting a microscopic thickness in contrast with the macroscopic extension of its surfaces and satisfying certain conditions of deposition and oxidation. From an ideal point of view the thin ferromagnetic film occurs as a many-body system of spins, which is finite in one direction, and which obeys some condition of uniformity and cohesion. A rapid progress of the technology together with a remarkable growth of the interest of physicists brought in the last 20 years the subject of films to a position of prominence in the physics of magnetism.

The interest of both engineers and physicists in thin films is mainly stimulated by the exciting fact that a typical thin ferromagnetic film would be the only known single ferromagnetic domain of macroscopic extension. This important point was stated clearly as early as in 1946 by

Kittel,[1] who succeeded in accounting for some basic types of magnetic domain configurations in thin films.

Following Kittel, it has been found that since the ferromagnetism is a cooperative phenomenon, any decrease in sizes or symmetry of a ferromagnetic sample ought to affect markedly its magnetic properties. In particular, one may assume that below some critical size the existence of domain boundaries becomes unfavorable energetically, so that the whole sample behaves as a single domain, magnetized to saturation in a certain direction.

The tendency to a single domain structure explains the rectangular hysteresis behavior of these specimens as well as the high values of the coercive force. Such qualities are very important in the technology of permanent magnets and, more recently, of magnetic memory devices; thus the increasing interest of technologists in the field is understandable.

Because of space limitations the present paper is restricted to topics which reveal the most striking features of the subject. An effort is made to show how such basic concepts as *surface defect* and *magnetic anisotropy* are used in building up a realistic model of thin ferromagnetic films, which can properly take account of factors like size, shape, structure, processing, and spoiling. The typical specimen considered is the single domain thin film magnetized to saturation in its plane.

Thus the work is focused mainly on *critical phenomena* and *phase transitions* on one hand, and *elementary excitations* on the other. We shall be interested in *saturation magnetization*, *Curie temperature*, *spin wave spectrum*, *dispersion laws*, and *resonant excitations*.

However, a discussion of the magnetic configurations in thin films with special emphasis on the domain structure as well as references to *ripple* are also given. It was not possible to include in this short text the topics pertaining mainly to the practical side, such as the theory of the magnetization curve, magnetization reversal, low and high frequency properties, galvanomagnetic and magnetooptic phenomena, or magnetic neutron diffraction.

II. Magnetic Energy

1. Macroscopic Description

There are many properties which justify the separate consideration of thin films. These properties are related to the existence of the surfaces, the importance of which might increase when the thin dimension of the

[1] C. Kittel, *Phys. Rev.* **70**, 965 (1946).

film is substantially reduced as compared with the usual dimensions of a bulk specimen.

For such thin films it might become energetically favorable, as was first proved by Kittel,[1,2] to disregard all domain boundaries if certain conditions are satisfied. Then the sample may be considered as being in a state of *uniform*—or at least nearly uniform—magnetization. Although significant differences between theory and experiment occur, which require consideration of the inhomogeneous structure of the film, we may begin by considering that the film is uniformly magnetized.

In this section quantitative expressions for the several types of energy that enter into the theory of ferromagnetic thin films will be given. Except when indicated otherwise, we shall assume that the thin film is a perfect single crystal, and regard it as a single domain in static equilibrium under a uniform strain system. Therefore, only the total energy of the system will be considered, although some of the expressions we shall write down may be employed also as energy densities. The magnetization **M** is considered to have a magnitude, called the saturation magnetization, determined by the temperature; its direction is to be determined by equilibrium criteria.

Although many expressions are valid also for bulk bodies, we shall use them here for the sake of consistency, but will point out essential differences which might appear between the behavior of the thin films and bulk materials.

The total energy is assumed to consist of: *exchange* energy, *anisotropy* energy, *magnetoelastic* energy, and *magnetostatic* energy.

In the thermodynamical treatment we shall choose as the system associated with the ferromagnetic thin film, the ferromagnet itself in which is included the inner uniform magnetic field. If the internal energy of the system is denoted by U then, following Carr,[3] we shall write

$$dU = T\,dS + \mathbf{H}\,d(V\mathbf{M}) + V_0\hat{\sigma} : d\hat{e}, \tag{1.1}$$

where $\mathbf{H}$ is the applied magnetic field, $V\mathbf{M}$ is the magnetic moment of the sample, $\hat{\sigma}$ is the symmetric stress tensor, $\hat{e}$ the symmetric strain tensor, V_0 is the unstrained volume,[4] and V is the actual volume.

As in thermostatics the convenient thermal variable is the temperature

[2] C. Kittel and J. K. Galt, *Solid State Phys.* **3**, 437 (1956).

[3] W. J. Carr, Jr., *in* "Handbuch der Physik" (S. Flügge, ed.), Vol. XVIII/2. Springer Verlag, Berlin and New York, 1966.

[4] $\hat{\sigma} : d\hat{e}$ denotes the scalar product $\sum_{ij} \sigma_{ij}\, de_{ij}$, the summation being performed over 1, 2, 3 for the x, y, z axes. For a full understanding of the notations see, however, Carr.[3]

T rather than the entropy, and the appropriate thermodynamic potential is not the internal energy U but the potentials

$$F = U - TS, \qquad G = F - V_0 \hat{\sigma} : \hat{e},$$
$$F_{\mathbf{H}} = F - (V\mathbf{M})\mathbf{H}, \qquad G_{\mathbf{H}} = G - (V\mathbf{M})\mathbf{H}. \tag{1.2}$$

The distinction between U and these potentials is important since for isothermal processes, these potentials play the role of energy functions. A detailed discussion on this subject can be found in the papers of Carr[3] and Brown.[5] (We shall only note that in the measurement of the anisotropy, for instance, we are concerned with the work which must be done in rotating the magnetization while maintaining the temperature constant, rather than maintaining the system adiabatically isolated.

To study the equilibrium and stability criteria, we must obtain the expressions of these potentials as functions of the internal coordinates.)

a. Exchange Energy

The exchange forces tend to keep the magnetization uniform. In a body of usual size, the exchange forces maintain an approximate uniformity only over short distances; over large distances they are in competition with long range magnetic forces. For films which are very thin, uniform magnetization appears. Therefore, the exchange contribution to the "free energy" is not essential in our static problem.[6]

b. Anisotropy Energy

Many properties of crystals are observed to be anisotropic, which while arising from a variety of causes, the anisotropy may be broadly classified according to the macroscopic symmetry. These anisotropies are observed in relation to some special axes in the material, not necessarily related to the crystal axes.

(*i*) *Magnetocrystalline Anisotropy.* At constant strain and constant temperature the free energy must depend on the components of $\mathbf{M}$, or upon the direction $\boldsymbol{\alpha}$ of the magnetization related to the crystal axes, in a manner consistent with the symmetry of the crystal.[7]

[5] W. F. Brown, Jr., "Micromagnetics." Wiley (Interscience), New York, 1963.

[6] The significant part of the exchange energy which depends on the derivatives of the point magnetization has to be taken into account in the cases of inhomogeneous magnetization problems (e.g., domain walls and ripple, the natural modes of precession of the magnetization).

[7] R. P. Birss, "Symmetry and Magnetism." Wiley, New York, 1964.

Formally, for a crystal which has a center of symmetry we may write

$$f \equiv F/V = K_0' + \sum K'_{mn}M_mM_n + \sum K'_{mnqp}M_mM_nM_qM_p + \cdots . \tag{1.3}$$

The terms of Eq. (1.3) have to be invariant when subject to the operations which characterize the symmetry properties of the crystal under investigation.[8] The requirements of symmetry impose severe restrictions on the number of independent components of the tensors K_{mn}, K_{mnqp}, etc. (see, however, Birss[7]).

For a cubic crystal, Eq. (1.3) becomes

$$f = K_0' + K_1'(\alpha_1^2\alpha_2^2 + \alpha_2^2\alpha_3^2 + \alpha_3^2\alpha_1^2) + K_2'\alpha_1^2\alpha_2^2\alpha_3^2 + \cdots . \tag{1.4}$$

and for a hexagonal crystal the magnetocrystalline anisotropy has the form

$$f = K_0' + K_1' \sin^2 \theta + K_2' \sin^4 \theta + (K_3' + K_4' \cos 6\varphi) \sin^6 \theta + \cdots, \tag{1.5}$$

where the hexagonal axis is the polar axis and θ and φ are the polar and the azimuthal angles, the latter measured from a hexagonal *a* axis.

It should be noted that in simplifying the form of the tensors appearing in Eq. (1.3), only the symmetry operators appropriate for a perfect bulk body lattice were used. The geometry of the thin film specimen will introduce additional symmetry features and the exclusion of any reference to the shape of the specimen can not be justified. It is our opinion that by simply adding a further term, i.e., the demagnetizing energy, to the free energy, the problem is not completely solved. The reason is that the demagnetizing energy is a volume energy; we actually require a surface free energy term taking account of the lower symmetry of the surface.[9]

(*ii*) *Surface anisotropy*. The surface anisotropy term was first introduced by Néel.[10] In his theory, the coupling energy of an atom pair

[8] The most general method of performing these operations is due to F. Seitz, *Z. Kristallogr.* **88**, 433 (1934).

[9] Consideration of microscopic models suggests that surface terms are also required for a realistic description of the materials, since atoms near the surface are in an unsymmetric situation. The introduction of the demagnetizing energy is not entirely satisfactory from this point of view since the demagnetizing energy is produced by long range forces rather than short ranged which are responsible for crystalline anisotropy.

[10] L. Néel, *J. Phys. Radium* **15**, 225 (1954).

can be expanded in a series of Legendre polynomials, the first term of which corresponds to the ordinary exchange coupling.

The pair energy $w(r, \phi)$ is dependent on the direction ϕ of the magnetic moment as measured from the bond direction, and the distance r between the two atoms

$$w(r, \phi) = g_0(r) + g_1(r)\, P_2(\cos\phi) + g_2(r)\, P_4(\cos\phi) + \cdots . \tag{1.6}$$

The second term of this series expansion necessarily contains one term corresponding to the dipolar coupling, which is long ranged and which yields the demagnetizing energy. The other terms are supposed to be related to the spin orbit coupling and therefore they are short ranged.

The second term, which takes account of the existence of the surfaces of the crystal, gives rise to an anisotropy energy *even in cubic crystals.* This anisotropy energy is generally uniaxial and is expressed by

$$f_{\mathrm{S}} = -K_{\mathrm{S}} \cos^2\theta, \tag{1.7}$$

where K_{s} is the surface anisotropy constant and θ is the angle of the spontaneous magnetization with the normal to the surface.

However the Néel work seems to be open to some criticism. As was suggested by Jacobs and Bean,[11] the coefficient of $P_4(\cos\phi)$ in Eq. (1.6) must be zero for a cubic crystal of constant dimensions and spin $\frac{1}{2}$. Moreover the second term does give rise in a quantum mechanical calculation to cubic anisotropy. We feel that these arguments do not invalidate Néel's theory since the given interpretation of the terms appearing in Eq. (1.6) is not very precise. Anyway, the concept of the surface anisotropy must be preserved as it follows from very general symmetry considerations.

Also, aside from the Néel surface anisotropy, there are still other terms contributing to the actual surface anisotropy which we shall discuss in connection with the spin-wave approach.

(*iii*) *Induced anisotropy.* We shall now study the induced anisotropy which appears in a magnetic annealed specimen. This anisotropy is found in certain *types of alloys* and it is normally attributed to directional ordering of the solute atoms.

By symmetry considerations the induced anisotropy of a magnetic annealed cubic crystal can be written as

$$\begin{aligned} g_{\mathrm{ind}} \equiv G_{\mathrm{ind}}/V = {} & -k_1({\alpha_1}^2\alpha_1'^2 + {\alpha_2}^2\alpha_2'^2 + {\alpha_3}^2\alpha_3'^2) \\ & -2k_2(\alpha_1\alpha_2{\alpha_1}'{\alpha_2}' + \alpha_2\alpha_3{\alpha_2}'{\alpha_3}' + \alpha_3\alpha_1{\alpha_3}'{\alpha_1}'), \end{aligned} \tag{1.8}$$

[11] I. S. Jacobs and C. P. Bean, *in* "Magnetism" (G. T. Rado and H. Suhl, eds.), Vol. III. Academic Press, New York, 1963.

where $\boldsymbol{\alpha}(\alpha_1, \alpha_2, \alpha_3)$ is the direction of the magnetization at a temperature T and $\boldsymbol{\alpha}'(\alpha_1', \alpha_2', \alpha_3')$ denotes the direction of magnetization at the temperature T' measured during the annealing treatment.[12] Note that if $k_1 = k_2 = k$, then

$$g_{\text{ind}} = -k \cos^2 \theta, \tag{1.8'}$$

where θ is the angle between the magnetization and the field applied during heat treatment. Thus the anisotropy becomes uniaxial. The induced anisotropy in a cubic crystal is sometimes described as uniaxial because it appears uniaxial in one plane of observation; it is truly uniaxial only if the condition $k_1 = k_2$ is satisfied.

The theory of the field induced magnetic anisotropy was proposed by Néel[13] and Taniguchi-Yamamoto.[14] According to this theory, the induced magnetic anisotropy energy is expressed by Eq. (1.8), where the constants k_1 and k_2 are both proportional to the product $c_A^2 c_B^2$ of the concentrations of the A and B atoms respectively. When the intrinsic magnetization lies in the plane $\{110\}$ Eq. (1.8) becomes

$$g_{\text{ind}} = -K_u \cos^2(\theta - \theta_0), \tag{1.9}$$

where K_u is the induced anisotropy constant, θ is the angle of magnetization, θ_0 is the angle of minimum energy direction, both measured from the direction $\langle 001 \rangle$ in the plane $\{110\}$; K_u and θ_0 depend on the values of k_1 and k_2 as well as on the direction of the applied field relative to the crystallographic axes.

When the direction of the annealing field does not coincide with the principal crystallographic axes, the minimum g_{ind} is not reached in the same direction as the annealing field but in a neighboring $\langle 111 \rangle$ direction; K_u has the greatest value for an annealing field along the $\langle 111 \rangle$ direction, a lesser value for the $\langle 110 \rangle$ direction, and the least value for the $\langle 100 \rangle$ direction. This behavior was explained in the case of permalloy specimens by Chikazumi,[15] who suggested that it is due to the formation of trigonal iron groups.

We shall also note that the Eq. (1.8′) is the proper expression for a polycrystal with random orientation of crystallites.

Another type of thermal treatment involves application of stresses at

[12] We observe that no consideration is given to the value of magnetic field **H** applied to orient the sample magnetization **M** during the anneal.

[13] L. Néel, *C. R. Acad. Sci. Paris* **237**, 1468, 1613 (1953).

[14] S. Taniguchi and M. Yamamoto, *Sci. Rep. Res. Inst., Tohoku Univ., Ser. A* **6**, 330 (1954).

[15] S. Chikazumi, *J. Phys. Soc. Japan* **11**, 551 (1956).

an elevated temperature. The phenomenological form of such an anisotropy shall be given later.

We conclude by noting that any uniaxial anisotropy is described by

$$g_{\mathrm{u}} = -K_{\mathrm{u}} \cos^2\theta, \tag{1.10}$$

where θ is the angle of the magnetization with the special axis characterizing the induced anisotropy.

(iv) Unidirectional anisotropy. A material in which a *single direction* rather than a single axis is preferred is said to possess unidirectional anisotropy. The phenomenon was discovered by Meiklejohn and Bean[16] and was attributed to an exchange mechanism (exchange anisotropy).

Vlasov and Mitsek[17] treated this problem thermodynamically by examining a system consisting of two subsystems one of which was ferromagnetic while the other was antiferromagnetic. The exchange and magnetic interactions between them were considered as perturbations to the state in which the subsystems were independent. It was considered that the exchange interaction was much greater than the magnetic interaction between the subsystems so that the crystalline anisotropy free energy of the system as a whole was the sum of the crystalline anisotropy free energies of the two subsystems. Under these conditions, a term might appear in the anisotropy energy of the whole system of the form

$$g_{\mathrm{ud}} \equiv G_{\mathrm{ud}}/V = -K_{\mathrm{ud}} \cos\theta, \tag{1.11}$$

where K_{ud} is the constant of the unidirectional anisotropy and θ is the angle between the magnetization and the easy axis of crystalline anisotropy.[18]

c. *Magnetoelastic Energy*

We shall define the magnetoelastic energy as the coupling energy of the elastic strains with the magnetization. Following Carr,[3] we shall expand the free energy F in powers of the strain components

$$F = F_0 + V_0\hat{B} : \hat{e} + (V_0/2)\hat{e} : \hat{C} : \hat{e} + \cdots, \tag{1.12}$$

[16] W. H. Meiklejohn and C. P. Bean, *Phys. Rev.* **105**, 904 (1957).

[17] K. B. Vlasov and A. I. Mitsek, *Fiz. Metal. Metalloved.* **14**, 487, 498 (1962).

[18] Subsystems with a single easy axis were examined. It was supposed that the easy axes of the subsystems are coincident. A treatment of the same problem for a system of cubic subsystems was done by Vlasov *et al.*[19]

[19] K. B. Vlasov, N. V. Volkenstein, C. V. Vonsovskii, A. I. Mitsek, and M. I. Turchinskaia, *Izv. Akad. Nauk SSSR* **28**, 423 (1964).

with F_0 independent of strains; $\hat{B}$ and $\hat{C}$ are respectively second- and fourth-rank tensors which may depend on temperature and magnetic moment. The term F_0 describes the *magnetocrystalline anisotropy* at *zero strain* and the following term represents the magnetoelastic energy

$$F_{\text{mel}} = V_0 \hat{B} : \hat{e}. \tag{1.13}$$

The components of the tensor $\hat{B}$ may be again determined if the invariance under the symmetry operations of the crystal is invoked.

(i) *Stress energy.* As we are concerned with the crystal upon which *stresses* are exerted we must consider the potential G rather than F. Then in the case of small stresses, the expansion of G is

$$G = G_0 - V_0 \hat{A} : \hat{\sigma} - (V_0/2)\hat{\sigma} : \hat{s} : \hat{\sigma}, \tag{1.14}$$

where the tensor s is the tensor of elastic compliance coefficients at constant temperature and magnetic moment, and G_0 contains the anisotropy energy at zero stress.

The stress energy[3] is defined by

$$G_{\text{stress}} = -V_0 \hat{A} : \hat{\sigma}. \tag{1.15}$$

It may be proved that the tensor $\hat{A}$ is given by

$$A = -\hat{s} : \hat{B} \tag{1.16}$$

and it may be interpreted as the strain tensor of magnetostriction, depending at constant temperature only on the magnetization. It can be seen that the coupling of stresses to magnetism is effected through magnetostriction. A more physical discussion on this point is given by Lee.[20]

Neglecting morphic effects we may write after Mason[21]

$$A_{ij} = \sum_{mn} T_{ijmn} M_m M_n + \cdots, \tag{1.17}$$

so that the Eq. (1.15) becomes

$$\begin{aligned} g_{\text{stress}} &\equiv G_{\text{stress}}/V_0 \\ &= -\sum_{ijmn} T_{ijmn} \sigma_{ij} M_m M_n + \text{higher terms}, \end{aligned} \tag{1.18}$$

[20] E. W. Lee, *Rep. Progr. Phys.* **18**, 184 (1955).
[21] W. P. Mason, *Phys. Rev.* **82**, 715 (1951).

where the independent components of the tensor may be determined from symmetry requirements.

The concrete expressions of T for cubic and hexagonal crystals are given by Mason[21,22]. Other expressions for stress anisotropy in some simple cases are given by MacDonald[23] who introduced the notion of *the internal field contribution of stresses* and defined it as follows:

$$(H_{\mathrm{st}}^{\mathrm{int}})_j = \sum_k D_{jk} M_k \,, \tag{1.19}$$

where the significance of the tensor D_{jk} is obvious.[24]

We shall now discuss some peculiar stress systems to which we shall often refer. First, we shall note the case of a *uniform stress system*, $\sigma_{ij} = \sigma\delta_{ij}$. It may be seen that the anisotropy is of the same type as the magnetocrystalline anisotropy. However the easy axis may be different depending of the sign of σ.

It seems that the most important cases are those of *linear* and *planar* stresses. A homogeneous *linear stress* of direction cosines γ_i is described as $\sigma_{ij} = \sigma\gamma_i\gamma_j$, and in the case of cubic crystals, the actual stress anisotropy is

$$\begin{aligned} g_{\mathrm{stress}} = {} & -\sigma h_2 \cos^2\theta - \sigma(h_1 - h_2)(\alpha_1^2\gamma_1^2 + \alpha_2^2\gamma_2^2 + \alpha_3^2\gamma_3^2) \\ & -\sigma(h_3 + 2h_4/3)(\alpha_1^2\alpha_2^2 + \alpha_2^2\alpha_3^2 + \alpha_3^2\alpha_1^2) \\ & -\sigma h_4(\alpha_1^4\gamma_1^2 + \alpha_2^4\gamma_2^2 + \alpha_3^4\gamma_3^2) + \cdots, \end{aligned} \tag{1.20}$$

where θ is the angle between the magnetization vector and the stress, while h_1, h_2, h_3, h_4 are the magnetostriction constants[3,27]. Also note that for $\gamma_2 = \gamma_3 = 0$, $\gamma_1 = 1$, Eq. (1.20) becomes

$$\begin{aligned} g_{\mathrm{stress}} = {} & -\sigma(h_1 + h_3 + 2h_4/3)\,\alpha_1^2 \\ & -\sigma(-h_3 + h_4/3)\,\alpha_1^4 - \sigma(h_3 + 2h_4/3)\,\alpha_2^2\alpha_3^2, \end{aligned} \tag{1.21}$$

and the last term on the right-hand side of Eq. (1.21) show that a biaxial anisotropy may be produced apart from the uniaxial anisotropy described by the first two terms of Eq. (1.21).

[22] W. P. Mason, *Phys. Rev.* **96**, 302 (1954).

[23] J. R. MacDonald, *Proc. Phys. Soc.* **A64**, 968 (1951).

[24] Sometimes the effective anisotropy field created by the stress is regarded as a demagnetizing field—see, for instance, MacDonald,[23] Griffiths,[25] Fraitova.[26]

[25] J. H. E. Griffiths, *Physica* **17**, 253 (1951).

[26] D. Fraitova, *Czech. J. Phys.* **B11**, 500 (1961).

[27] For polycrystalline materials and isotropic magnetostriction a uniaxial anisotropy appears.

Freedman[28] pointed out that a very important stress system in the physics of thin films seems to be the *isotropic planar stress*. Throughout this paper we shall denote by $0x$ the normal to the plane of the film. If we assume that an isotropic stress acts in this plane,

$$\sigma_{22} = \sigma_{33} = \sigma, \qquad \sigma_{11} = 0, \qquad \sigma_{ij} = 0, \qquad i \neq j,$$

then the stress anisotropy for cubic crystals is described by

$$\begin{aligned} g_{\text{stress}} &= \sigma(h_1 - 2h_3 + 2h_4/3)\,\alpha_1^2 + \sigma(2h_3 - h_4/3)\,\alpha_1^4 \\ &\quad + \sigma(-2h_3 + 2h_4/3)\,\alpha_2^2\alpha_3^2 \end{aligned} \tag{1.22}$$

It may be seen that aside from the uniaxial anisotropy there is a biaxial anisotropy in the stress plane and generally the easy axes of this system are not energetically equivalent. For hexagonal crystals with the c axis as the third coordination axis, the stress anisotropy is given in the same case by the expression

$$g_{\text{stress}} = \text{const. } \sigma\alpha_1^2 + \cdots, \tag{1.23}$$

and it may be seen that a uniaxial anisotropy can be developed.

(*ii*) *Magnetocrystalline anisotropy at zero stress.* We have seen that the magnetocrystalline anisotropy at zero stress is described by G_0. It may be proved that[3]

$$G_0 = F_0 - (V/2)\hat{B} : \hat{s} : \hat{B} \tag{1.24}$$

It has been found[2,7] that for both cubic and hexagonal crystals the only difference between G_0 and F_0 appears in the value of the anisotropy constants. Then for cubic crystals,

$$\begin{aligned} g_0 &\equiv G_0/V_0 \\ &= K_0 + K_1(\alpha_1^2\alpha_2^2 + \alpha_2^2\alpha_3^2 + \alpha_3^2\alpha_1^2) + K_2\alpha_1^2\alpha_2^2\alpha_3^2 + \cdots, \end{aligned} \tag{1.24$'$}$$

and for hexagonal crystals

$$g_0 = K_0 + K_1 \sin^2\theta + K_2 \sin^4\theta + \cdots. \tag{1.24$''$}$$

(*iii*) *Magnetic anisotropy in polycrystalline thin films.* In polycrystalline thin films with finite magnetostriction coefficients, the stresses present

[28] J. F. Freedman, *IBM J. Res. Develop.* **6**, 449 (1962); *J. Appl. Phys.* **33**, 1148S (1962).

in the film plane also can result in *magnetic anisotropy*. Let us consider for the purpose of analysis a simplified configuration consisting of a single domain thin film of cubic microcrystals, which are randomly oriented and do not interact, under the influence of an isotropic planar tension together with a linear stress. The average value of the effective anisotropy energy may then be obtained.[29] It was proved that magnetocrystalline anisotropy is averaged out. Also, it was concluded that the isotropic planar stress does not give rise *in the film plane* to a uniaxial anisotropy. For the stresses to result in magnetic anisotropy in the film plane, it is essential for the stresses themselves to be *anisotropic*.[30]

In a ferromagnetic thin film, a stress system may be caused through magnetostriction as induced by a magnetic field applied during deposition, the only role of which is to stabilize the direction of the magnetization. During and after deposition, but above a certain temperature T', the crystallites are free to deform by magnetostriction to an equilibrium strain-field which is fixed by forces imposed by the substrate. In the equilibrium state these forces cancel out totally the stresses caused through magnetostriction. After cooling the specimen below the temperature T', no significant changes in strain occur and then the deformations the system had in the neighborhood of T' are permanently fixed. Thus, due to the existence of the substrate that behaves as a constraint, the film becomes constrained.

To measure the uniaxial part of the magnetoelastic anisotropy, one has to rotate the magnetization out of its initial axis (the axis of the applied field during deposition), the result of which will be an increase of the magnetoelastic energy of the film. The energy related to this rotation of magnetization was called the "*constraint energy*." It may be thought that the constraint imposes upon the film a stress $\sigma = \lambda' E$ where E is Young's modulus and λ' is the value of the average saturation magnetostriction constant at the temperature T'. This stress is in the direction of the applied field during deposition and may be tensile or compressive. But we shall follow the work of West[31] and average the single crystal magnetoelastic energy over the ensemble of the randomly oriented crystallites considered above. The following expression for the anisotropy energy

$$\begin{aligned} \bar{g} &= \bar{G}/V \\ &= -(9/10)[(C_{11} - C_{12})\,\lambda_{100}\lambda'_{100} + 3C_{44}\lambda_{111}\lambda'_{111}]\cos^2\theta + \text{const.} \end{aligned} \tag{1.25}$$

[29] J. D. Blades, *J. Appl. Phys.* **30**, 260S (1959).
[30] M. Prutton, *Trans. 9th Nat. Vac. Symp. Amer. Vac. Soc.* 59 (1962).
[31] F. G. West, *J. Appl. Phys.* **35**, 1827 (1964).

is obtained for a cubic crystal; for a hexagonal polycrystal the anisotropy energy is

$$\bar{g} = -(2/15)\{(C_{11} - C_{12})(\lambda_a - \lambda_b) \times [(7/2)(\lambda_a' - \lambda_b') + \lambda_b'] + \cdots\} \cos^2 \theta + \text{const.}, \tag{1.26}$$

where the constants λ_i are the saturation magnetostriction constants and C_{ij} are the elastic constants. The primed values are referred to the constraint temperature T' and θ is the angle between magnetizations at temperature T' and at the temperature of measurement T.

We shall note that this theory works both for alloys[32] and single-metal films. In the case of alloys, we may therefore separate the uniaxial anisotropy into two parts: (i) the energy resulting from the short range directional ordering[13,14], and (ii) a magnetoelastic component described by the constraint model.[32] It is said that we are dealing with the pair-strain model of anisotropy. It seems that the *constraint model* is, however, more important for single-metal films where the theory of directional ordering can not predict any interesting effects. It was remarked by Ruske and Weber[34] that the same model can predict in certain circumstances a uniaxial anisotropy even in monocrystals.

Some improvements of the constraint model are to be noted. Although quantitative estimations are not given, Kneer and Zinn[35] have extended West's model to include the effects of the anisotropic deffects and impurities with a uniaxial component and dislocations (they have been included in the constraint stress field). The mechanism which could explain the appearence of anisotropic stresses was suggested by Prutton[30,36] to be the inverse effect to that described by Zener[37] in which directional stresses can cause directional ordering. Thus once again the constraint model is coupled with the directional ordering.[38] For permalloy films the most important mechanism seems to be the

[32] Permalloy films may be considered, although for some compositions no interesting results are expected; for a more detailed discussion see, for instance, Humphrey and Wilts.[33]

[33] C. H. Wilts and F. B. Humphrey, *J. Appl. Phys.* **39**, 1191 (1968).

[34] W. Ruske and P. Weber, *Phys. Status Solidi* **12**, 321 (1965).

[35] G. Kneer and W. Zinn, *Phys. Status Solidi* **17**, 323 (1966).

[36] M. Prutton, "Thin Ferromagnetic Films." Butterworths, London and Washington, D.C., 1964.

[37] C. Zener, *Phys. Rev.* **71**, 34 (1947).

[38] An excellent comment about these models and their role in the uniaxial anisotropy production for polycrystalline thin films (peculiarly Ni–Fe films) is given by Slonczewski.[39]

[39] J. C. Slonczewski, *IEEE Trans. Magn.* **4**, 15 (1968).

imperfection ordering as pointed by Soohoo.[40] As this mechanism works also for single metal films it might be considered as one of the most important sources of anisotropy in ferromagnetic thin films.

d. Magnetostatic Energy

For nonspherical crystals the long *range magnetic forces* produce an anisotropy which is determined by the *shape* of the specimen.

For constant strain the demagnetization self-energy contribution to the free energy F may be determined using a microscopic model. In our static model, the evaluation of the internal energy term is the first step. Since U and F coincide at $T = 0°K$, this contribution may be interpreted as a term in the free energy F at 0°K. A term of the same form, but with coefficients that are functions of temperature, T, may be introduced in the expression of F for arbitrary temperatures. Thus following Brown,[5,41] we may evaluate the demagnetizing energy of a ferromagnetic ellipsoid starting from the *dipolar energy*. We find for a cubic crystal

$$U/V = 2\pi(N_1M_1^2 + N_2M_2^2 + N_3M_3^2) - \tfrac{1}{2}(4\pi/3)\, M^2, \qquad (1.27)$$

where the second term describes the *Lorentz field* contribution to the internal energy. Note that this term does not depend on the direction of the magnetization vector **M** so that in a discussion of the anisotropy energy we may neglect it; N_1, N_2, N_3 are the demagnetizing factors of the ellipsoid.

For constant strain we shall define as *shape anisotropy* the function

$$f_{\text{shape}} \equiv F_{\text{shape}}/V = 2\pi \sum_\alpha N_\alpha M_\alpha^2, \qquad \sum_\alpha N_\alpha = 1. \qquad (1.28)$$

We note that for very thin films, which have two dimensions much greater than the third one, we may write

$$N_1 = 1, \qquad N_2 = N_3 = 0,$$

where the index 1 is related to the first coordination axis, normal to the film plane. Then, for such a film

$$f_{\text{shape}} = 2\pi M^2 \cos^2\theta, \qquad (1.29)$$

where θ is the angle between the magnetization vector and the film surface normal and M is the saturation magnetization which depends on tem-

[40] R. F. Soohoo, "Magnetic Thin Films." Harper and Row, New York, 1965.

[41] W. F. Brown, Jr., "Magnetostatic Principles in Ferromagnetism." North-Holland Publ., Amsterdam, 1962.

perature. It may be seen that this anisotropy makes the normal a hard axis of magnetization.[1,2]

2. Experimental Data on the Magnetic Energy of Single Crystals

As the anisotropy observations made on polycrystalline thin films were often and quite recently reviewed, e.g., by Malek and Schuppel,[42] Pugh,[43] Prutton,[30] and Wilts and Humphrey,[33] this section will be limited to anisotropy observations made on *monocrystalline films* only. Also it should be recognized that the study of anisotropy energy in connection with such complex systems as polycrystalline thin films requires first of all that it be established whether isolated crystallites possess the same anisotropy characteristics as bulk crystals. On the other hand, polycrystalline films give complicated magnetic properties which are not yet completely explained.

Single crystal films have been prepared mainly by epitaxial growth of the elements on heated single crystal substrates, the composition of which may be identical with that of the film or may differ from it. The first qualitative informations on the monocrystalline thin film anisotropy (iron films) were given by Unangst[44] and Elschner and Unangst.[45] They have found that the iron films grown on {100} rock salt plane, have two preferred perpendicular axes in the plane of the film. Later observations on iron, β-cobalt and nickel films which were generally deposited normal to the substrate and without a field present during deposition confirmed the observations of Elschner and Unangst. This may be understood if we remember that due to the shape anisotropy, the magnetization is generally forced to lie in the film plane. Then the cubic anisotropy energy in a {100} plane becomes

$$f = (K_1/4) \sin^2 \alpha \cos^2 \alpha, \tag{2.1}$$

which is biaxial, the two axes being energetically equivalent. Let us note that if the surface of the film is {110}, then the situation is somewhat more complicated[46] since the free energy is expressed as

$$f = -K_1 \cos 2\alpha - K_2 \cos 4\alpha, \tag{2.2}$$

which has two unequivalent minima at $\alpha = 0$ and $\alpha = \pi/2$.

[42] Z. Málek and W. Schüppel, *Phys. Status Solidi* **2**, 136 (1962).

[43] E. W. Pugh, *in* "Physics of Thin Films" (G. Hass, ed.), Vol. I. Academic Press, New York, 1963.

[44] D. Unangst, *Ann. Phys.* **7**, 280 (1961).

[45] B. Elschner and D. Unangst, *Z. Naturforsch.* **11a**, 98 (1956).

[46] L. V. Kirenskii and I. S. Edelman, *Fiz. Metal. Metalloved.* **18**, 340 (1964).

The magnetocrystalline anisotropy observed in the film plane would be characterized by the anisotropy constant K_1.

Aside from the observed anisotropy in the film plane, a perpendicular anisotropy was reported (see, among others, Chikazumi,[47] Kirenskii *et al*,[48] Mushailov *et al.*[49]) as well. It was mainly attributed to the magnetoelastic effects. We shall describe this anisotropy by the anisotropy constant $K_\perp$, defined as $K_\perp = 2\pi M^2 + K_1 + K_N$, where K_N is the anisotropy constant related to the magnetoelastic and other effects. In the absence of such anomalous effects the value of $K_\perp$ must equal $2\pi M^2 + K_1$. Taking account of the fact that K_1 is usually two orders of magnitude smaller than $2\pi M^2$ this anisotropy makes, in the simplest cases, a hard axis of the film surface normal.

Both polycrystalline and single-crystal films of magnetic materials are found to have a uniaxial anisotropy in the film plane, in addition to the magnetocrystalline effects.[50] In the case of single crystals, this anisotropy seems to be smaller than that of magnetocrystalline origin.[52] Nevertheless, this anisotropy is expected to be important for permalloy films.

Quantitative informations about the anisotropy of monocrystalline films were first reported by Boyd.[54] Many similar measurements were made after that.

As a general feature we may consider that if the films are removed from the substrate and their surfaces are not oxidized, their magnetocrystalline anisotropy constant K_1 has the same value as that of the corresponding bulk samples (some exceptions were however noted).

If still attached to the substrate the value of the constant K_1 is signi-

[47] S. Chikazumi, *J. Appl. Phys.* **32**, 81S (1961).

[48] L. V. Kirenskii, V. G. Pynko, G. P. Pynko, A. S. Komalov, N. I. Sivkov, G. I. Rusov, P. S. Galenov, M. A. Ovsyannikov, and S. G. Rusova, *Izv. Akad. Nauk SSSR* **31**, 716 (1967).

[49] E. S. Mushailov, V. G. Pynko, and G. P. Pynko, *in* "Physica Magnitnyh Plenok," p. 30. Irkutsk, 1968.

[50] Kirenskii *et al.*[48] have reported that many of their films (Fe, Ni, β-Co) present a uniaxial anisotropy in the film plane though the deposition was done at normal incidence and moreover no field was present during deposition. These findings seem to contradict earlier observations of Kirenskii *et al.*,[51] as well as those of other authors.

[51] L. V. Kirenskii, V. G. Pynko, R. V. Sukhanova, N. I. Sivkkov, G. P. Pynko, I. S. Edelman, A. S. Komalov, S. V. Kan, N. U. Synova, and A. G. Zveguinshev, *Phys. Metal. Metalloved.* **22**, 380 (1966).

[52] Anderson[53] has found that the induced uniaxial anisotropy of thin Ni films is of the same order of magnitude as the magnetocrystalline anisotropy. Moreover, this quantity is dependent on the film thickness and the pressure in the vacuum system during deposition.

[53] J. C. Anderson, *Proc. Phys. Soc.* **78**, 25 (1961).

[54] E. L. Boyd, *IBM J. Res. Develop.* **4**, 116 (1960).

ficantly different from the bulk constant. Moreover, it varies for the same film depending on the type of substrate on which the film was deposited. For Ni films it was reported[47,55] that K_1 is greater than the corresponding value of bulk Ni, whereas for Fe films the value of the magnetocrystalline constant K_1 is generally smaller than that of bulk Fe. It was found[47,49] that the perpendicular anisotropy contains in this case an important anomalous part K_N. Apart from magnetocrystalline and perpendicular anisotropies a uniaxial component in the film plane was generally observed.[48,53–55]

The explanation of *anomalous* values of the *magnetocrystalline anisotropy* constant K_1 for Ni films deposited on NaCl substrates was sketched both by Chikazumi[47] and Freedman.[28] It has been shown[28] that isotropic stresses of the magnitude expected from the difference in coefficients of thermal expansion of a NaCl substrate and a Ni film may account for the changes in the anisotropy constant K_1 (see Eq. (1.22)). Also it has been suggested[47] that the origin of the tensions acting in the film plane and which account for the changes of the anisotropy constant K_1 might be an epitaxial misfit between the substrate and the film. However, the anomalous part of the perpendicular anisotropy seems to be of different origin since the stress acting on the film, calculated from ΔK_1 (the anomalous part of K_1), is one order of magnitude too small and also gives the wrong sign.

It is not at all clear which are the origins of anomalous values of K_1 of other metals. Pynko *et al.*[55] have studied the influence of the uniaxial stresses on the magnetic properties of some films. It is to be noted that even uniaxial stresses may contribute to the biaxial anisotropy. Similar calculations to those of Freedman[28] show that this contribution may be expressed as follows

$$g_{\text{stress}} = -\sigma(h_3 + \tfrac{2}{3} h_4)\, \alpha_2{}^2\alpha_3{}^2 \tag{2.3}$$

if the uniaxial stress is considered to act along the third coordination axis and the magnetization is considered to lie in a plane perpendicular to it (see Eq. (1.20)). The reported observations[55] confirm this suggestion. If the films were compressed together with their substrate, important variations of the values of K_1, K_u where observed.

Without any doubt it may be stated that in thin films important tensions act, although the origins of these tensions are not completely understood. For a detailed discussion of this problem the reader is referred to the review papers of Malek and Schüppel[42] as well as to that

[55] G. P. Pynko, V. G. Pynko, and M. A. Ovsyannikov, *in* "Physica Magnitnyh Plenok," p. 291. Irkutsk, 1968.

of Hoffman.[56] These tensions may be isotropic and anisotropic as well (see, for instance, Prutton[36,57]).

The isotropic stress is characteristic of all evaporated films and may be made up of contributions due to several mechanisms. Among them we shall emphasize the differential contraction between the substrate and the film as the system cools from the deposition temperature to room temperature and the isotropic stress system related to the existence of a high number of uniformly distributed imperfections and vacancies trapped in a condensing film. We note that Boyd[54] has found that in his monocrystalline films there were imperfections with a density of roughly $10^{12}/cm^2$, uniformly distributed throughout the sample.

The mechanisms giving rise to anisotropic stresses are quite different. Directional ordering of imperfections and impurities in the presence of an external field or film contamination (for instance, Kirenski *et al.*[48] have considered that anisotropic stresses may appear due to water contamination of the NaCl substrates) may be considered. Supposing an inverse effect to that of Zener,[37] directional ordering could give rise to anisotropic stresses.

Thus generally, like MacDonald,[58] we shall consider that a *complex stress system*, which consists of isotropic plane tensions and anisotropic stresses, acts directly or indirectly upon a thin film. As proved in Section 1, these stresses have an important influence upon the values of K_1, $K_\perp$ and K_u. We may conclude that an important contribution to the explanation of the magnetic behavior of the monocrystalline thin films is given by the existence of such stress systems acting upon the specimens. However, this is not the single mechanism that is responsible for the magnetic anisotropy of the films.

One of the most important effects that determines the anisotropy in films of ferromagnetic alloys is the directional ordering of atom-like pairs. The induced anisotropy of Ni–Fe films was related by Smith[59] to the orientation of Fe–Fe pairs in a magnetic field, as was supposed in the original theory of Néel[13] and Taniguchi-Yamamoto.[14] We have already mentioned that the directional ordering of iron pairs alone is an insufficient description of the induced anisotropy even in Ni–Fe films (it is completely unsuitable for single-metal films). Therefore, the theory

[56] R. W. Hoffman, *in* "Physics of Thin Films" (G. Hass and R. E. Thun, eds.), Vol. 3. Academic Press, New York, 1966.

[57] M. Prutton, *in* "The Use of Thin Films in Physical Investigations" (J. C. Anderson, ed.). Academic Press, New York, 1966.

[58] J. R. MacDonald, *Phys. Rev.* **106**, 890 (1957).

[59] D. O. Smith, *J. Appl. Phys.* **32**, 70S (1961).

of the induced anisotropy had to be improved. Following Bozorth,[60] we may consider the directional ordering by lattice distributed vacancies and impurities, a mechanism which is belived to be among the most effective in producing uniaxial anisotropy in polycrystalline films too. The oxygen mechanism which we are going to discuss may be related to this ordering.

The possible effects of oxygen on giving rise to an antiferromagnetic oxide have been presented in Section 1 in connection with the exchange anisotropy. The interaction between the atoms at each side of a ferromagnetic–antiferromagnetic interface was proved to result in a *unidirectional* anisotropy. If, however, we suppose, as did Prutton,[30,36] that there are antiferromagnetic inclusions in the material a *uniaxial* anisotropy may appear. It was calculated that only 0.1 % of Ni0 in a Ni matrix could give rise to a uniaxial anisotropy of as much as 10^3 erg/cm^3. But the role of oxygen is not limited to the exchange anisotropy.

Heidenreich *et al.*[61] have shown that both bulk and thin film specimens of perminvar and permalloy contain oxygen faults in the {111} planes. They have observed that the heat treatment in a magnetic field is effective in orienting the faults if the concentration of oxygen is greater than 0.0014 %. The axes of individual faults are either in a {111} direction or in a direction neighboring that of the field. It was reported that the uniaxial anisotropy increases with increasing fault density. It may be supposed that the appearance of anisotropy is due either to an incomplete saturation of pseudodipolar interaction forces (which is dependent on the nonequivalence of the neighboring atoms—see Néel[62]) or to the shape anisotropy of fault aggregates (with an easy direction along the collective length of the faults).

Let us observe that this ordering may be effective in producing uniaxial anisotropy in monocrystalline films (not only permalloy films) and also gives rise to anisotropic stresses (and thus the magnetocrystalline anisotropy constant K_1 might be affected).

Many authors have tried to explain with this model appearence of the uniaxial anisotropy in their films. Thus Anderson[53] considered that the uniaxial anisotropy existing in his {100} Ni films[63] had to be related to

60 R. M. Bozorth, *Proc. Conf. Magn. Magn. Mater. Boston* p. 69 (1956).

61 R. D. Heidenreich, E. A. Nesbitt, and R. D. Burbank, *J. Appl. Phys.* **30**, 995 (1959).

62 L. Néel, *J. Appl. Phys.* **30**, 3S (1959).

63 We note that during deposition no field was applied. However, the mechanism was still working due to the appearance of spontaneous magnetization (the only role of the field consists in orienting the film magnetization in a certain direction) and then the subsequent deposition of metal may be regarded as taking place in a field created by this spontaneous magnetization.

the oxygen diffusion into a preferred direction (in this case, as expected, the ⟨110⟩ direction). The same explanation of the appearance of uniaxial anisotropy in thin Ni and β-Co films was advanced by Kirenski *et al.*,[51] Pynko *et al.*,[55] although other mechanisms are also invoked.

Whether or not this mechanism is operative even after the substrate is floating off, it remains to explain how it is possible for the magnetocrystalline anisotropy K_1 to attain the bulk value.

We shall finally mention another mechanism which was suggested as a possible source of uniaxial anisotropy in thin films and which is related to the substrate roughness. Kirenski *et al.*[48] and Pynko *et al.*[55] considered that the substrate irregularities could result in a shape anisotropy which may produce uniaxial anisotropy in the film plane. A simplified model of the substrate was proposed by Martinez[64] who showed that the demagnetizing field associated with such a substrate may introduce a uniaxial anisotropy. But this mechanism can not be the single one responsible for the uniaxial anisotropy since the temperature variation of K_1 could not be explained on this basis[55] alone. More realistic calculations and experiments in this range could, of course, establish the role played by the substrate in uniaxial anisotropy production.

At any rate, it is rather difficult to believe that there is only a single mechanism explaining the magnetic behavior of ferromagnetic thin films. Further experiments on monocrystalline thin films are expected to add new and important information on their fundamental peculiarities.

3. Microscopic Description

Thus far, we have centered our discussion on the various forms of energy which are encountered in analyzing, from a macroscopic viewpoint, the behavior of a ferromagnetic thin film. We shall now briefly examine the energy from an atomistic viewpoint, which is thought to be more useful in the understanding of what is going on physically.

The most important Hamiltonian term is the exchange Hamiltonian, but we shall not be too much concerned with it. We shall consider that this term is described by a Heisenberg-type Hamiltonian. The exchange interactions are evidently isotropic as they depend only on the angles between interacting spins, but are independent on the orientation of the sample magnetization relative to the crystal axes. The magnetic field equivalent to the exchange interaction is nearly 10^7 Oe. Also, we note that the exchange forces are short ranged, so that in many problems we may restrict ourselves to the so-called nearest-neighbors approxima-

[64] M. P. Martinez, *An. Real. Soc. Espan. Fis. Chim.* **A58**, 27 (1962).

tion. Sometimes two exchange integrals are used,[65,66] one describing the interactions of the spins in the same atomic layer, and the other describing the interaction between spins from different layers. However, the isotropic character of the exchange interactions is not destroyed, as this description has nothing to do with the orientation of the magnetization vector relative to the crystallographic axes.

On the other hand, we know that from a *macroscopic* point of view it is more difficult to orient the sample magnetization along certain directions which we have called hard axes as compared with the easy axes of magnetization. To orient the sample magnetization along the hard axes there is necessary a magnetic field which amounts to hundreds or, in particular cases, to thousands of oersteds. Therefore, in addition to the exchange coupling of spins in crystals, there occur certain other interactions which may be generally termed anisotropic interactions. The microscopic Hamiltonians which are responsible for the magnetic anisotropy should depend on the spin state and moreover they should reflect the atomic arrangement in the crystal. Perhaps the most plausible mechanism is the *interionic magnetic dipolar interaction*, described by the Hamiltonian term

$$\mathscr{H}_{\mathrm{d}} = \frac{1}{2} \sum_{i \neq j} (g\mu_{\mathbf{B}}{}^2/r_{ij}^3)\{\mathbf{S}_i\mathbf{S}_j - 3[(\mathbf{S}_i\mathbf{r}_{ij})(\mathbf{S}_j\mathbf{r}_{ij})/r_{ij}^2]\}, \tag{3.1}$$

where the notations are obvious.

But, as is well known, this equality is inadequate to describe the magnetic anisotropy in many crystals, being both too small and incapable of predicting magnetocrystalline anisotropy for bulk materials with cubic symmetry.[67] These considerations are not entirely valid for thin films. As it was pointed out by Néel[10] the dipolarlike coupling has to be relevant in the first approximation, even for cubic ferromagnetic thin films. These terms give rise to a term in the so-called "*surface anisotropy*," which we shall discuss later.

As it is generally recognized, the most important mechanism of anisotropy seems to be the coupling between the spin and the orbital motion of the electrons; the *spin-orbit coupling* makes the energy anisotropic since it makes the spin "well aware" of the dependence of the interatomic energy on how the orbital wave functions are oriented. By

[65] A. Corciovei and G. Ciobanu, *J. Phys. Chem. Solids* **26**, 1939 (1965).

[66] W. Brodkorb and W. Haubenreisser, *Phys. Status Solidi* **8**, K21 (1965).

[67] If, however, lower symmetries are considered, the dipolar or pseudodipolar interaction no longer vanishes in the first approximation, when all the spins are assumed nearly parallel; furthermore, even in cubic crystals the dipolarlike terms may contribute to the magnetocrystalline anisotropy if second-order effects are considered.

this coupling the spins "can see" the lattice. The microscopic energies arising from this source may be divided into two classes: the first class, which comprises the energies that are expressed in terms of the spin state of a single ion, and the second class, to which belong different couplings among the spins of two (or even more) ions. The interactions expressed in terms of the spin operator of one ion only are referred to as single ion anisotropy energy. This is an energy of an ion under the influence of the crystalline electric field to which the spin-orbit coupling contributes. The simplest terms are expressed as

$$C_x \sum_j (S_j^x)^2 + C_y \sum_j (S_j^y)^2 + C_z \sum_j (S_j^z)^2, \tag{3.2}$$

with a proper choice of the coordinate axes. This term may be important for $S \geqslant 1$. Such a term was considered for instance by Brodkorb[68] to describe the behavior of iron thin films. As far as the pair interactions are concerned, they may be described in the most general case by the *bilinear Hamiltonian*,

$$\mathscr{H}_{\mathrm{a}} = \sum_{i \neq j} \sum_{\alpha,\beta} P_{ij}^{\alpha\beta} S_i^{\alpha} S_j^{\beta}, \tag{3.3}$$

provided the magnitude of each spin is $\frac{1}{2}$. (In the case of $\frac{1}{2}$ spin, it is well known that any operator in the spin space can be expressed as a linear combination of S^x, S^y, S^z, and 1.) If the interaction is invariant with respect to the rotation around the axis joining the interacting ions, then the anisotropy energy is described by the "*pseudodipolar*" Hamiltonian first derived by Van Vleck,[69]

$$\mathscr{H}_{\mathrm{pd}} = \sum_{i \neq j} P_{ij}[\mathbf{S}_i \mathbf{S}_j - 3r_{ij}^{-2}(\mathbf{S}_i \mathbf{r}_{ij})(\mathbf{S}_j \mathbf{r}_{ij})], \tag{3.4}$$

where P_{ij} decreases very rapidly with increasing distance r_{ij} between ions.

Let us note that taking into account the magnetic interactions leading to anisotropy energy for a hexagonal Co crystal, Potapkov[70] has obtained the following expression

$$\mathscr{H}_{\mathrm{a}} = \sum_{i \neq j} P_{ij} S_i^x S_j^x \tag{3.5}$$

for the anisotropy Hamiltonian (the x axis is taken here along the c axis

[68] W. Brodkorb, *Phys. Status Solidi* **16**, 225 (1966).
[69] J. H. Van Vleck, *Phys. Rev.* **52**, 1195 (1937).
[70] N. A. Potapkov, *Dokl. Akad. Nauk SSSR* **144**, 297 (1962); **151**, 543 (1963).

of the crystal). In the case of cubic crystals[67] Keffer[71] suggested that the pseudodipolar Hamiltonian must be augmented by

$$\mathscr{H}'_{\text{pd}} = \sum_{i \neq j} P'_{ij} \sum_{\alpha} (S_i^{\alpha} S_j^{\alpha})(r_{ij}^{\alpha}/r_{ij})^2, \qquad \alpha = x, y, z. \tag{3.4'}$$

When the magnitude of each spin is larger than $\frac{1}{2}$, the bilinear interaction may be expanded in powers of spin operators, the first term of which is given by Eq. (3.3) which also can be used in this case. The next term will generally be a *quadrupole–quadrupole* interaction, which in principle is present only if $S \geqslant 1$. The quadrupole-quadrupole interaction

$$\mathscr{H}_{\text{q}} = \sum_{i \neq j} D_{ij}(\mathbf{r}_{ij}\mathbf{S}_i)^2 (\mathbf{r}_{ij}\mathbf{S}_j)^2 \tag{3.6}$$

can give rise to cubic anisotropy even in the first-order perturbation theory, whereas the pseudodipolar Hamiltonian contributes to it in the second order only.

These are the expressions with which we are dealing in describing the magnetocrystalline anisotropy (and that part of surface anisotropy related to it) of the thin ferromagnetic films, although they were derived to describe the magnetic behavior of bulk specimens. More detailed discussion about the physical significance and mathematical derivation of them may be found in the excellent expositions of Kanamori[72] and Bertaut.[73]

Although in the theory of ferromagnetic thin films the concept of "effective anisotropy field" was used in describing the volume anisotropy of the specimen we shall not discuss it here. It was pointed out[11,74] that it would be risky to approximate the actual anisotropy energy by it; its limitations were discussed by Corciovei.[74] However, it was stressed that it may be used provided certain precautions are taken.

We shall go further and discuss the significance of the dipolar interactions in the theory of ferromagnetic thin films. We have already seen that this term may contribute both to the surface anisotropy (see Néel[10]) and to the magnetocrystalline anisotropy if there is no cubic symmetry. The most important effect of the dipolar interactions is that connected

[71] F. Keffer, *in* "Handbuch der Physik" (S. Flügge, ed.), Vol. XVIII/2, p. 1. Springer Verlag, Berlin and New York, 1966.

[72] J. Kanamori, *in* "Magnetism" (G. T. Rado and H. Suhl, eds.), Vol. I. Academic Press, New York, 1963.

[73] E. F. Bertaut, *in* "Magnetism" (G. T. Rado and H. Suhl, eds.), Vol. III. Academic Press, New York, 1965.

[74] A. Corciovei, *IEEE Trans. Magn.* **MAG-4**, 6 (1968).

with their long range character. They are shape-dependent and thus give rise to shape anisotropy, which we have already discussed. It was proved that the shape anisotropy may be written as

$$f_{\text{shape}} = 2\pi \sum N_\alpha M_\alpha^2, \qquad \sum N_\alpha = 1, \quad N_\alpha \geqslant 0, \tag{3.7}$$

if the crystal is considered to be an ellipsoid and the local field contribution is neglected.

The shape of the external boundary of a sample, considered as a discrete medium, explicitly enters the thermodynamic properties[75] through the dipole sums

$$n_\alpha(i) = [V/(4\pi N_0)] \sum_{j(\neq i)} r_{ij}^{-3}(1 - 3(r_{ij}^\alpha/r_{ij})^2), \qquad \alpha = x, y, z, \tag{3.8}$$

where V is the volume of the crystal and N_0/V is the number of spins per unit volume. Considering only large ellipsoidal specimens, we may decompose the dipolar sums into a sum about an origin (i) and a remainder which is accounted for by the classical demagnetizing factors just introduced:

$$n_\alpha = n_\alpha^0 - N_\alpha^0 + N_\alpha , \tag{3.9}$$

where

$$n_\alpha^0 = V_p/(4\pi N_p) \sum_{j\in V_p(j\neq i)} r_{ij}^{-3}(1 - 3(r_{ij}^\alpha/r_{ij})^2) \tag{3.10}$$

is the lattice sum over a volume V_p close to the origin (i). N_α^0 is the demagnetizing factor for the surface enclosing this volume, and N_α the demagnetizing factor for the external boundary of the sample. Usually V_p is taken as a small sphere centered on the origin. Then we may arrive at the definition of the demagnetizing factors given by Van Kranendonk[76] and Van Vleck

$$N_\alpha = V/(4\pi N_0) \sum_{j\in V-V_p} r_{ij}^{-3}(1 - 3(r_{ij}^\alpha/r_{ij})^2) + \tfrac{1}{3}. \tag{3.11}$$

It was proved by Benson and Mills[77] that for thin films, the most important contribution to a dipolar sum is given by the atoms which are situated either in the same plane as the origin (i) or in the planes nearest to it. This was the condition that was used in the earlier calculations of Fraitova,[26] Brodkorb[68] or Corciovei and Vamanu.[78] Since the

[75] P. Levy, *Phys. Rev.* **170**, 595 (1968).
[76] J. Van Kranendonk and J. H. Van Vleck, *Rev. Mod. Phys.* **30**, 1 (1968).
[77] H. Benson and D. Mills, *Phys. Rev.* **178**, 839 (1969).
[78] A. Corciovei and D. Vamanu, *Rev. Roum. Phys.* **12**, 629 (1967).

microscopic approach has imposed the idea of nonuniform magnetization of the thin films, it is believed that no unique definition of demagnetizing factors can be given in the range of microscopic theories. There were, however, attempts[26,68] to give a definition of these factors such as would comply best with their classical values. It has been found[68] that for a bcc lattice the values of the new demagnetizing factors $\tilde{N}_\alpha$ are different from the classical values ($\tilde{N}_2 = \tilde{N}_3 = 0.040$, $\tilde{N}_1 = 0.920$). Calculations similar to those of Brodkorb carried out for sc lattices[78] seemed to reinforce the discrepancy between the values of the new and classical factors (thus it has been found that $\tilde{N}_2 = \tilde{N}_3 = 0.083$, $\tilde{N}_1 = 0.834$). But using the reliable results for sc lattices of Benson and Mills, it may be shown that the new demagnetizing factors are almost equal to the classical factors,[79] their values being $\tilde{N}_2 = \tilde{N}_3 = 0.002$ and $\tilde{N}_1 = 0.996$. Therefore, in many problems the values of the classical demagnetizing factors may be used.

Let us observe that the anisotropy energy density at zero temperature can be considered as an "intrinsic" or microscopic anisotropy energy. Therefore, we may take account of the dipolar interaction using the shape anisotropy energy written in the form

$$f_{\text{shape}} = 2\pi \sum_{\alpha} (N_\alpha - \tfrac{1}{3}) M_\alpha^2. \tag{3.12}$$

Thus Luttinger and Kittel,[80] Oguchi,[81] and Tyablikov[82] observed that for deriving the ferromagnetic resonance condition, it suffices to calculate the eigenvalues of the Hamiltonian,

$$\mathscr{H} = (2\pi/V)(g\mu_{\mathrm{B}})^2 \sum_{\alpha} \sum_{ij} n_\alpha S_i^\alpha S_j^\alpha, \tag{3.13}$$

rather then those of the much more complicated Hamiltonian employed by Holstein and Primakoff.[83] The correspondence formula in deriving Eq. (3.13) is obvious by

$$M_\alpha \to (g\mu_{\mathrm{B}}/V) \sum_{i} S_i^\alpha. \tag{3.14}$$

This procedure involves errors of unknown magnitude and it is presumed

[79] It seems that the difference between the reported calculations of demagnetizing factors is due to either an overestimation of the nearest-planes contribution[68] or an underestimation of the in-plane contribution.[78]

[80] J. M. Luttinger and C. Kittel, *Helv. Phys. Acta* **21**, 480 (1948).

[81] T. Oguchi, *Progr. Theor. Phys.* **17**, 659 (1957).

[82] S. V. Tyablikov, *Fiz. Tverd. Tela* **2**, 361 (1960).

[83] T. Holstein and H. Primakoff, *Phys. Rev.* **58**, 1098 (1940).

that they are negligible at low temperatures.[84] It may be that this Hamiltonian can be used to describe the shape-dependence of the thermodynamical quantities of a thin film at low enough temperatures provided that the film could be supposed in a state of nearly uniform magnetization.

One of the most difficult problems in the quantum description of the behavior of a ferromagnet is the finding of the most proper way of expressing the coupling between stresses and magnetization. This coupling plays an important role in the theory of ferromagnetic thin films. It is known that the tensor which performs this coupling is related to the *magnetostriction tensor* (see Eq. (1.17)). The quantum mechanical study of the magnetoelastic energy can start with the treatment of Callen and Callen[85,86]. However, it seems that there are other possibilities as well. Thus we may use the classical expressions (see Eq. (1.19)) and follow MacDonald,[23] Griffith,[25] or Fraitova[26] in deriving the concept of "*effective field*." By defining new *demagnetizing* factors analogous to the true factors involved in Eq. (1.28), we may proceed by giving a quantum-mechanical significance to this effective field. We think that we can go further and start from the very beginning with the classical expression (1.17) and then use arguments analogous to those of Kittel and Luttinger[80] or to those of Tyablikov.[82] It may be supposed that the quantum expression of the magnetoelastic energy is obtained if relation (3.14) is used. We have to ensure that the constants entering the theory are temperature independent.[87]

(Thus different situations which appear in the physics of thin films may be described. As we have already noted, the most tractable physical situations are related to the *planar* and *linear* stress systems. A quantum description of such systems may be easily done if the above considerations would be employed. The lack of quantitative studies of such problems prevents us from knowing the limits of the procedure we have proposed.)

Actually, we have tried thus far to give some physical motivations for different terms entering the thin film Hamiltonian. The theory tends to describe the actual thin film taking account of the particular circumstances like technology and aging. With a view approaching this ambitious aim, we try to see what is to be expected from simplest models.

[84] At the Curie temperature, these errors might become quite important.

[85] E. R. Callen and H. B. Callen, *Phys. Rev.* **129**, 578 (1963).

[86] E. R. Callen and H. B. Callen, *Phys. Rev.* **139A**, 455 (1965); E. R. Callen, *J. Appl. Phys.* **39**, 519 (1968).

[87] This is due to the fact that the classical expression is considered at $T = 0°K$. The temperature dependence of the constants may be obtained if pertinent statistical calculations are made.

The most widely used is the model of *planar sublattices* introduced by Néel[88] in treating bulk magnetism and subsequently adapted by Valenta[89–91] for thin ferromagnetic films. In the case of the thin ferromagnetic films the natural way of choosing Néel-like sublattices consists in taking them as crystallographic planes parallel to the film surface (see Fig. 1). Indirect experimental evidence of crystallographic planes

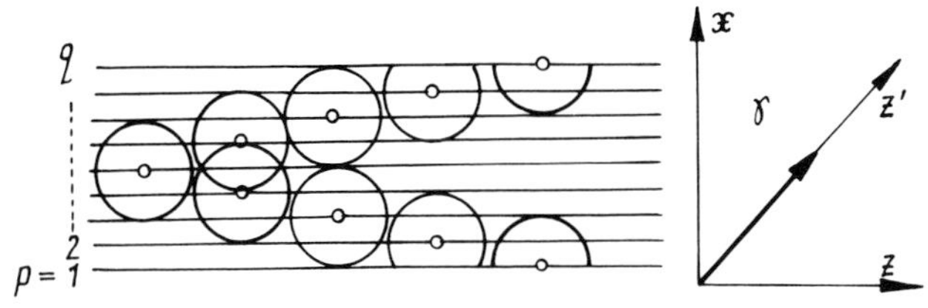

FIG. 1. The thin film in the sublattices model; γ is the magnetization direction. The *natural* surface defect is suggested by intersecting the first sphere of coordination with the film surfaces.

parallel to the surfaces, were found in X-ray diffraction experiments by Croce *et al.*[92] Using such a model it becomes possible to investigate a wide class of thin films, with various types of surface lattices, starting from a given type of bulk lattice. For instance, by cutting the three types of the cubic lattices (sc, fcc, bcc) along some preferential planes one finds thin specimens showing quadratic, tetragonal, rhomboedral, or hexagonal surface lattice (see Fig. 2). The only requirement which has to be satisfied by this model is that in the limiting case of the bulk body built by increasing the number of sublattices (monolayers) indefinitely, the thin film theory has to become identical with its bulk correspondent.

A first step in the theory consists, in distinguishing between the "*thin film system of coordinates*" and the "*local system of coordinates*" of each localized spin. In fact the axial symmetry of the film recommends a system of coordinates xyz with two axes, say y, z, in the plane of the film and x perpendicular to the film surface (Fig. 1). The spin operator and the unit vector of the quantization axis are in this system $S^{\alpha}_{\mathrm{f}p}$ and $\gamma^{\alpha}_{\mathrm{f}p}$, $\alpha = x, y, z$, respectively. Here $p = 1, 2, \ldots, q$ labels the planar

[88] L. Néel, *J. Phys.* **8**, 5, 241, 265 (1944).
[89] L. Valenta, *Czech. J. Phys.* **7**, 127, 136 (1957).
[90] L. Valenta, *Izv. Akad. Nauk SSSR, Ser. Fiz.* **21**, 879 (1957).
[91] L. Valenta, *Phys. Status Solidi* **2**, 112 (1962).
[92] P. Croce, G. Devant, et M. F. Verhaeghe, *Proc. Int. Symp. Basic Probl. Thin Film Phys.* (R. Niedermayer and H. Mayer, eds.), p. 381. Vandenhoeck & Ruprecht, Göttingen, 1966.

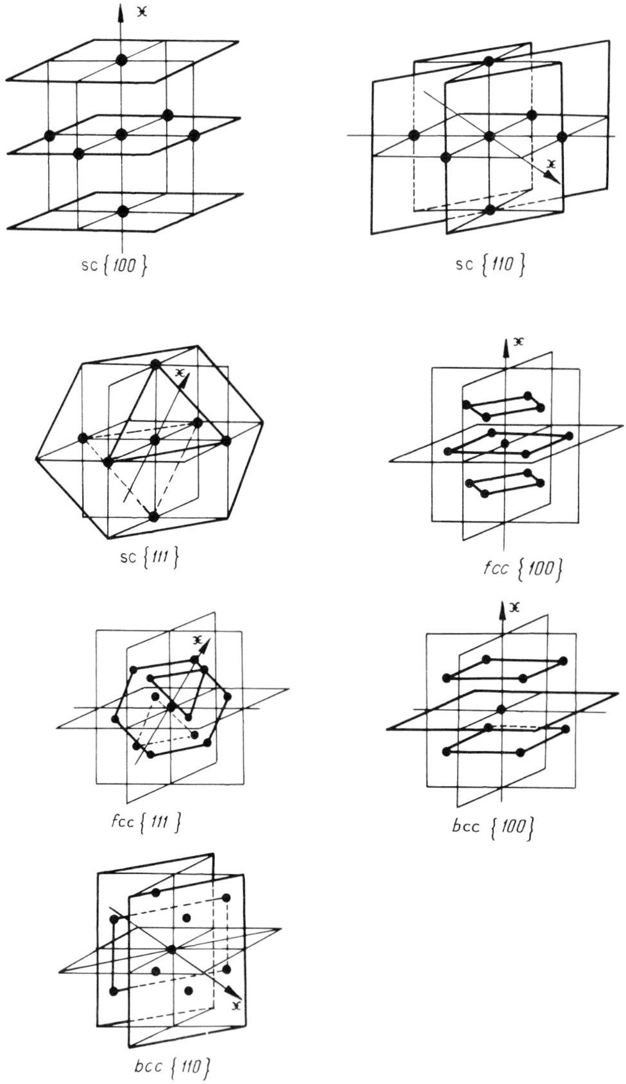

FIG. 2. Typical orientations of the film surfaces in cubic lattices. The x axis is normal to the film surfaces. Only first-order neighbors of the central atom are denoted.

sublattices (monolayers) and $\mathbf{f}$ is a lattice vector in the plane of the sublattice. On the other hand it is convenient to describe the states of the spin $\mathbf{f}p$ in a system $0x'y'z'$ with the $0z'$ axis along the unit vector $\gamma_{\mathrm{f}p}$ so that $S_{\mathrm{f}p}^{z'}$ is a diagonal operator.

The *canonical transform* relating the dynamical variables $S^{\alpha}_{\mathbf{f}p}$ and $S^{\alpha'}_{\mathbf{f}p}$ is the well-known rotation[93]

$$S^{\alpha}_{\mathbf{f}p} = \gamma^{\alpha}_{\mathbf{f}p} S^{z'}_{\mathbf{f}p} + \mathscr{A}^{\alpha}_{\mathbf{f}p} S^{+'}_{\mathbf{f}p} + \mathscr{A}^{\alpha *}_{\mathbf{f}p} S^{-'}_{\mathbf{f}p}, \tag{3.15}$$

where the coefficients $\mathscr{A}^{\alpha}_{\mathbf{f}p}$ and $\mathscr{A}^{\alpha *}_{\mathbf{f}p}$ obey the following rules

$$\begin{aligned} i\mathscr{A}_{\mathbf{f}p} &= \boldsymbol{\gamma}_{\mathbf{f}p} \times \mathscr{A}_{\mathbf{f}p}, &\quad \mathscr{A}_{\mathbf{f}p} \cdot \mathscr{A}^{*}_{\mathbf{f}p} &= \tfrac{1}{2}, \\ i\boldsymbol{\gamma}_{\mathbf{f}p} &= 2\mathscr{A}_{\mathbf{f}p} \times \mathscr{A}^{*}_{\mathbf{f}p}, &\quad \mathscr{A}_{\mathbf{f}p} \cdot \mathscr{A}_{\mathbf{f}p} &= 0, \quad \boldsymbol{\gamma}_{\mathbf{f}p} \cdot \mathscr{A}_{\mathbf{f}p} = 0. \end{aligned} \tag{3.16}$$

One defines as the trial ground state of our system the state of maximum projection $S_{\mathbf{f}p}$ of all the spins along their own quantization axes.[94] Accordingly any deviation from the fundamental state is described by the set of displacements $S_{\mathbf{f}p} - S^{z'}_{\mathbf{f}p}$. Following the usual commutation rules for $S^{\pm'}_{\mathbf{f}p} = S^{x'}_{\mathbf{f}p} \pm i\, S^{y'}_{\mathbf{f}p}$, namely

$$[S^{z'}_{\mathbf{f}p}, S^{\pm'}_{\mathbf{g}p'}] = \pm S^{\pm'}_{\mathbf{f}p}\, \delta_{\mathbf{f}p,\mathbf{g}p'},$$

where $\delta_{\mathbf{f}p,\mathbf{g}p'}$ is the Kronecker symbol, one sees that $S^{-'}_{\mathbf{f}p}$ increases the displacement $S_{\mathbf{f}p} - S^{z'}_{\mathbf{f}p}$ while $S^{+'}_{\mathbf{f}p}$ reduces it.

According to the Eq. (3.15), one may write

$$S^{\alpha}_{\mathbf{f}p} = \gamma^{\alpha}_{\mathbf{f}p} S_{\mathbf{f}p} + \delta S^{\alpha}_{\mathbf{f}p}, \tag{3.17}$$

where $\delta S^{\alpha}_{\mathbf{f}p}$ is a comprehensive form of the spin deviation from the fundamental state

$$\delta S^{\alpha}_{\mathbf{f}p} = \mathscr{A}^{\alpha}_{\mathbf{f}p} S^{+'}_{\mathbf{f}p} + \mathscr{A}^{\alpha *}_{\mathbf{f}p} S^{-'}_{\mathbf{f}p} - \gamma^{\alpha}_{\mathbf{f}p}(S_{\mathbf{f}p} - S^{z'}_{\mathbf{f}p}). \tag{3.18}$$

The thin-film energy may be presented now as a series expansion with respect to the deviations (3.18) as follows:

$$\begin{aligned} \mathscr{H}(S^{\alpha}_{\mathbf{f}p}) &= \mathscr{H}(\gamma^{\alpha}_{\mathbf{f}p} S_{\mathbf{f}p}) + \sum_{\alpha} \sum_{\mathbf{f}p} [\partial \mathscr{H} / \partial \gamma^{\alpha}_{\mathbf{f}p} S^{-1}_{\mathbf{f}p} \delta S^{\alpha}_{\mathbf{f}p}] \\ &\quad + \tfrac{1}{2} \sum_{\alpha,\beta} \sum_{\mathbf{f}p,\mathbf{g}p'} [\partial^2 \mathscr{H} / (\partial \gamma^{\alpha}_{\mathbf{f}p}\, \partial \gamma^{\beta}_{\mathbf{g}p'})\, S^{-1}_{\mathbf{f}p} S^{-1}_{\mathbf{g}p'}\, \delta S^{\alpha}_{\mathbf{f}p}\, \delta S^{\beta}_{\mathbf{g}p'}] + \cdots . \end{aligned} \tag{3.19}$$

The coefficients of this expansion depend on the composition of the model Hamiltonian $\mathscr{H}$, which ought to be built up by contributions of the types discussed above.

[93] S. V. Tyablikov, "Metody kvantovoi teorii magnetizma." Izd. "Nauka," Moscow, 1965.

[94] The Tyablikov's approach[93] covers satisfactorily some basic requirement of the involved problem of the ground state energy of magnetics, even for antiferromagnets and ferrites. It seems definite that it works for ferromagnetic systems, as described in this paper.

Since the mean value of $S^{\alpha}_{\mathbf{f}p}$ in the trial ground state is

$$\langle \cdots S_{\mathbf{f}p} \cdots | S^{\alpha}_{\mathbf{f}p} | \cdots S_{\mathbf{f}p} \cdots \rangle = \gamma^{\alpha}_{\mathbf{f}p} S_{\mathbf{f}p}\,, \tag{3.20}$$

it is clear that $\mathscr{H}(\gamma^{\alpha}_{\mathbf{f}p} S_{\mathbf{f}p}) = \mathscr{E}_0$ denotes the *mean energy* of the trial ground state. It is a function of the magnetic configuration, by the direction cosines $\gamma^{\alpha}_{\mathbf{f}p}$.

The set of directions $\boldsymbol{\gamma}_{\mathbf{f}p}$ leading to the *lowest mean energy* of the *ground state* may be found, as usual, by a system of equations involving the Lagrange factors $\lambda_{\mathbf{f}p}$:

$$\partial\mathscr{E}_0/\partial\gamma^{\alpha}_{\mathbf{f}p} - 2\lambda_{\mathbf{f}p} S_{\mathbf{f}p} \gamma^{\alpha}_{\mathbf{f}p} = 0, \qquad \sum_{\alpha} (\gamma^{\alpha}_{\mathbf{f}p})^2 = 1. \tag{3.21}$$

Thus the Hamiltonian takes the form

$$\begin{aligned} \mathscr{H}(S^{\alpha}_{\mathbf{f}p}) = {} & \mathscr{E}_0(\gamma^{\alpha}_{\mathbf{f}p} S_{\mathbf{f}p}) - 2 \sum_{\mathbf{f}p} \lambda_{\mathbf{f}p} n_{\mathbf{f}p} \\ & + \tfrac{1}{2} \sum_{\alpha,\beta} \sum_{\mathbf{f}p,\mathbf{g}p'} [\partial^2 \mathscr{H}/(\partial\gamma^{\alpha}_{\mathbf{f}p} \partial\gamma^{\beta}_{\mathbf{g}p'}) \\ & \times S^{-1}_{\mathbf{f}p} S^{-1}_{\mathbf{g}'p} (T^{\alpha}_{\mathbf{f}p} - \gamma^{\alpha}_{\mathbf{f}p} n_{\mathbf{f}p})(T^{\beta}_{\mathbf{g}p'} - \gamma^{\beta}_{\mathbf{g}p'} n_{\mathbf{g}p'})] + \cdots \end{aligned} \tag{3.22}$$

where $\gamma_{\mathbf{f}p}$ now corresponds to the actual magnetic configuration and the contributions in Eq. (3.22) are noted as

$$T^{\alpha}_{\mathbf{f}p} = \mathscr{A}^{\alpha}_{\mathbf{f}p} S^{+'}_{\mathbf{f}p} + \mathscr{A}^{\alpha *}_{\mathbf{f}p} S^{-'}_{\mathbf{f}p}\,, \qquad n_{\mathbf{f}p} = S_{\mathbf{f}p} - S^{z'}_{\mathbf{f}p}\,. \tag{3.23}$$

This standard expansion is a good starting point for the approaches reported here.

III. Domains and Ripple

The question of the magnetic configurations in a ferromagnetic thin film may be approached from two different points of view, corresponding respectively to *micromagnetics* and to *domain theory*. The background of these theories and their criticism may be found in the excellent papers of Brown,[5,95] Döring,[96] Shtrikman and Treves [97] for micromagnetics, and Kittel,[98] Kittel and Galt,[2] and Brown[5] for the domain theory.

[95] W. F. Brown, Jr., *J. Appl. Phys.* **30**, 62S (1959).

[96] W. Döring, *in* "Handbuch der Physik" (S. Flügge, ed.), Vol. XVIII/2, p. 341. Springer Verlag, Berlin and New York, 1966.

[97] S. Shtrikman and D. Treves, *in* "Magnetism" (G. T. Rado and H. Suhl, eds.), Vol. III, p. 395. Academic Press, New York, 1965.

[98] C. Kittel, *Rev. Mod. Phys.* **21**, 541 (1949).

Both approaches have been proved particularly successful in the study of the magnetization direction in thin ferromagnetic films, their success being naturally facilitated by the improvement of the observation methods of the magnetization directions. Many experimental facts and theoretical attempts in this realm may be found in pertinent monographs and review articles such as those of Prutton[36] and Soohoo[40] for domains and domain walls, of Middelhoek[99] for walls, of Lever[100] and Cohen[101] for the magnetization ripple.

4. Magnetization in Thin Single Domain Specimens

Let us start with the simplest model of thin films[102] as a micromagnet, the single domain film in a form of a flat ellipsoid of revolution with perfect homogeneity. Geometrically, such a specimen is described by the equation

$$x^2/a^2 + (y^2 + z^2)/b^2 = 1, \qquad a \ll b. \tag{4.1}$$

Physically it may be described satisfactorily by the following composition of the magnetic energy:

$$\mathscr{H} = \int d\mathbf{r} \left\{ f(M^2) - \tfrac{1}{2} \sum_{ij} \alpha_{ij} \mathbf{M}[\partial^2 \mathbf{M}/(\partial x_i \, \partial x_j)] - \tfrac{1}{2} \rho(\mathbf{M} \cdot \mathbf{n})^2 + 2\pi \mathbf{M} \hat{N} \mathbf{M} - \mathbf{M} \cdot \mathbf{H} \right\}. \tag{4.2}$$

The first term, of density $f(M^2)$ includes all contributions which depend only on the *square* of the magnetizetion. It is mainly of exchange and dipolar origin. The second term is the significant part of the exchange contribution describing the *inhomogeneity* of the magnetization.[103] The third term carries the contribution of an *anisotropy* of uniaxial type along the axis of unit vector **n**. The fourth term denotes the *demagnetizing*

[99] S. Middelhoek, *J. Appl. Phys.* **34**, 1054 (1963).

[100] K. D. Leaver, *Thin Solid Films* **2**, 149 (1968/1969).

[101] M. S. Cohen, *in* "Thin Film Phenomena" (K. L. Chopra, ed.), Chapter 10. McGraw-Hill, New York, 1969.

[102] This section follows closely A. I. Akhiezer, B. G. Bar'yakhtar, and S. V. Peletminskii, "Spinovye volny." Izd. "Nauka," Moscow, 1967.

[103] Since ideal single domain film of ellipsoidal shape does not admit static inhomogeneities of the magnetization, originating in domain structure or magnetization ripple, it is clear that in the present problem the inhomogeneities of the magnetization may be only of dynamic origin (natural or induced modes of precession waves, in the sense of Brown[95]).

energy of the specimen, introduced by the demagnetizing tensor $\hat{N}$. In our case the demagnetizing tensor is diagonal

$$N_1 = N_{xx} = N_{\perp}, \qquad N_2 = N_{yy} = N_3 = N_{zz} = N_{\parallel}, \tag{4.3}$$

the components $N_{\perp}$ and $N_{\parallel}$ being given by

$$N_{\perp} = [(1 + e^2)/e^3](e - \arctan e), \qquad N_{\parallel} = \tfrac{1}{2}(1 - N_{\perp}),$$

and

$$e = ((b/a)^2 - 1)^{1/2}. \tag{4.4}$$

Finally, the fifth term gives the energy of the specimen in an external field **H**.

The dynamics of the magnetization is described by the gyromagnetic equation,[104] whose simplest form is

$$\partial \mathbf{M}/\partial t = \gamma \mathbf{M} \times \mathbf{H}_{\text{eff}}, \tag{4.5}$$

where $\gamma = g\mu_B/\hbar$ is the gyromagnetic ratio. The effective field $\mathbf{H}_{\text{eff}}$ corresponding to the composition (4.2) of the energy is

$$\mathbf{H}_{\text{eff}} = (\mathbf{H} - 4\pi\hat{N}\mathbf{M}) + \sum_{ij} \alpha_{ij}[\partial^2\mathbf{M}/(\partial x_i\, \partial x_j)]$$

$$+ \rho\mathbf{n}(\mathbf{M} \cdot \mathbf{n})^2 - 2\mathbf{M}f'(M^2). \tag{4.6}$$

The term in parentheses is known as the "internal field" of the specimen. Equation (4.5) denotes a *precession* of the magnetization around a certain equilibrium direction. Due to the various kinds of coupling, the precession of the magnetization has a wavelike character throughout the sample.

In order to get the *equilibrium* direction of the magnetization one assumes the torque exerted by the effective field on the magnetization to vanish. Assuming the magnetization as *uniformly* distributed one finds in this manner the equation

$$\mathbf{M}_0 \times (\mathbf{H} - 4\pi\hat{N}\mathbf{M}_0 + \rho\mathbf{n}(\mathbf{M}_0 \cdot \mathbf{n})) = 0, \tag{4.7}$$

[104] The gyromagnetic form of the equation of movement of the magnetic moment originates in the quantum mechanical equation

$$i\hbar\, \partial M/\partial t = [\mathbf{M}, \mathscr{H}].$$

The effective field, which occurs as a functional derivative of the energy with respect to the magnetization, $\mathbf{H}_{\text{eff}} = -\delta\mathscr{H}/\delta\mathbf{M}$ appears when passing from the noncommutative operators $\mathbf{M}(\mathbf{r}) = g\mu_B \sum_f S^z\, \delta(\mathbf{r} - \mathbf{r}_f)$ to their macroscopic averages $\langle\mathbf{M}(\mathbf{r})\rangle$, i.e., to the mean values over a small volume centered on $\mathbf{r}$ of the statistical averages.

which generally permits many solutions. The physical solution corresponds to the effective minimum of the *energy density*

$$w = 2\pi \mathbf{M}_0 \hat{N} \mathbf{M}_0 - \tfrac{1}{2}\rho(\mathbf{M}_0 \cdot \mathbf{n})^2 - \mathbf{M}_0 \cdot \mathbf{H}. \tag{4.8}$$

To point out the main tendencies of the magnetization direction in a single domain specimen the following example may be helpful: *easy axis* and external field "*perpendicular* to the film" (along the small axis of the ellipsoid), i.e., $\mathbf{n} = (1, 0, 0)$ and $\mathbf{H} = (H, 0, 0)$. Now, it is easy to show, using Eqs. (4.7) and (4.8), that for values of the external perpendicular field exceeding a certain threshold, $|[4\pi(N_\perp - N_\parallel) - \rho]M_0|$, there is only one solution of Eq. (4.7), namely $M_0 = (M_0, 0, 0)$, i.e., the magnetization is "perpendicular" to the film. As the external field has values below the same threshold we have to compare two solutions of Eq. (4.7) leading to different values of the energy density (4.8), namely $\mathbf{M}_0 = (H/[4\pi(N_\perp - N_\parallel) - \rho], M_0^y, M_0^z)$, corresponding to the energy density w_{I}, and $\mathbf{M}_0 = (M_0, 0, 0)$, with the energy density w_{II}. Thus, when $w_{\mathrm{II}} < w_{\mathrm{I}}$, the magnetization is "perpendicular," while if $w_{\mathrm{I}} < w_{\mathrm{II}}$, the magnetization exhibits an oblique incidence, with a "perpendicular" component proportional to the external perpendicular field. It is easy to see that the first situation (perpendicular magnetization) occurs as the perpendicular uniaxial anisotropic constant exceeds a certain value depending on the shape anisotropy (see Section 1, d), namely,[105]

$$\rho > 4\pi(N_\perp - N_\parallel), \tag{4.9}$$

while the second situation (oblique incidence) occurs as

$$\rho < 4\pi(N_\perp - N_\parallel). \tag{4.10}$$

In this second situation, if the external field is absent, the magnetization lies in the film plane.

We can conclude that the shape anisotropy tends to keep magnetization in the plane of the film[106] while the uniaxial anisotropy perpendicular to the film opposes the shape anisotropy.

[105] Remember that $0 < N_\perp - N_\parallel \lesssim 1$.

[106] One gets a straightforward expression of this tendency taking in the initial equations (4.7) and (4.8), $\rho = 0$, $H = 0$. Then the following alternative appears:

$$\text{(I)} \quad \mathbf{M}_0 = (M_0\, 0, 0)$$

and

$$\text{(II)} \quad \mathbf{M}_0 = (0, M_0^y, M_0^z), \qquad (M_0^y)^2 + (M_0^z)^2 = M_0^2.$$

The actual solution, which minimizes the energy density obviously corresponds to magnetization lying in the plane of the film.

Note that, although the shape anisotropy keeps the magnetization in the plane, it is *not able to point a certain axis* in this *plane*. To fix the equilibrium direction of the magnetization in the film plane it is necessary for there to be also an *uniaxial anisotropy* along an axis lying in the plane of the film. In fact, by adding a contribution of the form $-\frac{1}{2}\rho_{\parallel}(\mathbf{M}_0 \cdot \mathbf{n}_{\parallel})^2$, where $\mathbf{n}_{\parallel}$ is a unit vector of components (0, 0, 1), in the energy density of the specimen, a detailed examination of the Eq. (4.7) leads to the following alternative: As $H > |[4\pi(N_{\perp} - N_{\parallel}) - (\rho_{\perp} - \rho_{\parallel})]M_0|$ we have $\mathbf{M}_0 = (M_0, 0, 0)$, i.e., the magnetization lies along the external, perpendicular field. As $H < |[4\pi(N_{\perp} - N_{\parallel}) - (\rho_{\perp} - \rho_{\parallel})]M_0|$ we have

(1) perpendicular magnetization $M_0 = (M_0, 0, 0)$, if the perpendicular anisotropy is strong enough:

$$\rho_{\perp} - \rho_{\parallel} > 4\pi(N_{\perp} - N_{\parallel}), \tag{4.11}$$

(2) oblique incidence in the xz-plane,

$$\mathbf{M}_0 = (H/[4\pi(N_{\perp} - N_{\parallel}) - (\rho_{\perp} - \rho_{\parallel})], 0, M_0^z)$$

if the parallel anisotropy is strong enough:

$$\rho_{\perp} - \rho_{\parallel} < 4\pi(N_{\perp} - N_{\parallel}), \tag{4.12}$$

(2′) magnetization along z axis, if, in addition to the condition (4.12), the external field vanishes.

It is also note worthy, that, for given values of the anisotropy constants $\rho_{\perp}$ and $\rho_{\parallel}$ the magnetization direction depends on the thickness of the film. In this case, one can use as a rude estimate, $e \simeq b/a$, so that we have approximately for the demagnetizing factors

$$N_{\perp} \approx 1 - (\pi a)/(2b), \qquad N_{\parallel} \approx (\pi a)/(4b).$$

In this manner the condition (4.12) becomes

$$a/b < [4\pi - (\rho_{\perp} - \rho_{\parallel})]/(3\pi^2) = (a/b)_c\,. \tag{4.13}$$

Assuming the absence of the external field, this result means that thin films of sufficiently small thickness exhibit a "parallel" magnetization, while films with a thickness such that $a/b > (a/b)_c$ exhibit a "perpendicular" magnetization.

Let us now look at the *dynamics* of the magnetization in our model. According to the gyromagnetic equation (4.5) the magnetization vector performs a *precession* around an equilibrium direction. This precession introduces *local dynamic inhomogeneities* of the magnetization, which have a wavelike character throughout the sample[107] and are coupled mainly by exchange forces.

One takes the magnetization to be of the form

$$\mathbf{M} = \mathbf{M}_0 + \mathbf{m} \tag{4.14}$$

and assumes $|\mathbf{m}| \ll |\mathbf{M}_0|$, $\mathbf{m} \cdot \mathbf{M}_0 = 0$. Neglecting higher-order terms in (4.5), the so-called "linearized" gyromagnetic equation is obtained:

$$\partial\mathbf{M}/\partial t = \gamma\mathbf{M}_0 \times \Big[\mathbf{h} + \sum \alpha_{ij}[\partial^2\mathbf{m}/(\partial x_i\, \partial x_j)] + \rho\mathbf{n}(\mathbf{m}\cdot\mathbf{n}) - (1/M_0^{\,2})[\mathbf{M}_0 \cdot \mathbf{H}_0^{(i)} + \rho(\mathbf{M}_0 \cdot \mathbf{n})^2]\mathbf{m}\Big]. \tag{4.15}$$

Here $\mathbf{H}_0^{(i)} = \mathbf{H} - 4\pi\hat{N}\mathbf{M}_0$ is the equilibrium value of the internal field $\mathbf{H}^{(i)} = \mathbf{H}_0^{(i)} + \mathbf{h} = \mathbf{H} - 4\pi\hat{N}\mathbf{M}$, where $\mathbf{h} = -4\pi\hat{N}\mathbf{m}$ denotes displacement of the internal field from its equilibrium value due to the magnetization displacements $\mathbf{m}$. In the following, we assume the simplest form of the exchange term namely $\alpha_{ij} = \alpha\delta_{ij}$, which corresponds to crystals of cubic symmetry.

The question of the boundary conditions required by the second-order differential equation (4.15) was widely discussed in the literature (see, e.g., Rado and Weertman[108]). The most manageable in occur the following three cases:

(*i*) *"free" surfaces*, requiring the vanishing of the normal derivatives of the magnetization displacement on the surfaces,

$$\partial\mathbf{m}/\partial x|_{x=0} = 0, \qquad \partial\mathbf{m}/\partial x|_{x=L} = 0; \tag{4.16}$$

(*ii*) *completely "pinned"* surfaces, meaning the vanishing of the displacement itself on the surfaces

$$\mathbf{m}(x = 0) = 0, \qquad \mathbf{m}(x = L) = 0; \tag{4.17}$$

(*iii*) *one "pinned" / one "free"* surface—a mixture of the above cases, namely,

$$\partial\mathbf{m}/\partial x|_{x=0} = 0, \qquad \mathbf{m}(x = L) = 0. \tag{4.18}$$

[107] For a detailed discussion about the connection between the static and dynamic problem in micromagnetism see Brown.[95]

[108] G. T. Rado and I. Weertman, *Phys. Rev.* **94**, 1386 (1954).

Let us focus our attention on a very thin single domain film which may be imagined as practically infinite in extension. Its essential parameters are: the *thickness* L the demagnetizing tensor $\hat{N}$ of the approximate components $N_\perp \simeq 1$, $N_\parallel \simeq 0$, the perpendicular uniaxial anisotropy of constant $\rho_\perp$ and the parallel uniaxial anisotropy of constant $\rho_\parallel$.

Looking for the natural modes of precession of the spontaneous magnetization in the sense of Brown[95] we assume the absence of any external field. In this case, according to the previous discussion, the equilibrium magnetization occurs either perpendicular ($\rho_\perp - \rho_\parallel > 4\pi$) or parallel ($\rho_\perp - \rho_\parallel < 4\pi$) to the film. In both cases, the small displacements of the magnetization are superpositions of normal modes of the form

$$\mathbf{m}(\mathbf{r}, t) = m(\mathbf{r}) \exp[i(E/\hbar)t]. \tag{4.19}$$

We may expect that, due to the presence of the surfaces, the propagation of the precession in thin films would have a *running* character in the plane but a *standing* character in a direction perpendicular to the plane of the film. Accordingly, the following form of the spatial part $\mathbf{m}(\mathbf{r})$ is suggested:

$$\mathbf{m}(\mathbf{r}) = \mathscr{M}(x) \exp[i(\nu_y y + \nu_z z)], \tag{4.20}$$

where $\boldsymbol{\nu} = (\nu_y, \nu_z)$ is the propagation vector of a wave running in the plane of the film, obeying the usual periodic boundary conditions.

Denoting generally by $u(x)$ and $v(x)$ the two components of the amplitude $\mathscr{M}(x)$ in the plane perpendicular to $\mathbf{M}_0$ we have now equations of the form

$$d^2u/dx^2 = au + bv, \qquad d^2v/dx^2 = cu + dv, \tag{4.21}$$

where the concrete values of the coefficients a, b, c, d depend on the magnetization direction.

These equations must be solved with appropriate boundary conditions. Assuming for instance $ad - bc \neq 0$ and $b \neq 0$ the results may be summarized as follows.

(*i*) *Film with* free *surfaces*. The boundary conditions (4.16) give $u'(0) = u'(L) = 0, v'(0) = v'(L) = 0$. The components of the amplitude $\mathscr{M}(x)$ are found to be

$$u(x) = -\mathscr{N} b/(s^2 + a) \cos(sx), \qquad v(x) = \mathscr{N} \cos(sx) \tag{4.22}$$

where $\mathscr{N}$ is a certain normalizing constant. The momentum s of the standing wave (4.22) has the spectrum

$$s_\tau = \pi\tau/L, \qquad \tau = 0, 1, \ldots . \tag{4.23}$$

The value $s = 0$ corresponds to a mode of constant amplitude along the normal to the film plane.

(*ii*) *Film with completely* pinned *surfaces*. The boundary conditions (4.17) require now the vanishing of the amplitude itself on the surfaces $u(0) = u(L) = 0$, $v(0) = v(L) = 0$. The amplitudes are

$$u(x) = -\mathcal{N}b/(s^2 + a)\sin(sx), \qquad v(x) = \mathcal{N}\sin(sx), \tag{4.24}$$

where the spectrum of the momentum s is given by

$$s_\tau = \pi\tau/L, \qquad \tau = 1, 2, 3,\ldots . \tag{4.25}$$

This denotes standing waves with nodes on both surfaces. Note that the momentum never vanishes in the case of pinned surfaces, i.e., the amplitudes depend essentially on the perpendicular coordinate x.

(*iii*) *The mixed case of a film with* one free/one pinned *surface*. Assuming the surface $x = 0$ as free and the surface $x = L$ as pinned, the boundary conditions (4.18) become $u'(0) = v'(0) = 0$, $u(L) = v(L) = 0$. Accordingly, the amplitude is given by

$$u(x) = -\mathcal{N}b/(s^2 + a)\cos(sx), \qquad v(x) = \mathcal{N}\cos(sx). \tag{4.26}$$

It is only apparently identical with the amplitude of the wave in a thin film with both surfaces free, since, in this case, the momentum spectrum is given by

$$s_\tau = (2\tau + 1)\,\pi/(2L), \qquad \tau = 0, 1, 2,\ldots, \tag{4.27}$$

which denotes standing waves with a node on the pinned surface.

In all cases described above the energy of the precession wave appearing in (4.19) may be found in a form of a dispersion law

$$s^2 = \tfrac{1}{2}\{[(a - d)^2 + 4bc]^{1/2} - (a + d)\}. \tag{4.28}$$

Let us now consider the peculiarities of the two main cases, of ***perpendicular*** and ***parallel*** magnetization. In both cases it is easy to find the concrete form of the coefficients a, b, c, d of Eq. (4.21).

In this case of perpendicular magnetization ($\rho_\perp - \rho_\parallel > 4\pi$), the dispersion law is found as

$$E_k = g\mu_B M_0\alpha\{[k^2 + (\rho_\perp - 4\pi)/\alpha][k^2 + (\rho_\perp - \rho_\parallel - 4\pi)/\alpha]\}^{1/2}, \tag{4.29}$$

where k measures the momentum of the precession wave,

$$k^2 = \nu^2 + s^2$$

and s has a spectrum which follows from the boundary conditions, as shown above. We emphasize that in the case of perpendicular magnetization, the dispersion law, exhibits always a gap

$$E_0 = g\mu_B M_0\{(\rho_\perp - 4\pi)(\rho_\perp - \rho_\parallel - 4\pi)\}^{1/2}, \tag{4.30}$$

which increases with the increasing perpendicular anisotropy and is diminished by the demagnetizing effect and, eventually, by the parallel anisotropy. The fact that even the excitation of the mode of momentum $k = 0$ requires a certain energy E_0 corresponds to the existence of an equilibrium direction of the magnetization, stressed by the perpendicular anisotropy and securing the stability of the ferromagnetic phase.

Let us compare this situation with those appearing when the magnetization lies in the film plane ("parallel" magnetization). The dispersion law is

$$Ek = g\mu_B M_0\alpha\{[k^2 + (\rho_\parallel/\alpha)][k^2 + (4\pi - (\rho_\perp - \rho_\parallel))/\alpha]\}^{1/2} \tag{4.31}$$

One sees that in this case the energy gap

$$E_0 = g\mu_B M_0\{\rho_\parallel[4\pi - (\rho_\perp - \rho_\parallel)]\}^{1/2} \tag{4.32}$$

results only from the parallel anisotropy: as the parallel anisotropy vanishes the gap itself vanishes and we must imagine that the ferromagnetic phase of a film with free surfaces disappears.

Going further let us write down the form of the precession wave amplitudes $\mathscr{M}_k(x)$. In the case of *perpendicular* magnetization we have

$$\begin{aligned} u_k(x) &= \mathscr{M}_k^y(x) \\ &= -i\mathscr{N}\{[k^2 + ((\rho_\perp - \rho_\parallel) - 4\pi)/\alpha]/[k^2 + (\rho_\perp - 4\pi)/\alpha]\}^{1/2}\, w_k(x), \\ v_k(x) &= \mathscr{M}_k^z(x) \\ &= \mathscr{N} w_k(x), \end{aligned} \tag{4.33}$$

where $w_k(x)$ is a function of the type sine or cosine, according to the boundary conditions (see Eqs. (4.22), (4.24), and (4.26)). One sees that as long as the parallel anisotropy exists ($\rho_\parallel \neq 0$), the precession wave occurs elliptically polarized in the yz-plane, meaning that during the precession $|\mathscr{M}_k^y(x)| < |\mathscr{M}_k^z(x)|$. As the parallel anisotropy vanishes the precession wave becomes circularly polarized.

In the case of *parallel* magnetization along the z-axis we have

$$\begin{aligned} u_k(x) &= \mathscr{M}_k^x(x) \\ &= -i\mathscr{N}\{[k^2 + (\rho_\parallel/\alpha)]/[k^2 + (4\pi - (\rho_\perp - \rho_\parallel))/\alpha]\}^{1/2}\, w_k(x) \\ v_k(x) &= \mathscr{M}_k^y(x) \\ &= \mathscr{N} w_k(x). \end{aligned} \tag{4.34}$$

Here, even when the perpendicular anisotropy is neglected, the precession waves occur elliptically polarized in the xy-plane due to the demagnetizing effect. Intuitively, we may say that the precession of the magnetization around an axis lying in the plane of the film requires a periodic demagnetizing effort which increases the energy each time as the magnetization comes out of the film plane (we have indeed $|\mathscr{M}_k^x| < |\mathscr{M}_k^y|$).

5. Domain Structure

As early as 1946, Kittel[1] showed that the domain structure of a thin ferromagnetic specimen depends essentially upon its *thickness*. In order to illustrate this, he chose an easy axis perpendicular to the film plane. Subsequent experimental facts, accumulated especially during the last ten years, revealed convincingly that the distribution of the anisotropy field is by far the more complicated, so that at present a more complete picture of the influence of the anisotropy field on the domain structure can be drawn. A combination of the shape anisotropy with magnetocrystalline and stress anisotropies may result in one of the following typical situations: thin monocrystalline film with *biaxial anisotropy* field in its plane, thin film with *uniaxial anisotropy* situated in the plane, and thin film with *uniaxial anisotropy* perpendicular to its plane.[109]

The first case of *biaxial anisotropy* in the plane is obtained if a monocrystalline film which exhibits a cubic structure (e.g., Fe) is grown on a carefully prepared and oriented substrate. Results obtained using Bitter powder and Lorentz microscopy techniques. are reported by Sato *et al.*,[110–112] Gondo,[113] Gondo and Funatogawa,[114] Sato *et al.*[112] summarized the following features of the observed domain structures:

(1) As the thickness of the film decreases, the domain structure with magnetizations in the film plane becomes smaller in size and more sensitive to the imperfections of the substrate.

(2) Closure domains over the entire film are usually not formed; however, they can be obtained by careful demagnetization. In such a case a *checkerboard* domain pattern is obtained.

[109] The two last categories include monocrystalline and polycrystalline films as well. It must be remarked that in the polycrystalline case, it is not important what anisotropy individual crystallites exhibit since on the macroscopic scale of interest here, anisotropies other than uniaxial ones average out.

[110] H. Sato, R. S. Toth, and R. W. Astrue, *J. Appl. Phys.* **33**, 1113 (1962).

[111] H. Sato, R. S. Toth, and R. W. Astrue, *J. Appl. Phys.* **34**, 1064 (1963).

[112] H. Sato, R. S. Toth, and R. W. Astrue, *in* "Single Crystal Films" (M. H. Francombe and H. Sato, eds.), p. 395. Pergamon Press, Oxford, 1964.

[113] Y. Gondo, *J. Phys. Soc. Jap.* **17**, 1129 (1962).

[114] Y. Gondo and Z. Funatogawa, *J. Phys. Soc. Japan Suppl. BI* **17**, 621 (1962).

(3) 90°-walls are more frequent compared to the 180°-walls; they exhibit a fine structure which is very sensitive to imperfections.

A film with *uniaxial* anisotropy in its plane is represented typically by a permalloy (80 Ni–20 Fe) film. In this case, the study of the domain structure was incidental in a study of memory applications. The main experimental results obtained in this field may be summarized as follows. The films of usual thicknesses (up to 10000 Å) do not exhibit walls parallel to the film plane. Closure domains or open structure are present, depending on the anisotropy intensity.[115] If the film is a monodomain, then edge domains of irregular shape appear, in which the magnetization direction is inverted so that the demagnetizing energy of the specimen is reduced. These edge domains play a major role in the magnetization reversal processes. The main experimental results as well as simple theoretical models are reviewed in the monographs of Prutton[36] and Soohoo.[40]

Let us now consider thin films with *uniaxial perpendicular anisotropy*. A series of theoretical investigations due to Kittel,[1] Málek and Kamberský,[116] Kaczér *et al.*,[117] Silcox[118] and Jacubovics[119] lead to the prediction that the domain structure in these specimens depends upon the ratio $p = K_\perp'/(2\pi M^2)$ where $K_\perp'$ is the phenomenological constant of the anisotropy energy ($K_\perp' = K_1 + K_N$; see Section 2).

For $p > 1$ (e.g., for films MnBi,[120] $p \simeq 4$, and for films of magnetoplumbite,[121] $p \simeq 3.5$) Málek and Kamberský[116] showed that below a certain thickness (100 Å for MnBi) the film is a monodomain with the magnetization perpendicular to the film plane (Fig. 3a). For greater thicknesses the lowest free energy state is that of domain strips magnetized alternatively up and down perpendicular to the film surfaces (open structure, Fig. 3c). The case $p < 1$ corresponds to that imagined by Kittel[1] (for Co thin foils $p \simeq 0.36$) and afterwards was treated more accurately by Kaczér *et al.*[117,122] For thicknesses less than a certain threshold the sample is a monodomain, the magnetization lying in the film plane (Fig. 3b).

[115] K. Kuwahara, T. Goto, A. Nishimura, and Y. Ozaki, *J. Appl. Phys.* **35**, 820 (1964).

[116] Z. Málek and V. Kamberský, *Czech. J. Phys.* **8**, 416 (1958).

[117] J. Kaczér, M. Zelený, and P. Šuda, *Czech. J. Phys.* **B13**, 579 (1963).

[118] J. Silcox, *Phil. Mag.* 8, 7, 1395 (1963).

[119] J. P. Jacubovics, *Phil. Mag.* **14**, 881 (1966).

[120] H. J. Williams, R. C. Sherwood, F. G. Foster, and E. M. Kelley, *J. Appl. Phys.* **28**, 1181 (1957).

[121] J. Kaczér and R. Gemperle, *Czech. J. Phys.* **B10**, 505 (1960).

[122] We ignore the possibility of existence of a certain metastable structure with perpendicular magnetization, which may exist as p is less but approaching unity (see Kaczér)

For greater thicknesses, the domain structure may be an open (Fig. 3c) or closed (Fig. 3d) structure.[123]

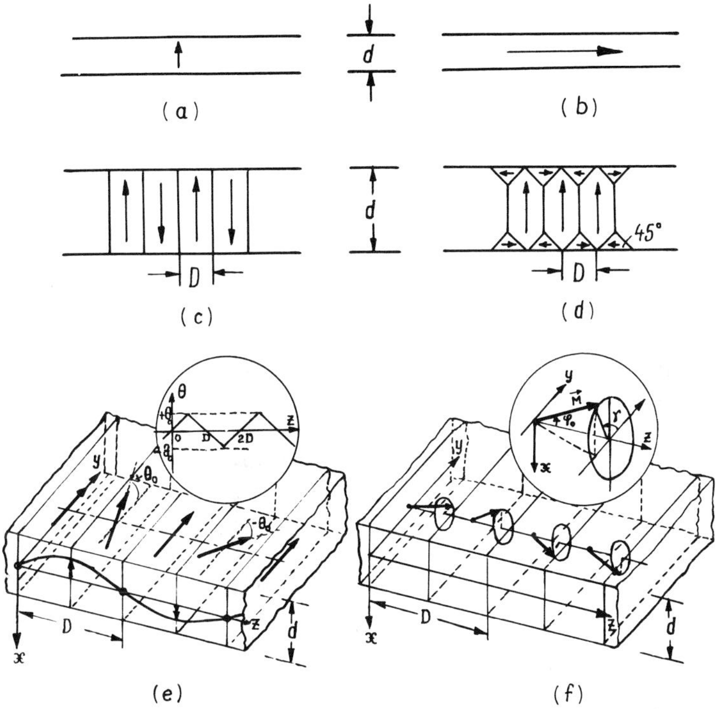

FIG. 3. Types of domain structures existing in thin films which exhibit uniaxial anisotropy, perpendicular to the film plane: (a) monodomain magnetized perpendicular to the film plane; (b) monodomain magnetized in the film plane; (c) open structure; (d) closure structure; (e) longitudinal stripe domain structure; (f) transverse stripe domain structure. In Fig. 3e the curve in the (x, z) plane shows the variation of the normal component of the magnetization along the z axis. The inset of the Fig. 3e illustrates the "saw-toothed" model for the variation of the normal component of the magnetization along the z axis. The inset of Fig. 3f explains the meaning of the angles φ_0 and γ.

In Kittel's approximation[1] an abrupt transition from the monodomain to the multidomain structure is assumed, but later Kaczér *et al.*[117] predicted the existence of a transition structure, with the magnetization oscillating out of the film plane. Such a structure was indeed observed

[123] Silcox[118] and Jacubovics[119] revealed the possibility that an open, or closure structure inclined from the normal easy axis does exist.

experimentally[124] and was called *stripe domain* structure[125,126] due to its appearance in the Bitter patterns[124,125] and in electron microscope micrographs.[127] (Earlier workers observed *mottled* thin films. Due to an insufficient resolution, the Bitter patterns presented a spotted or a mottled appearance.[128]) Well-resolved stripe domains were first observed on Ni rich permalloy films (Ni content larger than 81 %) which exhibit negative magnetostriction;[124–127] subsequently, careful experiments proved their existence in other films (e.g., Fujiwara *et al.*[129] for Fe films, Jacubovics[119] for Co foils, De Blois[130] for Ni single crystal platelets).

More detailed investigations due to Ferrier and Puchalska[131,132] and Baltz[133] showed that in obliquely evaporated Ni rich permalloy films two different configurations of magnetization distribution in stripe domains may exist: (i) a *longitudinal stripe domain structure* in which the normal component of the magnetization oscillates alternatively up and down, while the tangential one parallel to the in-plane component remains parallel to a given direction (the stripes are parallel to this direction) (Fig. 3e); and (ii) a *transverse stripe domain structure* in which the magnetization has a constant component in the direction of the mean magnetization while the other component spirals as it passes through stripe domains (Fig. 3e) so that the stripes are perpendicular to the easy axis in the plane.

Originally[125,134] the stripe domain structure was attributed to the perpendicular anisotropy caused by magnetoelastic effects (i.e., either

[124] R. J. Spain, *Appl. Phys. Lett.* **3**, 208 (1963).
[125] N. Saito, H. Fujiwara, and Y. Sugita, *J. Phys. Soc. Japan* **19**, 1116 (1964).
[126] The *stripes* are a specific magnetic structure which appears in thin films for a definite range of thickness. The magnetization direction varies continuously, almost periodically, everywhere in the sample but with great amplitudes, in contrast with another quasiperiodical phenomenon, *magnetization ripple* (see Sec. 6), which denotes the small variations of the magnetization direction *inside a monodomain.* Apart from the common quasiperiodical variation of the magnetization, the two phenomena have no common incidences, their origins being different. Thus, the appearance of stripe domains is purely the effect of the geometry of the specimen, for a certain *macroscopic* distribution of the anisotropy field, whereas the magnetization ripple is a phenomenon which appears at *any* thickness, its appearance being conditioned by the existence of *microscopic* variations of the anisotropy field inside the specimen.
[127] T. Koikeda, K. Suzuki, and S. Chikazumi, *Appl. Phys. Lett.* **4**, 160 (1964).
[128] E. E. Huber, Jr. and D. O. Smith, *J. Appl. Phys.* **30**, 267S (1959).
[129] H. Fujiwara, Y. Sugita, and N. Saito, *J. Phys. Soc. Jap.* **19**, 782 (1964).
[130] R. W. DeBlois, *J. Appl. Phys.* **36**, 1647 (1965).
[131] I. B. Puchalska and R. P. Ferrier, *Thin Solid Films* **1**, 437 (1967/1968).
[132] R. P. Ferrier and I. B. Puchalska, *Phys. Status Solidi* **28**, 335 (1968).
[133] A. Baltz, *J. Appl. Phys.* **37**, 1485 (1966).
[134] R. J. Spain, *Appl. Phys. Lett.* **6**, 8 (1965).

to tensions or compressions in films, see Sec. 1). Subsequent investigations[131–133,135] also revealed the importance of the angle of incidence effects[136] in films obtained by evaporation in vacuum. Different stripe domain structures at various angles of incidence were observed by Lorentz microscopy.[131,132] At small oblique incidences of the vapor beams, the stripe structure was observed to correspond to the longitudinal one, while the stripe domains obtained at larger angles of incidence show a transverse structure.

The theory of the stripe structures obtained by standard methods may be found in the paper of Saito *et al.*[125] for the longitudinal structure and in the paper of Sukiennicki[137] for the transverse structure. These treatments assume a film of uniform thickness d, a uniaxial anisotropy in the plane determining the mean direction of the magnetization and a normal component which cause the magnetization to deviate from the in-plane direction. Throughout this section, the modulus of magnetization is assumed to be constant in the sample.

For the longitudinal structure the coordinate system is shown in Fig. 3e. Assuming a saw-toothed dependence of the angle which describes the magnetization deflection,

$$\theta = (-1)^n \, 2\theta_0(2z - n\lambda)/\lambda, \qquad n = 0, 1, 2, \ldots, \tag{5.1}$$

where θ_0 denotes the maximum deflection angle, and λ the wavelength of the stripe structure. The free energy density, f, of the thin film, which includes the exchange, the perpendicular anisotropy, and the magnetostatic contributions may be calculated. By minimizing it with respect to λ, the wavelength of the longitudinal structure is found[125] to be

$$\lambda = 4(Ad/M^2)^{1/3}\, \theta_0^{2/3}\{(\cos \theta_0) \times [((\pi/2) - \theta_0)^{-1} - ((\pi/2) + \theta_0)^{-1}]\}^{-2/3}, \tag{5.2}$$

where A denotes the exchange constant. One sees that λ increases monotonically with θ_0.

The longitudinal stripe structure is stable if the usual conditions $(\partial f/\partial\theta_0) = 0$ and $(\partial^2 f/\partial\theta_0{}^2) \geqslant 0$ are fulfilled. Thus in the limit of small deflection angles, $\theta_0 \ll 1$, it may be shown[125] that there is a critical thickness d_c above which this structure is stable

$$d_c = 27(8/\pi^2)^2\,(AM^4/(K_\perp')^3)^{1/2}. \tag{5.3}$$

[135] D. S. Lo and M. M. Hanson, *J. Appl. Phys.* **38**, 3 (1967).
[136] M. S. Cohen, *J. Appl. Phys.* **32**, 87S (1961).
[137] A. Sukiennicki, *Phys. Status Solidi* **29**, 417 (1968).

In the case of the *transverse stripe* domain structure, we ought to take into account the contribution of the in-plane anisotropy energy too. The wavelength λ of the transverse stripe structure and the maximum deflection angle φ_0 may be found as[137]

$$\lambda = 2(2\pi^2 Ad/M^2)^{1/3} \tag{5.4}$$

and

$$\cos\varphi_0 = K_{\parallel}\{K_{\perp}' - 2[(2\pi/\lambda)^2 A + M^2\lambda/(2d)]\}^{-1}, \tag{5.5}$$

where $K_{\parallel}$ is the phenomenological constant of the parallel anisotropy. From the obvious condition $0 \leqslant \varphi_0 \leqslant \pi/2$, it follows that

$$K_{\perp}' \geqslant K_{\parallel} + 2[(2\pi/\lambda)^2 A + M^2\lambda/(2d)] \tag{5.6}$$

or equivalently

$$d \geqslant d_c = (54\pi^2 AM^4)^{1/2}(K_{\perp}' - K_{\parallel})^{-3/2}, \tag{5.7}$$

i.e., the minimum of the energy density is realized if λ and φ_0 satisfy Eqs. (5.4) and (5.5) only for thicknesses greater than a certain critical thickness d_c or if the perpendicular anisotropy constant fulfills Eq. (5.6).

The foregoing results show that the existence of the stripe structure requires certain conditions for the film thickness and anisotropy. Unfortunately, quantitative agreement between the predictions of these simplified models and experiment is poor. A micromagnetic approach[138] which allows more realistic variations of the magnetization inside the specimen gives improved agreement between experiment and theory.

Domain Walls

Thus far we have seen that a thin ferromagnetic film may exhibit a domain structure under certain circumstances. Between two adjacent domains there is a transition region of finite thickness, i.e., *domain wall.* It is expected that the domain walls in thin films, as well as the domain configuration, would exhibit many peculiarities. The width and the magnetization distribution inside the domain wall (wall structure) will depend essentially on the film thickness as well as on the relative orientations of magnetization in the adjacent domains.

The actual wall structure corresponds to the minimum of the total free energy of the wall, composed of the exchange, anisotropy, and magnetostatic contribution. A rigorous approach to this problem would be beset by insurmountable difficulties. Therefore, it is usually supposed

[138] Y. Murayama, *J. Phys. Soc. Jap.* **21**, 2253 (1966); **23**, 510 (1967).

that (i) the magnetization is forced to vary only *along the normal* to the wall and (ii) the problem is split into two parts, one concerned with the wall *width* and the other with the magnetization *distribution* inside the wall. Thus we confine ourselves to unidimensional models which are fully characterized by the total angle of the magnetization rotation, the width, and the magnetization distribution inside the wall.

Two types of 180°, one-dimensional walls may exist in thin films,[139–141] *Bloch walls* in thicker films, and *Néel walls* in thinner films. Inside the Bloch walls, which are the common walls in the bulk ferromagnets,[2] the magnetization keeps its normal to the wall component constant so as to avoid magnetic poles on the wall surfaces (Fig. 4a). Then at the intersection of the wall with the film surfaces, magnetic poles arise and a stray field normal to the film plane is created. Inside the Néel wall[142] the magnetization always lies parallel to the film plane to avoid magnetic poles on the film surfaces. Then magnetic poles on the wall surfaces arise, and hence a stray field is created in the plane of the film (Fig. 4b).

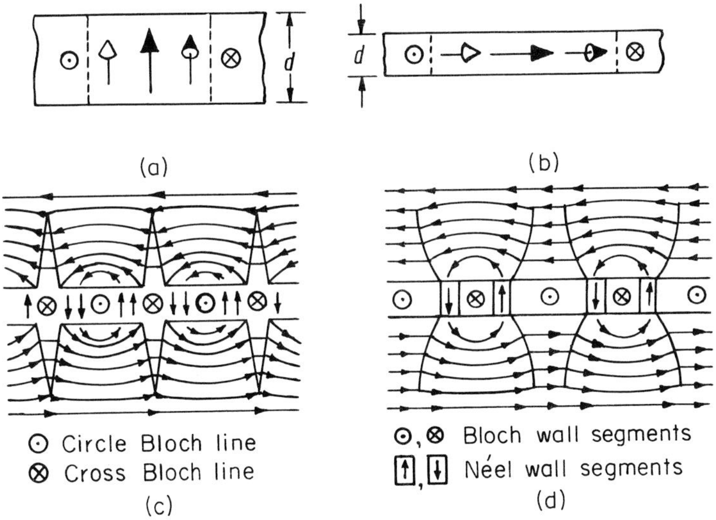

FIG. 4. Domain wall structures in permalloy thin films. (a) 180° Bloch wall; (b) 180° Néel wall; (c) cross-tie wall; (d) Bloch–Néel wall.

[139] W. F. Brown, Jr. and A. E. LaBonte, *J. Appl. Phys.* **36**, 1380 (1965).
[140] A. Aharoni, *J. Appl. Phys.* **37**, 3271 (1966).
[141] A. L. Olson, H. N. Oredson, E. J. Torok, and R. A. Spurrier, *J. Appl. Phys.* **38**, 1349 (1967).
[142] L. Néel, *C. R. Acad. Sci. Paris* **241**, 533 (1955).

In films thick enough to support 180° Bloch walls, Middelhoek[99] pointed out that a critical value, smaller than 180°, of the total angle of the magnetization rotation can be defined, so that for walls of smaller angle the energy of a Néel wall falls below the energy of the Bloch wall, while for walls of larger angle the situation would be reversed. Later on, Torok *et al.*[143] showed that the structure of the walls of wall angle falling between the critical angle and 180° is rather of an intermediate type, involving both Bloch and Néel components.

Although the previously mentioned models give the general features of a real domain wall, they are too simplified to comply with the actual circumstances. Thus in addition a *fine structure* of the domain walls in thin films was experimentally observed. A Bloch wall which exhibits fine structure consists of Bloch segments alternatively magnetized up and down, separated by narrow regions in the center in which the magnetization lies in the film plane as in Néel walls. According to Stein and Feldtkeller,[144] such a separation region appears as a Néel line. In a Néel wall which exhibits fine structure, the oppositely magnetized segments are separated by narrow regions with the magnetization *perpendicular* to the film plane which are called Bloch lines.

A special case of fine structure of a Néel wall is the so-called *cross-tie wall*[145,146] which presents a great density of Bloch lines of two distinct types named by Feldtkeller[148] "circle" and "cross" Bloch lines. In the case of a circle Bloch line the magnetization surrounds the Bloch line, and in the case of a cross line the magnetization causes a crosslike pattern (see Fig. 4c).

Experimental investigation by the Bitter powder method (see, e.g., Methfessel *et al.*[149]) showed that in a permalloy film, Néel walls exist in films thinner than 200 Å, cross-tie walls in the range of thicknesses from approximately 200–900 Å, and Bloch walls in thicker films. The number of cross ties per unit length of wall was reported to vary drastically with the film thickness,[149] with a maximum at approximately 750 Å and decreasing monotonically both for lesser and greater thicknesses.

[143] E. J. Torok, A. L. Olson, and H. N. Oredson, *J. Appl. Phys.* **36**, 1394 (1965).

[144] K. U. Stein and E. Feldtkeller, *J. Appl. Phys.* **38**, 4401 (1967).

[145] Because it appears at intermediate thicknesses, this type of wall was treated initially as a transition structure between Néel and Block structures.[36,40,146,147] But the fact that the main wall consists of alternatively magnetized Néel segments led Middelhoek[99] to point out that the cross-tie wall is rather a lower energy mode of a Néel wall than a distinct one.

[146] E. E. Huber, Jr., D. O. Smith, and J. B. Goodenough, *J. Appl. Phys.* **29**, 294 (1958).

[147] R. M. Moon, *J. Appl. Phys.* **30**, 92S (1959).

[148] E. Feldtkeller, *Symp. Elec. Magn. Properties Thin Magn. Layers, Leuven*, 1961, p. 100.

[149] S. Methfessel, S. Middelhoek, and H. Thomas, *IBM J. Res. Develop.* **4**, 96 (1960).

Feldtkeller and Fuchs[150] found by Lorentz microscopy methods that in the thickness range of about 900 Å, sections of Bloch and Néel walls may exist in the same film; even the same wall may consist of segments of the two types (Fig. 4d), so that the transition thickness from one structure to another one is not precisely delimited. Two main theoretical concepts are encountered in one-dimensional wall models. One goes further with some severe approximations, predicting only the energy and the wall width. Examples of such models are Néel-like theories[99,142] and also models applying the Ritz method.[150–153] Such approaches are not entirely satisfactory since they do not provide any information about the wall structure.

These shortcomings are overcome by the second type of approach which consists in solving either the variational problem for the wall energy,[139,154,155] or the torque equations for the magnetization direction inside the wall[156] using computer programming. In this way theoretical predictions on the wall structure are obtained in agreement with experiment.[157,158]

6. Magnetization Ripple

In Section 1 it was supposed that in the case of a perfect monocrystalline thin film regarded as a monodomain, the static equilibrium is obtained if the magnetization is purely uniform. Let us consider now a polycrystalline thin film with an in-plane easy axis, exhibiting a monodomain structure which is in static equilibrium. Direct observations on the magnetization direction by Lorentz microscopy of very high resolution[159] showed that in the film plane the vector **M** exhibits static periodical local angular deviations of several degrees from a mean direction. These quasi-periodical nonuniformities of the magnetization direction which occur on a microscopic scale are known as the *magnetization ripple*.[159] Though the interpretation of the Lorentz micrographs is compli-

[150] E. Feldtkeller and E. Fuchs, *Z. Angew. Phys.* **18**, 1 (1964).
[151] H. D. Dietze and H. Thomas, *Z. Phys.* **163**, 523 (1961).
[152] R. E. Behringer and R. S. Smith, *J. Franklin Inst.* **272**, 14 (1961).
[153] F. F. Y. Wang, *J. Appl. Phys.* **39**, 865 (1968).
[154] R. Collette, *J. Appl. Phys.* **35**, 3294 (1964).
[155] R. Kirchner and W. Döring, *J. Appl. Phys.* **39**, 855 (1968).
[156] H. N. Oredson and E. J. Torok, *IEEE Trans. Magn.* **MAG-4**, 44 (1968).
[157] E. Feldtkeller, *Z. Angew. Phys.* **15**, 206 (1963).
[158] E. Fuchs, *Z. Angew. Phys.* **14**, 203 (1962).
[159] H. W. Fuller and M. E. Hale, *J. Appl. Phys.* **31**, 238 (1960).

cated,[159–162] there is general agreement that this phenomenon is not a result of the method of observation used.[163]

Two possible configurations of ripple structure, *longitudinal* and *transverse* (Fig. 5) have been reported. In the case of longitudinal ripple

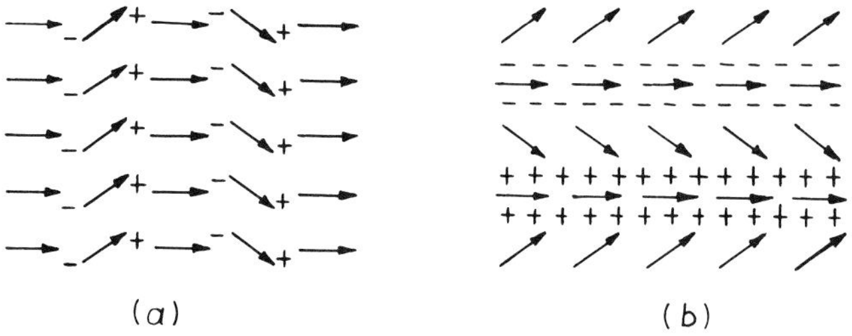

FIG. 5. The two types of magnetization ripple in thin films and their volume poles (schematic representation). (a) Longitudinal ripple; (b) transverse ripple.

the magnetization direction varies from point to point along the mean direction of magnetization of the monodomain, being constant in a direction perpendicular to it. In contrast, for the transverse ripple the variations of the magnetization direction are registered only in a direction perpendicular to the mean magnetization. A qualitative analysis[159] of the two types of ripple shows that they are equivalent with respect to exchange, local anisotropy, uniform anisotropy, and external magnetic forces, but not with respect to the volume poles density, which is much *larger* for the transverse ripple than for the longitudinal one. On this basis Fuller and Hale concluded that the *longitudinal ripple is favored* by the magnetostatic interaction, so that the observed ripple should be mainly longitudinal.

The ripple structure originates from *local* variations of the *anisotropy* forces, which appear in polycrystalline thin films. Local variations of the anisotropy were revealed by Fuchs[167] to be caused by various factors, among which the most important seem to be the crystalline anisotropy

[160] D. Wohlleben, *Phys. Lett.* 22, 564 (1966).
[161] D. Wohlleben, *J. Appl. Phys.* 38, 3341 (1967).
[162] M. S. Cohen, *J. Appl. Phys.* 38, 4966 (1967).
[163] Direct observation of ripple may be made by Lorentz microscopy and magneto—optical high—resolution methods. For recent developments in these fields see Cohen,[164] Lambeck,[165] and Leaver.[166]
[164] M. S. Cohen, *IEEE Trans. Magn.* **MAG-4**, 48 (1968).
[165] M. Lambeck, *IEEE Trans. Magn.* **MAG-4**, 51 (1968).
[166] K. D. Leaver, *J. Appl. Phys.* 39, 1157 (1968).
[167] E. Fuchs, *Z. Angew. Phys.* 13, 157 (1961).

forces which vary randomly in direction from crystallite to crystallite, and stress anisotropy forces, as offered by Fujii *et al.*,[168] due to inhomogeneous tensions between crystallites as pointed by Leaver[100] and Harte.[169] The relative contributions of these mechanisms to the local variations of the easy axis are not firmly established experimentally, so that this local anisotropy field must be treated on a statistical basis, as generally as possible.[170]

Consistent theoretical results on ripple may be obtained if a micromagnetic approach is used. Owing to mathematical difficulties the theory is restricted to a bidimensional model based on the assumption that the magnetization direction is *independent* of the coordinate *normal* to the film plane. Moreover no normal component of the magnetization is assumed. It is supposed that a uniform anisotropy field exists in the film plane subjected to an external field H. Both these fields determine the mean direction of the magnetization. On these uniform fields is superimposed a *randomly distributed anisotropy field* which causes the nonuniformities in the magnetization direction. These nonuniformities result in magnetic poles and hence a stray field is present everywhere in the film.

Therefore, the free energy density of the film results from the following contributions: exchange f_{ex}, uniform anisotropy $f_{K_{\text{u}}}$, local anisotropy f_K, stray field $f_{\text{m.s.}}$, and external field f_{H} energies. The total free energy is then

$$\int f\, d\mathbf{r} = \int (f_{\text{ex}} + f_{K_u} + f_K + f_H + f_{\text{ms}})\, dr. \tag{6.1}$$

If the magnetization direction in the film is described by the angle $\phi(\mathbf{r})$, between the local magnetization and the easy axis then the above expression is a functional with respect to $\phi(\mathbf{r})$. Therefore, the actual state of the film corresponds to a solution of the variational problem

$$\delta \int f\, d\mathbf{r} = 0, \tag{6.2}$$

where the variation is taken with respect to $\phi(\mathbf{r})$.

168 T. Fujii S. Uchiyama, E. Yamada, and Y. Sakaki, *J. Phys. Soc. Japan* **6**, 1 (1967).

169 K. J. Harte, *J. Appl. Phys.* **39**, 1503 (1968).

170 The earlier *anisotropy dispersion model* which supposed that the film is a collection of noninteracting crystallites (see, e.g., Prutton[36]), or the more sophisticated *complex biaxial anisotropy model* proposed by Torok *et al.*,[171,172] may explain many experimental facts, but they are in disagreement with the values of some experimentally determinable parameters (see among others, Leaver[100] for the values of α_{90}).

There is still another procedure which may be followed based on the fact that the torque which acts on the magnetization is everywhere zero

$$\mathbf{M}(\mathbf{r}) \times \mathbf{H}_{\text{eff}}(\mathbf{r}) = 0. \tag{6.3}$$

Since the effective field may be written immediately, the problem may be solved by taking the Fourier components of the total torque as vanishing everywhere. In both methods difficulties arise from local anisotropy and stray field terms.

In the following we discuss briefly main outlines of the theories of Hoffmann[173–176] and Harte[169,177,178], which illustrate respectively the two methods sketched above.

Using arguments furnished partly by Lorentz micrographs and partly by qualitative considerations which prove the existence of exchange and magnetostatic coupling between individual crystallites, Hoffmann[173–176] showed that for finding the direction $\phi(\mathbf{r})$ of the magnetization at the point $\mathbf{r}$ inside the film, only magnetostatic and exchange interactions for film regions situated inside a small "coupling region" which surrounds the point $\mathbf{r}$ are important. Thus the magnetic state of the coupling region may be found as if it were isolated from the remaining part of the film. This approximation considerably simplifies the micromagnetic problem.

The angle $\phi(\mathbf{r})$ which gives the direction of the magnetization in the point $\mathbf{r}$ (see Fig. 6) is expressed as

$$\phi(\mathbf{r}) = \varphi_0 + \varphi(\mathbf{r}), \tag{6.4}$$

where φ_0 is the angle which gives the deviation of the mean magnetization $\mathbf{M}_0$ from the easy axis due to the external field $\mathbf{H}$, and $\varphi(\mathbf{r})$ is the angle which denotes the deviation at the point $\mathbf{r}$ from the mean magnetization. Lorentz micrographs show that $\varphi(\mathbf{r}) \ll 1$, so that all energies included in Eq. (6.2) may be expanded in powers of φ restricted to the second

[171] E. J. Torok, H. N. Oredson, and A. L. Olson, *J. Appl. Phys.* **35**, 3469 (1964).
[172] E. J. Torok, *J. Appl. Phys.* **36**, 952 (1965).
[173] H. Hoffmann, *J. Appl. Phys.* **35**, 1790 (1964).
[174] H. Hoffmann, *in* "Basic Problems in Thin Film Physics" (R. Niedermayer and H. Mayer, eds.), p. 386. Vandenhoeck and Ruprecht, Göttingen, 1966.
[175] H. Hoffmann, *IEEE Trans. Magn.* **MAG-2**, 566 (1966).
[176] H. Hoffmann, *IEEE Trans. Magn.* **MAG-4**, 32 (1968).
[177] K. J. Harte, *Proc. Int. Conf. Magn. Nottingham* (1964).
[178] K. J. Harte, *J. Appl. Phys.* **37**, 1295 (1966).

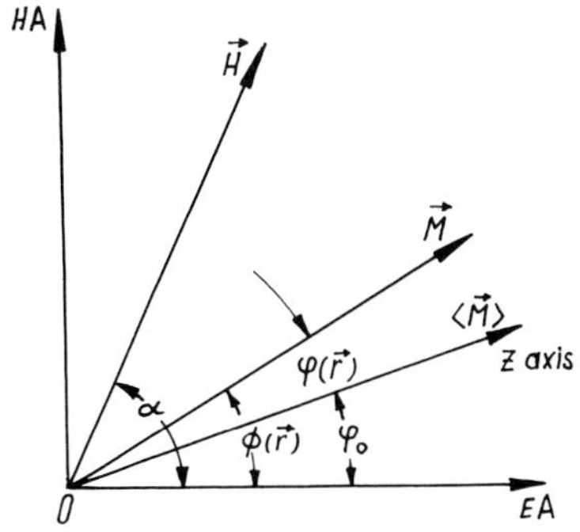

FIG. 6. Coordinate system and notations used in the variational treatment of ripple. The x axis is perpendicular to the film plane. (Letters with arrows correspond to boldface in the text.)

order. In particular, after some approximations, the local anisotropy energy f_K and the stray field energy $f_{\text{m}\cdot\text{s}\cdot}$ may be written as

$$f_K = Kf_0(\varphi_0, y, z) + K\varphi f_1(\varphi_0, y, z) + \tfrac{1}{2} K\varphi^2 f_2(\varphi_0, y, z) \tag{6.5a}$$

and

$$f_{\text{ms}} = \tfrac{1}{2} M^2 c_z[\varphi\, \partial^2\varphi/\partial z^2 + (\partial\varphi/\partial z)^2 - (\partial^2\varphi/(\partial y \partial z)) - \varphi^2(\partial\varphi/\partial z)^2 - \tfrac{2}{3}\varphi^3\partial^2\varphi/\partial z^2] - \tfrac{1}{2}M^2 c_y \varphi\, \partial^2\varphi/\partial y^2. \tag{6.5b}$$

Following Hoffmann[173] the expression for f_K holds over a volume V_1 inside which the exchange and magnetostatic couplings between individual crystallites affect seriously the magnetization direction in V_1. This volume is supposed to be centred around r; K denotes an "effective" local anisotropy density energy, constant over V_1. The expression for $f_{\text{m}\cdot\text{s}\cdot}$ holds inside another volume V^* in which the variation of $\varphi(r')$ from $\varphi(r)$ is small, so that it may be expanded in a power series; c_z and c_y are two magnetostatic constants which depend on the volume V^*.

After carrying out the calculations needed by the variational procedure, an Euler equation which describes the ripple effect is obtained. This equation is nonlinear, and its general solution has not yet been written down. Neglecting the nonlinear terms, the following linear equation is found:

$$\nabla^2\varphi + (M^2 c_y/(2A))(\partial^2\varphi/\partial y^2) - (K_u/A)\, h(\alpha)\varphi = Kf_1/(2A), \tag{6.6}$$

and its solution may be expressed in terms of a modified Bessel function

$$\varphi(\varphi_0, y, z) = -(K/(4\pi AW)) \iint_{\text{film}} f_1(\varphi_0, \xi, \eta)\, K_0(\zeta)\, d\xi\, d\eta, \qquad (6.7)$$

where

$$\zeta = [(K_u/A)\, h(\alpha)]^{1/2}\, [(z - \xi)^2 + (y - \eta)^2/W^2]^{1/2}, \qquad (6.7a)$$

$W = 8dM^2[AK_u h(\alpha)]^{-1}$, d is the film thickness, and

$$h(\alpha) = h \cos(\alpha - \varphi_0) + \cos 2\varphi_0, \qquad h = H/H_K. \qquad (6.7b)$$

The solution shows that the coupling region V^* is an ellipse centered on r with a semiminor axis

$$R_{ec} = [A/(K_u h(\alpha))]^{1/2} \qquad (6.8a)$$

along the mean magnetization and a semimajor axis

$$R_{mc} = W[A/(K_u h(\alpha))]^{1/2} \qquad (6.8b)$$

perpendicular to the mean magnetization; R_{ec} is the coupling length of the exchange forces, while R_{mc} represents the coupling length of the transverse stray field, hence these dimensions may be accepted for V_1 as well. Recalling that a transverse stray field is characteristic for the transverse ripple, it must be concluded that the results of the linear ripple theory may describe the transverse ripple configuration only. Furthermore in the linear approximation the thin film must be regarded as being composed of "generalized crystallites" of equal dimensions, over which the local anisotropy forces have an effective value.[179]

The experimentally determinable quantities given by Hoffmann's theory[173,176] are: the ripple wavelength, defined as

$$\lambda_{TR} = 2\pi R_{ec} = 2\pi[A/(K_u h(\alpha))]^{1/2}, \qquad (6.9)$$

the magnetization dispersion, given by

$$\delta = (\langle\varphi^2\rangle)^{1/2} = 0.17[S/(Md^{1/2})^{1/2}][AK_u h(\alpha)]^{-3/8} \qquad (6.10)$$

in the case of randomly distributed uniaxial anisotropies (S is a structure

[179] This is true only when the real crystallites have equal dimensions, i.e., they may be characterized by the same mean diameter D. This is possibly achieved when all skew effects due to oblique incidence during the film deposition are avoided.

constant), and the incremental susceptibility (dc bias field and ac tickle field parallel to each other)

$$\chi_i = (M/H_K)(\delta^2/h(\alpha)) \propto h(\alpha)^{-7/4}. \tag{6.11}$$

Due to the experimental difficulties in preparing films the properties of which would comply with the requirements of the theoretical model[179] only few determinations agreed with the predicted values of the ripple wavelength and magnetization dispersion,[180] as well as with the predicted values of the incremental susceptibility.[166,181]

A possible source of disagreement between the predicted values and the experimental data given by Baltz and Doyle[182] might be related both to the linear character of the theory and to the subsequent approximations which were made in the framework of the linear theory. This may be understood if we recall that Hoffmann's linear theory, though simple, has the disadvantage that being built up by replacing long-range magnetostatic forces by short-range forces inevitably introduces errors of unknown magnitude. To asses a magnitude of these errors we shall follow Brown[183] in his criticism of Hoffmann's approach.

Brown observed that there existed already[184] an exact linearized equation derived by him as early as in 1940 (see also Brown,[5] p. 61) namely

$$\nabla^2\varphi + (M/2A)\, h_y - (K_{\mathrm{u}}/A)\, h(\alpha)\varphi = Kf_1/(2A), \tag{6.12}$$

which becomes identical to that of Hoffmann (6.6) if the stray field perpendicular to the mean magnetization h_y is approximated by

$$h_y = c_y(\partial^2 M_y/\partial y^2) = c_y M(\partial^2\varphi/\partial y^2). \tag{6.13}$$

By solving the exact equation (6.12) assuming a concentrated inhomogeneity, Brown succeeded in finding the magnetization dispersion as well as the angular displacement $\varphi(\mathbf{r})$. He concludes that although the value of δ could be made comparable with that of Hoffmann the values of $\varphi(\mathbf{r})$, especially in the regions far from the point at which the inhomo-

[180] T. Suzuki and C. H. Wilts, *J. Appl. Phys.* **39**, 1151 (1968).

[181] C. S. Comstock, Jr., A. C. Sharp, R. L. Samuels, and A. V. Pohm, *IEEE Trans. Magn.* **MAG-4**, 39 (1968).

[182] A. Baltz and W. D. Doyle, *J. Appl. Phys.* **25**, 1814 (1964).

[183] W. F. Brown, Jr., *J. Appl. Phys.* **40**, 1214 (1969).

[184] W. F. Brown, Jr., *Phys. Rev.* **58**, 736 (1940).

geneity is located does not agree with the values of Hoffman's theory.

Another problem to be investigated is the *stability* of the ripple structure. In the framework of the approximations already mentioned Hoffmann[175] has given a micromagnetic approach similar to the Stoner–Wohlfarth theory[185] of a monodomain particle.[186] One shows that the stability of the ripple structure is determined essentially by the concurrent influences of the uniform effective field (see Eq. (6.7b)) and two stray fields originating in the local magnetostatic interaction and directed respectively parallel and antiparallel to the magnetization. Over the regions in which the local deviations of the magnetization tend to their extreme values the stray field parallel to the magnetization is important, as compared with the antiparallel one, and acts to stabilize the ripple structure. Over the region on which $\varphi(\mathbf{r})$ vanishes, the antiparallel stray field is dominant and the ripple shows its maximum instability. As the antiparallel stray field exceeds the uniform effective field the ripple structure tends to the so-called "blocked structure".[175,176]

On the other hand, Harte[169,177,178] developed a micromagnetic theory of the ripple structure based on the condition of local equilibrium of the magnetization which requires the vanishing at each point of the total torque of the forces acting on the magnetization (Eq. (6.3)). The local demagnetizing field was treated by Harte in a straightforward way by solving the Maxwell equations, which permitted a better approach to the long-range effects as compared with Hoffmann's treatment.

To introduce the local anisotropy field the following model is used: the film is supposed to consist of cylindrical cells perpendicular to the film plane (e.g., crystallites, clumps of crystallites, or larger regions) over which the anisotropy is uniform. One assumes random cross-sectional areas and shapes, with the condition that the cell dimensions are on the average isotropic, and cell sizes and shapes are on the average homogeneous. Each cell is a generalized crystallite having a randomly oriented magnetic anisotropy.

The Fourier transform of the torque equation permits an approach to the nonlinear effects on the magnetization dispersion. One finds that the magnetization dispersion shows two coupling lengths related to exchange and magnetostatic interactions, similarly to those appearing in Hoffman's theory. The magnetization dispersion is very sensitive to the inhomogeneity scale. In the case of fine scale inhomogeneities, the results of the linear approximation coincide with Hoffmann's results.

[185] E. C. Stoner and E. P. Wohlfarth, *Phil. Trans. Roy. Soc.* **A240**, 74 (1948).

[186] Detailed discussion of the Stoner-Wohlfarth theory for thin monodomain films may be found, e.g., in the monographs of Prutton,[36] Soohoo,[40] or Cohen.[101]

IV. Saturation Magnetization

All the thermodynamic properties of a system can be obtained by finding the eigenvalues of the Hamiltonian, constructing the partition function and taking the appropriate partial derivatives of the free energy. In this chapter we shall focus our attention only on a simple but fundamental problem, namely that of the saturation magnetization of a *monodomain thin film*. Unfortunately, with the models commonly used in the theory of ferromagnetic films, although much simplified and hence not too realistic, exact theoretical results are hardly to be expected. In practice, the problem of the saturation magnetization is much too difficult to be solved rigorously, and various approximate methods have been proposed. However certain features of the solution of the Heisenberg model are known and these provide both physical guides in the choice of approximations and criteria to judge the adequacy of these approximate methods. As is well known, it was rigorously proved by Mermin and Wagner[187] that a *unidimensional isotropic* Heisenberg system with short range interactions cannot be ferromagnetic.

Some rigorous results valid for the isotropic *two-dimensional* Heisenberg system with short range interactions are to be noted. First as Wortis[188] demonstrated, the existence of bound magnon states causes the breakdown of simple spin wave theory for this system. Second, Mermin and Wagner[187] have proved that such a system exhibits no spontaneous magnetization at all positive temperatures. Quite recently it was argued by Kats[189] that a two-dimensional system described by an anisotropic Heisenberg Hamiltonian may exhibit long-range order for some positive temperatures. Hence *phase transitions* in *two dimensions probably depend on the very existence of anisotropy*.[190] The exact results obtained both in the study of Ising systems and those obtained for the Heisenberg models suggested that the problem of *the existence of long range order* is closely connected with the *dimensionality of the system* as well. Thus Kats[189] proved that an anisotropic Heisenberg unidimensional system does not become ferromagnetic. Since the simple spin wave theory does take account of the dimensionality of the system we shall often use it as an

[187] N. D. Mermin and H. Wagner, *Phys. Rev. Lett.* **17**, 1133 (1966).

[188] M. Wortis, *Phys. Rev.* **132**, 85 (1963).

[189] E. I. Kats, *JETF* **56**, 2043 (1969).

[190] A similar result was obtained in the frame of spin wave theory but we note that this statement, until the paper of Mermin and Wagner,[187] was of questionable validity.

heuristic approach to the statistical mechanics of thin films, provided some conditions for its applicability are fulfilled.[191]

7. Molecular Field and High Temperature Expansions

To investigate our system theoretically, we need statistical approximations which, at the very least, do not conflict with the prior findings.

Other features of the solutions can be derived from rigorous series expansion of the free energy. The reason is that reliable information regarding the true thermodynamical properties of a given model may be obtained from a series of successive approximations which enable us to assess the value of a particular property and hence to estimate the error involved in stopping at a particular stage.

Two different procedures have been employed in the theory of ferromagnetic thin films. These two methods usually begin with similar Hamiltonians and arrive at predictions of the Curie temperature by alternate routes; in one procedure the high temperature expansion of the partition function is found without further assumption whereas the other uses a physical approximation to treat a simplified problem in a semiclosed form. The first method is useful only if the expansion may be stopped at a given stage. In the second class of approximate methods a small section of the film is treated exactly and the exchange interactions of this small section with the remainder of the crystal are treated in an approximate way. As the exchange interactions affect the alignment of the magnetic moments, a physically challenging approximation is to replace them by an effective field the properties of which are to be determined in some consistent way. The advantage of this method is that *the nature* of the approximation and of the terms ignored is known precisely at each step. An unfortunate aspect of both methods is that their predictions do not depend on the dimensionality of the system, but on the number of spins in the range of interaction. Although the exact results could undermine our confidence in the validity of these approximations, we shall also see how it might restore it.

To begin with, let us refer to the breakdown of the simple spin wave theory in two dimensions. Conventional spin wave theory is valid at low temperatures for actual systems exhibiting an ordered phase, provided that bound-magnon complications are not significant. There are already indications, as shown by Lines,[192a] that bound magnon states are fairly

[191] Though some of the predictions of the spin wave theory are not well grounded, this kind of theory belongs to those which, to some extent, make use of the translational invariance of the lattice structure and can therefore relate the existence of long-range order to the dimensionality of the system under study.

[192a] M. E. Lines, *J. Appl. Phys.* **40**, 1352 (1969).

ineffective, even if present, for the weakly anisotropic ferromagnet in two dimensions. Then for such a system we may employ the conventional spin-wave picture. The spin-wave divergencies which cause the instability of the two-dimensional Heisenberg ferromagnet could be removed by the introduction of even small amounts of *anisotropy*, and therefore the long-range order in two dimensions exists for positive temperatures, as proved by Robinson.[192b]

On the other hand, a reasonable approach to the Heisenberg model would be to consider the Ising Hamiltonian as an unperturbed Hamiltonian and the noncommutative transverse coupling terms as a perturbing interaction. The free energy of the Heisenberg model is equal to that of the Ising system plus an integral of the transverse correlation terms, as shown by Callen.[192c] Now at sufficiently high temperatures, we may expect the transverse correlations to be vanishingly small. Thus the free energy of the Ising and Heisenberg models are very nearly equal in this region. Therefore, we may expect that *the predictions of molecular field*[193] become much more reasonable *for anisotropic spin systems with short range interactions, as compared to isotropic Heisenberg systems.*

Finally we note that the molecular field theory is now recognized as the limiting form of the rigorous solution, for long-range exchange interaction. This interpretation is meaningful since the basic assumption of the molecular field theory is that the internal field resulting from interactions between neighboring atoms is steady. However, the field is fluctuating and only in the limit in which the number of the neighbors interacting with a given atom is increasing does the significance of these fluctuations diminish, i.e., the field becomes steady. We may therefore expect *that the predictions of the molecular field* become more exact as the *dimensionality* of the crystal *increases.*[194] We conclude that we have to employ the previously mentioned approximations for anisotropic spin systems only. Since our physical system, although closely approximating isotropic Heisenberg systems, necessarily contains anisotropic terms too, it is highly probable that a *true long-range order* will be effected.

Before discussing various approximate methods applied in the theory

[192b] D. W. Robinson, *Commun. Math. Phys.* **14**, 195 (1969).

[192c] H. B. Callen, *in* "Physics of Many Particle Systems: Methods and Problems" (E. Meeron, ed.), Vol. I. Gordon and Breach, New York, 1966.

[193] The mean field approximation of the Heisenberg ferromagnet might be considered the solution to an Ising Hamiltonian.

[194] Note that this observation applies equally for isotropic and anisotropic Heisenberg systems, and does not contradict the theorem of Mermin and Wagner,[187] which, in fact, can not be applied to long-range exchange interactions.

of ferromagnetic thin films, let us consider the concept of spontaneous magnetization. Let the Hamiltonian of the system be

$$\mathscr{H} = \mathscr{H}_0 - \mathscr{M}\mathscr{H}, \tag{7.1}$$

$$\mathscr{M} = g\mu_B \sum_{\mathbf{f}p} \mathbf{S}_{\mathbf{f}p}, \tag{7.2}$$

where $\mathscr{H}_0$ is a spin dependent Hamiltonian including the Heisenberg Hamiltonian and anisotropy terms; H is an external applied field directed along the α axis (chosen so that the Zeeman term commutes with $\mathscr{H}_0$). *Spontaneous magnetization* is present if with $H = 0$, $\mathscr{M}$ attains in some sense a nonzero value. For bulk ferromagnets $\mathscr{M}/V$ has to be an *intensive* thermodynamical quantity, that is the value of $\mathscr{M}$ has to be proportional to the size of the system. This is no longer valid for very thin films, although $\mathscr{M}$ does not cease to be a thermodynamically defined quantity. Due to the existence of surfaces, the value of the spontaneous magnetization has to depend to a certain extent on the thickness of the film. We note that for bulk samples the "surface" energies are negligible compared to "volume" energies.

Let us consider the average magnetization per atom,

$$m_{Nq}(H) = (Nq)^{-1}\langle\mathscr{M}\rangle = (Nq)^{-1} \ln \operatorname{tr}(\mathscr{M}e^{-\beta\mathscr{H}})/\operatorname{tr} e^{-\beta\mathscr{H}} \tag{7.3}$$

and the free energy per atom

$$f_{Nq}(H) = -(\beta Nq)^{-1} \ln \operatorname{tr}(e^{-\beta\mathscr{H}}) \tag{7.4}$$

for a thin film containing N atoms in each of the q planes parallel to the surface. Evidently $m_{Nq}(H) = [\partial f_{Nq}(H)/\partial H]_T$. The *spontaneous magnetization* $m(q)$ is the limit as $H \to 0^+$ of the average magnetization per atom, in the limit of an infinite *plane* system of finite thickness

$$m(q) = \lim_{H\to 0+} \lim_{N\to\infty} m_{Nq}(N) = -\lim_{H\to 0+} \lim_{N\to\infty} [\partial f_{Nq}(H)/\partial H]_T. \tag{7.5}$$

Actually, in the theory of ferromagnetic thin films the existence of this limit has not been proved and it is only presumed that it exists.[195]

Sometimes another definition of the spontaneous magnetization is used. If one calculates the free energy per atom now as a function of magnetization, for zero applied external field, this function, in the

[195] The problem of a similar limit entering the theory of bulk ferromagnetic bodies was studied by R. Griffiths, *Phys. Rev.* **152**, 240 (1966). Its existence was proved under fairly general conditions. That is why it is belived that the limit in Eq. (7.5) exists.

limit of the infinite system may possess for $T < T_c$ two minima at the values $\pm m_0$. The spontaneous magnetization is defined exactly as m_0. It was proved that this definition is reasonable and is equivalent to that given by Eq. (7.5) for a long ranged Ising system. Therefore, it is expected that in the *molecular field approximation such a definition is allowed.* For Heisenberg systems with short range interactions, this definition is generally questionable when no magnetic field is present. A detailed discussion on this subject is given by Nakano.[196]

a. Molecular Field

When applying the classical Weiss picture of the molecular field to thin ferromagnetic specimens, the Valenta model of *planar sublattices* acquires an important physical meaning. As is well known, the essence of the molecular field method consists in replacing the direct exchange interaction of the spin pairs by an interaction of each spin with an effective field, which is taken as due to the neighboring spins. As only nearest neighbors are taken to be responsible for the effective field, one deals with the usual form of the molecular field method.

It is obvious that the effective field as viewed in the molecular field picture depends essentially on the distribution of neighboring spins. Thus, while in an infinite bulk ferromagnet the effective field is uniform throughout the sample, in finite specimens the effective field must exhibit a certain distribution, in the sense that the intensity of the effective field at a given point inside the specimen should depend essentially on the position with respect to the *boundaries* of the point under consideration. In a thin ferromagnetic film it is natural to assume that the *planar sublattices* are planes in which the intensity of the effective field depends only on the position of the plane in the film.

Let us assume in the following the simplest composition of the model thin film Hamiltonian, including only exchange isotropic term and a Zeeman term, namely

$$\mathscr{H} = -\sum_{\mathbf{f}p,\mathbf{g}p'} A_{\mathbf{f}p,\mathbf{g}p'} \mathbf{S}_{\mathbf{f}p} \cdot \mathbf{S}_{\mathbf{g}p'} - g\mu H \sum_{\mathbf{f}p} S^z_{\mathbf{f}p}, \tag{7.6}$$

although the procedure may be easily extended to Hamiltonians which take account of the anisotropy terms required both by physical considerations and the plausibility arguments for employing the molecular field methods as given above. One splits the thin film Hamiltonian into two parts, $\mathscr{H}_0$ and $\mathscr{H}_1$ so that $\mathscr{H}_0$ contains the ground state energy $\mathscr{E}_0$ and the contribution of the effective field, built up according to the

[196] H. Nakano, *Progr. Theor. Phys.* **9**, 403 (1953).

picture developed above (see Section 3), while $\mathscr{H}_1$ contains the remaining part of the Hamiltonian

$$\mathscr{H} = \mathscr{H}_0 + \mathscr{H}_1 \tag{7.7}$$

where

$$\mathscr{H}_0 = \mathscr{E}_0 + \sum_p \rho_p \sum_{\mathbf{f}\in p} n_{\mathbf{f}p} . \tag{7.8}$$

The ground state energy $\mathscr{E}_0$ depends on the magnetic configuration through the direction cosines $\gamma_{\mathbf{f}p}$. The equivalence of all sites of a plane monolayer suggests that the same direction γ_p may be assumed for all spins lying in the same monolayer, so that

$$\mathscr{E}_0 = -S^2 \sum_{p,p'} \gamma_p \cdot \gamma_{p'} \sum_{\mathbf{f}\in p,\mathbf{g}\in p'} A_{\mathbf{f}p,\mathbf{g}p'} - g\mu_{\mathrm{B}} HN \sum_p \gamma_p{}^z. \tag{7.9}$$

The second term in (7.8) introduces the contribution of the effective field by the set of undetermined factors ρ_p, $p = 1, 2,\ldots, q$, $n_{\mathbf{f}p}$ being defined by (3.23). It is easily found, using the standard expansion (3.19), that

$$\mathscr{H}_1 = \sum_p \sum_{\mathbf{f}\in p} (R_p - \rho_p)\, n_{\mathbf{f}p} - \sum_{p,p'} \gamma_p \cdot \gamma_{p'} \times \sum_{\mathbf{f}\in p,\mathbf{g}\in p'} A_{\mathbf{f}p,\mathbf{g}p'} n_{\mathbf{f}p} n_{\mathbf{g}p'} + \mathscr{H}'. \tag{7.10}$$

Here we have separated the terms which are diagonal in spin deviations and have let $\mathscr{H}'$ include the remaining part.

$$R_p = 2S \sum_{p'} \gamma_p \cdot \gamma_{p'} A^0_{pp'} + g\mu_{\mathrm{B}} H\gamma_p{}^z, \tag{7.11}$$

where $A^0_{pp'}$ denotes Fourier transforms in the plane of the film of the exchange integral, of the form

$$A^{\nu}_{pp'} = \sum_{\mathbf{g}\in p'} A_{\mathbf{f}p,\mathbf{g}p'} \exp[-i(\mathbf{f} - \mathbf{g}) \cdot \boldsymbol{\nu}]. \tag{7.12}$$

Now, to answer fully the question how the effective field must be chosen, the set of factors ρ_p must be determined. It is convenient to use a *variational procedure* due to Peierls[197] and Bogoljubov.[198] One proves

[197] R. Peierls, *Phys. Rev.* **54**, 918 (1938).

[198] This part follows closely the Tyablikov's approach to ferrites.[93] The variational procedure was applied in deriving Valenta's molecular field equations in thin films by H. Koppe and R. J. Jelitto, *Phys. Status Solidi* **9**, 357 (1965).

that for an *arbitrary* splitting[199] of the Hamiltonian of the system into two parts (7.6) the free energy can be overestimated by the right-hand side of the following inequality,

$$F(\mathscr{H}) \leqslant F(\mathscr{H}_0) + [\mathrm{tr}(\mathscr{H}_1 e^{-\beta\mathscr{H}_0})]/[\mathrm{tr}\, e^{-\beta\mathscr{H}_0}], \tag{7.13}$$

where

$$F(\mathscr{H}_0) = -\beta^{-1} \ln \mathrm{tr}\, e^{-\beta\mathscr{H}_0}. \tag{7.14}$$

In our case the right-hand side of the inequality (7.13) occurs as a *trial free energy* depending on the factors ρ_p which introduced the effective field Hamiltonian. One chooses the factors ρ_p so as to furnish the closest approach to the actual free energy of the ferromagnetic film. That is, one has to minimize

$$F_{\mathrm{trial}} = F(\mathscr{H}_0) + [\mathrm{tr}(\mathscr{H}_1 e^{-\beta\mathscr{H}_0})]/[\mathrm{tr}\, e^{-\beta\mathscr{H}_0}] \tag{7.15}$$

with respect to ρ_p, $p = 1, 2,\ldots, q$.

Noticing that the spin deviations $S - S^{z'}_{\mathrm{f}p}$ take only the values $0, 1, 2,\ldots, 2S$, the traces in the expression (7.15) of the trial free energy can be easily calculated. Using

$$[\mathrm{tr}(n_{\mathrm{f}p} e^{-\beta\mathscr{H}_0})]/[\mathrm{tr}\, e^{-\beta\mathscr{H}_0}] = \sum_{n=1}^{2S} [n e^{-\beta\rho_p n}] \Big/ \Big[\sum_{n=0}^{2S} e^{-\beta\rho_p n}\Big] \equiv \overline{n_p}, \tag{7.16}$$

the trial free energy is

$$F_{\mathrm{trial}} = \mathscr{E}_0 - \beta^{-1} N \sum_{p=1}^{q} \ln \Big(\sum_{n=0}^{2S} e^{-\beta\rho_p n}\Big) + N \sum_{p=1}^{q} (R_p - \rho_p)\, \overline{n_p}$$
$$- N \sum_{p,p'=1}^{q} A^0_{pp'} \gamma_p \cdot \gamma_{p'} \overline{n_p}\, \overline{n_{p'}}\,. \tag{7.17}$$

Now, there are generally two sets of *extremal conditions* to be imposed. The first is concerned with the parameters ρ_p of the molecular field:

$$\partial F_{\mathrm{trial}}/\partial \rho_p = 0, \qquad p = 1, 2,\ldots, q, \tag{7.18}$$

[199] The Bogolyubov's formulation emphasizes the *arbitrary* character of the splitting (4.6) of the Hamiltonian. In particular $\mathscr{H}_1$ may be *not* a small perturbation of $\mathscr{H}_0$, and the choice of $\mathscr{H}_0$ may be determined by the actual needs of the problem.

whereas the other involves the parameters of the magnetic configuration

$$\partial\left[F_{\text{trial}} + (N/2)\sum_{\alpha} \lambda_p(\gamma_p{}^{\alpha})^2\right]\Big/\partial\gamma_p{}^{\alpha} = 0, \qquad \sum (\gamma_p{}^{\alpha})^2 = 1, \tag{7.19}$$

λ_p being Lagrange multipliers. Writing these equations explicitly, one finds

$$\rho_p = g\mu_{\text{B}}H\gamma_p{}^z + 2S\sum_{p'} A^0_{pp'}\boldsymbol{\gamma}_p\cdot\boldsymbol{\gamma}_{p'}\sigma_{p'} \tag{7.20}$$

and

$$\sum_{p'} A^0_{pp'}S\sigma_p S\sigma_{p'}\gamma^{\beta}_{p'} + \lambda_p\gamma_p{}^{\beta} = -S\sigma_p g\mu_{\text{B}}H\delta_{\beta z}\,. \tag{7.21}$$

Here we have $\sigma_p = 1 - (1/S)\overline{n_p}$. According to the definition of the Brillouin function

$$B_S(x) = 1/(2S)[(2S+1)\coth((2S+1)\,x/2) - \coth(x/2)] \tag{7.22}$$

the quantities σ_p become

$$\sigma_p = B_S\left[\beta\left(g\mu_{\text{B}}H\gamma_p{}^z + 2S\sum_{p'} A^0_{pp'}\boldsymbol{\gamma}_p\cdot\boldsymbol{\gamma}_{p'}\sigma_{p'}\right)\right]. \tag{7.23}$$

The spontaneous magnetization of the sample, σ, is found from the free energy to be

$$\sigma = \lim_{H\to 0+}\lim_{N\to\infty}[(\partial F(H)/\partial H)_T/(\partial F(H)/\partial H)_{T=0}]. \tag{7.24}$$

Then, using the trial free energy one finds

$$\sigma = (1/q)\sum_{p}\sigma_p(H=0). \tag{7.25}$$

Obviously, the quantity σ_p is the relative spontaneous magnetization of the pth layer and it is a thermodynamic quantity within the framework of molecular field theory. In fact, the operators $\sum_{f\in p} S^z_{\text{f}p}$ do not commute with the total Hamiltonian of our system, but they commute with the Hamiltonian of the molecular field theory. We note that the quantities ρ_p have been found by minimizing F_{trial} and thus we have rederived the equations of the *molecular field* theory. But this minimum free energy is still higher than the true free energy. That is why we cannot suppose that we have proved that the spontaneous magnetizations σ_p are generally thermodynamic quantities.

Now let us focus our attention on the problem of the magnitude of the spontaneous magnetization in a single domain film. For one-half spins, the specific magnetization (7.22) is

$$\sigma_p = \tanh\Big((\beta/2)(g\mu_B H + \sum_{p'} A^0_{pp'}\sigma_{p'}\Big). \tag{7.26}$$

Taking into account only the contribution of the *first-order neighbors* the Fourier transforms of the isotropic exchange integrals can be written as $A^0_{pp'} = A\Gamma^0_{p'-p}$. In general, we have $A^\nu_{pp'} = A\Gamma^\nu_{pp'}$, where

$$\Gamma^\nu_{pp'} = \sum_{\mathbf{f}\in p'} \exp(i\mathbf{f}\cdot\boldsymbol{\upsilon}) \tag{7.27}$$

the sum being performed only over *nearest neighbors* $f \in p'$ of an arbitrary atom from the pth monolayer. These quantities depend on the lattice type and surface orientation and act as "effective numbers of nearest neighbors." In the present case $\Gamma_0{}^0$ is just the number of nearest neighbors of an atom lying in the same monolayer as the atom itself, while $\Gamma_1{}^0$ denotes the number of nearest neighbors belonging to a neighboring monolayer. In view of the qualitative nature of the discussion we restrict ourselves to the high temperature range. Consequently we have

$$\sigma_p = \tfrac{1}{2}\beta\Big(g\mu_B H + A\sum_{t=-\infty}^{\infty} \Delta(p,t)\,\Gamma_t{}^0\sigma_{p+t}\Big), \tag{7.28}$$

where we use the notation (Jelitto[200])

$$\Delta(p,t) = \begin{cases} 1 & \text{for } 1 \leqslant p+t \leqslant q, \\ 0 & \text{otherwise.} \end{cases} \tag{7.29}$$

Thus one sees that the simplest form of the molecular field approach leads to the following linear system for the relative spontaneous magnetization of the film

$$\sigma_p = (\beta A/2)\sum_t \Delta(p,t)\,\Gamma_t{}^0\sigma_{p+t}\,, \qquad p = 1, 2, \ldots, q. \tag{7.30}$$

The *molecular field* in the pth monolayer is in this model

$$H_p^{\text{mol}} = A/(g\mu_B)\sum_t \Delta(p,t)\,\Gamma_t{}^0\sigma_{p+t}\,. \tag{7.31}$$

Complying with the traditional picture it is proportional to the magnetization. Here it depends also on the local distribution of spins.

[200] R. J. Jelitto, *Z. Naturforsch.* **19a**, 1567 (1964).

We note that the system of equations (7.30) represents one of the most typical problems for the sublattices model. In the following we will have to face such systems frequently. Therefore, we solve this system in some detail, as an example.

Explicitly, the system (7.30) can be written

$$-2x\sigma_1 + \sigma_2 = 0, \tag{7.32}$$

$$\sigma_{p-1} - 2x\sigma_p + \sigma_{p+1} = 0, \qquad p = 2, 3, \ldots, q-1, \tag{7.33}$$

$$\sigma_{q-1} - 2x\sigma_q = 0, \tag{7.34}$$

where

$$2x = 2/(\beta A \Gamma_1^0) - \Gamma_0^0/\Gamma_1^0 \tag{7.35}$$

Looking for solutions of the form $\sigma_p = r^p$, one finds from (7.33)

$$r = x \pm (x^2 - 1)^{1/2}. \tag{7.36}$$

Thus the general solution of (7.33) is the linear combination

$$\sigma_p(x) = c_1(x + (x^2 - 1)^{1/2})^p + c_2(x - (x^2 - 1)^{1/2})^p. \tag{7.37}$$

Assuming $|x| < 1$ we can put $x = \cos k$ and the general solution takes the form

$$\sigma_p(k) = c_1 e^{ikp} + c_2 e^{-ikp} \tag{7.38}$$

It must be suited also for Eqs. (7.32) and (7.34), from which the constants c_1 and c_2 are found as solutions of a linear and homogeneous system

$$\begin{aligned} c_1 + c_2 &= 0, \\ e^{i(q+1)k} c_1 + e^{-i(q+1)k} c_2 &= 0. \end{aligned} \tag{7.39}$$

There are nontrivial solutions only for

$$k_\tau = \pi\tau/(q+1), \qquad \tau = \text{integer}. \tag{7.40}$$

Now, from Eq. (7.35) one finds the *eigentemperatures*

$$T^\tau = A/(2k_B)(\Gamma_0^0 + 2\Gamma_1^0 \cos(\pi\tau/(q+1)) \tag{7.41}$$

and from Eqs. (7.38) and (7.39),

$$\sigma_p = \text{const.} \sin(p\pi\tau/(q+1)). \tag{7.42}$$

Assuming $|x| > 1$ and taking $x = \cosh k$, one finds $\sigma_p = c_1 e^{kp} + c_2 e^{-kp}$.

Such a combination is a trivial solution of the whole system (7.32)–(7.34); if introduced in Eqs. (7.32) and (7.34), it gives $\sigma_p \sim \sinh k$, but the corresponding compatibility condition requires $k = 0$, so that $\sigma_p = 0$ for all p.

Let us consider the general method of solving systems of equations like (7.32)–(7.34). Such systems occur as a natural consequence of the planar sublattices model. Equations (7.32) and (7.34) must be considered as boundary conditions, which determine the actual trend of the general solution. *An alternative*, pointed out by Jelitto,[200] is to assume that all the inner equations are of the same type, so that the *ideal periodicity* of the bulk body is artificially imposed. It is quite natural that plane waves in the perpendicular direction would be the particular solutions of this *uniform* system. In the simplest case, the problem takes the following form

$$\begin{aligned} \sigma_0 - 2x\sigma_1 + \sigma_2 &= 0, \\ \sigma_1 - 2x\sigma_2 + \sigma_3 &= 0, \\ &\vdots \\ \sigma_{q-2} - 2x\sigma_{q-1} + \sigma_q &= 0, \\ \sigma_{q-1} - 2x\sigma_q + \sigma_{q+1} &= 0, \end{aligned} \tag{7.43}$$

with the new boundary conditions

$$\sigma_0 = 0, \qquad \sigma_{q+1} = 0. \tag{7.44}$$

One sees easily that, assuming a general solution of the form

$$\sigma_p = c_1 e^{ipk} + c_2 e^{-ipk}, \tag{7.45}$$

these boundary conditions lead to the same results as the correct end equations. This *simpler variant* is often preferred in similar problems.

Another variant of practical use consists, obviously, in diagonalizing directly the secular matrix which corresponds to the system (7.32)–(7.34). In our case we have

$$\begin{vmatrix} -2x & 1 & 0 & \cdots & & & 0 \\ 1 & -2x & 1 & & & & \\ 0 & 1 & -2x & & & & \vdots \\ & & & \ddots & & & \\ \vdots & & & & -2x & 1 & 0 \\ & & & & 1 & -2x & 1 \\ 0 & & \cdots & & 0 & 1 & -2x \end{vmatrix} = 0. \tag{7.46}$$

This determinant represents the Gegenbauer polynomial of qth degree[201] $C_q^1(-x)$. Using $x = \cos k$ when $|x| < 1$, and using $x = \cosh k$ when $|x| > 1$, one finds

$$C_q^1(\cos k) = (\sin(q+1)k)/\sin k = 0$$

and

$$C_q^1(\cosh k) = (\sinh (q+1)k)/\sinh k = 0$$

respectively. From the first equation we have $k_\tau = \pi\tau/(q+1)$, τ-integer, and alternatively from the second $k = 0$, which are exactly the previous results.

Let us return to Eq. (7.41) which leads to many values of the temperature corresponding to transition temperatures. We shall analyze the *maximal eigentemperature*. We only assert that this maximal eigentemperature is not degenerate and to this eigenvalue corresponds the eigenvector $\sigma_1, \ldots, \sigma_q$ with all coordinates positive. Hence the solution, being both nondegenerate and positive, has a *homogeneous ferromagnetic* character; i.e., the type of ordering compatible with the highest eigenvalue is *ferromagnetic*. The other transition temperatures correspond in principle to other types of ordering.

We may ask how it is possible to decide which of the different kinds of ordering will actually occur. If we imagine cooling the material from a high temperature, then it appears obvious that the type of ordering with the highest transition temperature will occur first. We have then to find out if there will be transitions to other types of ordering when the temperature is further reduced. It may be proved, as pointed by Smart,[202] that for normal systems such as the Heisenberg system this does not happen.

The above mentioned eigentemperatures are functions of the number of monolayers (film thickness) as well as of the lattice type and surface orientation via Γ_0^0 and Γ_1^0. Valenta[89–91] has stated that the maximum eigentemperature can be taken as the Curie temperature of the ferromagnetic thin film. Using numerical methods Valenta solved in a better approximation the system (7.23) and reported the dependence of the Curie temperature, as well as of the relative magnetization (Eq. (7.25)), on the film thickness, lattice type and surface orientation (Fig. 7). One sees that there are many cases in which even a single planar sub-

[201] We use the notation of L. Brillouin and M. Parodi, *in* "Propagation des ondes dans les milieux périodiques," p. 97. Masson and Cie, Dunod, Paris, 1956.

[202] I. S. Smart, "Effective Field Theories of Magnetism." Saunders, Philadelphia, Pennsylvania, 1966.

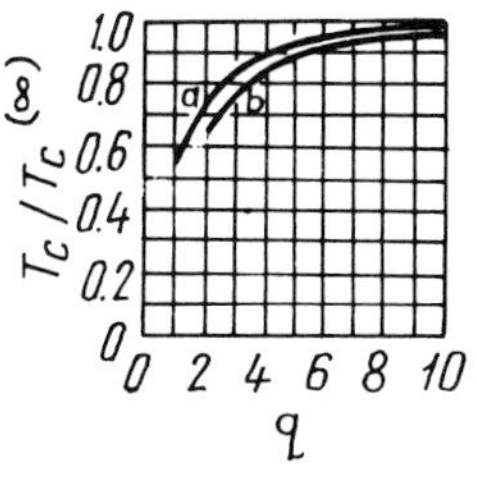

FIG. 7. Ni and Co Curie temperature versus film thickness. Curve (a) Valenta's method [L. Valenta, *Izv. Akad. Nauk., SSSR, Ser. Fiz.* **21**, 879 (1957) (Russian)]; Curve (b) Corciovei's method [A. Corciovei, *Czech. J. Phys.* **B10**, 568 (1960)].

lattice seems to support the ferromagnetic ordering of spins. As emphasized previously this is only a result of the molecular field approximation.

The Valenta model of *planar sublattices* provides a natural interpretation of the molecular field approach to thin films. The following picture is illuminating: each spin of an inner monolayer is acted upon by a *molecular field* given by the interaction within this monolayer with neighbouring monolayers. Since the surface atoms have fewer nearest neighbors than the interior ones, the Curie temperature and hence the magnetization at a given temperature are reduced. The molecular field diminishes from the interior to the exterior of the film and, therefore, the effect of the spin alignment itself is diminished in the external sublattices, which thus posses a smaller relative magnetization.

b. The constant coupling approximation

Certain improved approximations can be made within the framework of the molecular field approach. Various refinements introducing short-range order have been devised. These refinements are related to the various *cluster* approximations which may be derived as a cluster series from a direct expansion of the partition function so that the zero order results in the molecular field theory.[203] The two-spin cluster result is identical to the constant coupling approximation.

In a series of papers, Wojtczak[204,205] has applied the *constant coupling approximation* in the variant of Kasteleijn and Kranendonk[206] and its form as modified by Oguchi and Ono[207] to the study of ferromagnetic

[203] For such an approach see B. Stried and H. B. Callen, *Phys. Rev.* **130**, 1798 (1963).
[204] L. Wojtczak, *Bull. Acad. Pol. Sci.* **16**, 535 (1968).
[205] L. Wojtczak, *J. Phys.* **30**, 578 (1969).
[206] P. N. Kasteleijn and J. Van Kranendonk, *Physica* **22**, 317 (1956).
[207] T. Oguchi and I. Ono, *J. Phys. Soc. Japan* **21**, 2178 (1966).

thin films. It was supposed that the Hamiltonian of the system is

$$\mathscr{H} = - \sum_{\mathbf{f}p,\mathbf{g}p'} A_{\mathbf{f}p,\mathbf{g}p'}[S^z_{\mathbf{f}p}S^z_{\mathbf{g}p'} + \eta_{pp'}(S^x_{\mathbf{f}p}S^x_{\mathbf{g}p'} + S^y_{\mathbf{f}p}S^y_{\mathbf{g}p'})] - g\mu_{\mathrm{B}}H \sum_{\mathbf{f}p} S^z_{\mathbf{f}p}, \tag{7.47}$$

where $\eta_{pp'}$ is an anisotropy parameter.[208] Taking account of the fact that the constant coupling may be considered as an improvement of the molecular field theory in which the interaction between a pair of spins is treated exactly while the interaction with the remainder of the film is accounted for by an effective field, we may use the model of physical sublattices of Valenta.[89] Then the partition function of the film may be written as $Z = \prod_p Z_p$ where Z_p is calculated in a representation in which the z component of the total spin of the pth layer is diagonal. This component may be related to the long range order parameter σ_p. The free energy is given by[206]

$$\beta F = \sum_p \int_0^\beta E_p(\beta', \sigma_p)\, d\beta' - \sum_p w(\sigma_p), \tag{7.48}$$

where $w(\sigma_p)$ is the number of states corresponding to a given value of σ_p. The mean value of the energy of the system of spins from the pth layer $E_p(\beta, \sigma_p)$ can be related to the partial density matrix $\rho_{pp'}$ as follows

$$E_p(\beta, \sigma_p) = (N/2) \sum_{p'} \Gamma^0_{pp'} \operatorname{tr}(\rho_{pp'}\mathscr{H}_{pp'}), \tag{7.49}$$

where $\mathscr{H}_{pp'}$ describes the interaction (of the type (7.47)) between two spins situated in the planes p and p', and $\rho_{pp'}$ is the projection of the total density matrix into a two-dimensional subspace, being given by

$$\rho_{pp'} = \exp(-\beta\mathscr{H}^{\mathrm{e}}_{pp'})/\operatorname{tr}\exp(-\beta\mathscr{H}^{\mathrm{e}}_{pp'}). \tag{7.50}$$

Here $\mathscr{H}^{\mathrm{e}}_{pp'}$ is an *effective Hamiltonian* for a pair of spins situated in the planes p and p'. It is, of course, this operator which is used to attempt to compute the cluster methods. In the case of $\frac{1}{2}$ spins this Hamiltonian has the general form

$$\mathscr{H}^{\mathrm{e}}_{ij} = -A_1\mathbf{S}_i\mathbf{S}_j - A_2S_i^zS_j^z - g\mu_{\mathrm{B}}A_3(S_i^z + S_j^z). \tag{7.51}$$

[208] It is considered that the z-axis, situated in the film plane, is an easy axis of magnetization.

In the theory of ferromagnetic thin films the *constant coupling approximation* consists in taking for $\mathscr{H}^{\mathrm{e}}_{pp'}$ the operator

$$\mathscr{H}^{\mathrm{e}}_{pp'} = - A_{\mathbf{f}p,\mathbf{g}p'}[S^z_{\mathbf{f}p}S^z_{\mathbf{g}p'} + \eta_{pp'}(S^x_{\mathbf{f}p}S^x_{\mathbf{g}p'} + S^y_{\mathbf{f}p}S^y_{\mathbf{g}p'})]$$

$$- g\mu_{\mathrm{B}}H_p\, S^z_{\mathbf{f}p} - g\mu_{\mathrm{B}}H_{p'}\, S^z_{\mathbf{g}p'}\,. \qquad (7.52)$$

H_p denotes the molecular field in the pth layer which has to be determined in some consistent way (see formula (7.31)). It remains, therefore, to calculate the eigenvectors and eigenvalues of Hamiltonians $\mathscr{H}^{\mathrm{e}}_{pp'}$ and $\mathscr{H}_{pp'}$.

Supposing that this calculation has been made, we can obtain the spontaneous magnetizations σ_p from the equation

$$\sigma_p = \mathrm{tr}[\rho_{pp}(S^z_{\mathbf{f}p} + S^z_{\mathbf{g}p})]. \qquad (7.53)$$

Wojtczak[205] has proved that

$$\sigma_p = \sinh(\beta g\mu_{\mathrm{B}}H_p)/(\cosh(\beta g\mu_{\mathrm{B}}H_p) + e^{-\beta A/2}\cosh(\beta\eta A/2)), \qquad (7.54)$$

where η and A are respectively the anisotropy constant and the exchange integral taken for nearest neighbors. The molecular fields H_p are then determined from the equation $\partial F/\partial\sigma_p = 0$. It may be seen that in the case of zero coupling ($A = 0$) the system (7.54) becomes the molecular field system (7.26).

The solution of the system (7.54) may be found as shown in Section 7a) and is given by

$$\sigma_p = X\cos[\alpha(p - (q+1)/2], \qquad p = 1, 2, \ldots, q, \qquad (7.55)$$

where α and X are constants. The Curie temperature may be determined requiring again a nontrivial solution for the system (7.54).

Comparing the Eqs. (7.55) giving the variation of the spontaneous magnetization across the film with similar calculations of Pearson[209] and Valenta *et al.*[210] it may be noted that this variation is much more sensitive to the position of the layer in the film than was predicted by Pearson[209] and Valenta.[91] Comparing the predicted values of the Curie temperature with the experimental results of Gradmann and Müller,[211] it may be stated that the constant coupling approximation reproduces well enough these data (Fig. 8).

[209] J. Pearson, *Phys. Rev.* **138A**, 213 (1965).

[210] L. Valenta, W. Haubenreisser, and W. Brodkorb, *Phys. Status Solidi* **26**, 191 (1968).

[211] U. Gradmann and J. Müller, *J. Appl. Phys.* **39**, 1379 (1968).

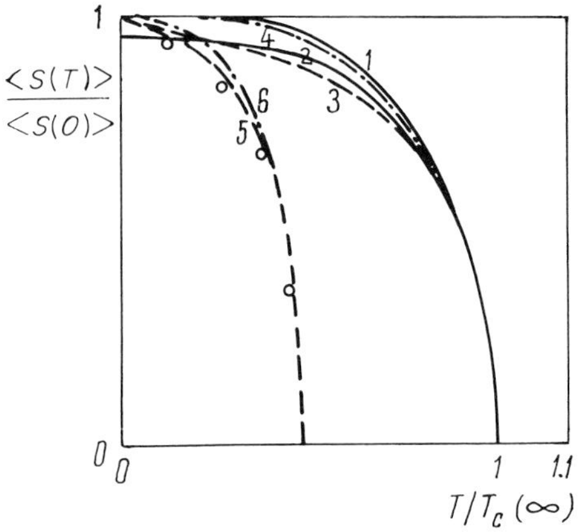

FIG. 8. Spontaneous magnetization versus temperature. (1) the mean value of the magnetization in the molecular field approximation; (2) the mean value of the magnetization in the constant coupling approximation [after P. N. Kastelejn and J. Van Kranendonk, *Physica* **22**, 317 (1956)]; (3, 4) the mean value of the magnetization in the constant coupling approximation [after L. Wojtczak, *J. Phys.* **30**, 578 (1969)]; (5, 6) the mean value of the magnetization for $q = 6$, and $q = 3$ (after L. Wojtczak *ibid.*); (7) the experimental points for $q \approx 3$ [after U. Gradmann and J. Müller, *J. Appl. Phys.* **39**, 1379 (1968)].

c. The Bethe–Peierls–Weiss Method

In the previous paragraph we have seen that the constant coupling method gives somewhat different and presumably better results for thin films than does the molecular field method. A possible next step in improving the approximation would be to consider more than two spins in a *cluster*. By further enlarging the *cluster* of spins with which a given spin interacts, one may expect that the approximation becomes better and better since more interactions within the cluster are treated exactly. However successive cluster approximations are not expected to converge rapidly for a dense ferromagnet.

The Bethe–Peierls–Weiss method remains one of the approaches, which though using a large cluster of spins,[212] is manageable. In fact Weiss[213] estimated that higher approximations would give only a negligible improvement. Pearson[209] has extended the Bethe–Peierls–Weiss approximation to the case of thin films. As usual since the theory is of an effective field type, the Valenta model of sublattices was used. Thus a different molecular field was defined for each plane of the film which leads to a theory closely resembling that of Valenta. The molecular

[212] The cluster consists of an arbitrary central atom and its nearest neighbors.
[213] P. R. Weiss, *Phys. Rev.* **74**, 1493 (1948).

fields are self-consistently determined by requiring that the average spin projection of the central atom and one of its nearest neighbors in the same plane are equal. The cluster Hamiltonian in the Pearson formulation is given by

$$\mathscr{H}_p = -AS_{0p}\left(S_{p-1} + \sum_{f\neq 0} S_{fp} + S_{p+1}\right)$$

$$-H_{p-1}S^z_{p-1} - H_p \sum_{f\neq 0} S^z_{fp} - H_{p+1}S^z_{p+1}, \qquad (7.56)$$

where S_{0p} is the spin of the central atom of the cluster in the plane p of the film whereas S_{fp} is one of its nearest neighbors in the same plane; $S^z_{p\pm 1} = \sum_{f\in p\pm 1} S^z_{fp\pm 1}$ the sum being performed only over the nearest neighbors of the atom $(0, p)$. The spontaneous relative magnetization in the layer p is calculated thus:

$$\sigma_p = S^{-1}\,\mathrm{tr}(S^z_{0p}e^{-\beta\mathscr{H}_p})/(\mathrm{tr}\, e^{-\beta\mathscr{H}_p}). \qquad (7.57)$$

The molecular fields H_p are chosen such that

$$\mathrm{tr}(S^z_{fp}e^{-\beta\mathscr{H}_p}) = \mathrm{tr}(S^z_{0p}e^{-\beta\mathscr{H}_p}). \qquad (7.58)$$

Although Pearson started with an isotropic Heisenberg Hamiltonian the theory can be equally well formulated so as to take account of the existence of the anisotropy, imposed both by physical arguments and theoretical criteria introduced in the introduction of Section 7.

Pearson has calculated the variation of the magnetization through the film thickness. It was found that the rise of the magnetization to its central value is very little dependent on the film thickness. The spatial decrease of the magnetization at the surfaces was even more abrupt than in the molecular field theory. Thus for a film of twenty layers, the magnetization falls at the surface to 60 % of its value in the interior and almost the entire *collapse* occurs within four layers (Fig. 9). Unfortunately there is no direct experimental evidence of this effect which is expected in the neighborhood of the Curie temperature. With decreasing temperature, the spatial variation of the magnetization has to become less important since all the layers of the film approach saturation. Therefore, it is hard to believe that spin wave resonance measurements could give reliable information about the magnitude of this effect.

d. *The Kirkwood Method*

The method developed by Kirkwood[214] in the study of order–disorder phenomena and later applied to the thermodynamics of the spin systems

[214] J. G. Kirkwood, *J. Chem. Phys.* **6**, 70 (1938).

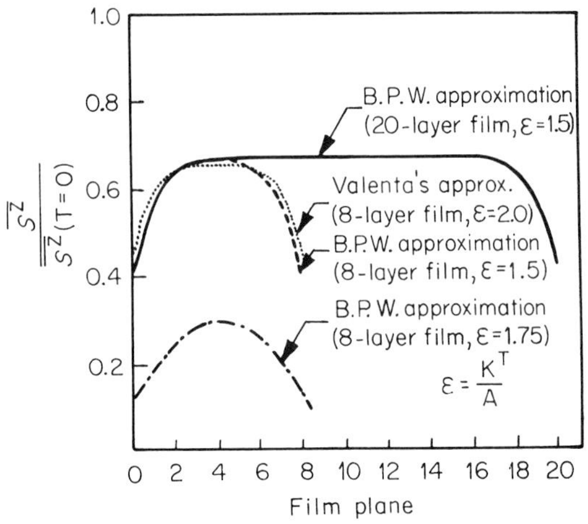

FIG. 9. Spatial distribution of the magnetization across the film [after J. Pearson, *Phys. Rev.* **138A**, 213 (1965)].

is based upon rigorous statistical mechanics and gives the molecular field theory as a leading approximation. We shall discuss here only the principal features of this theory employed in the study of ferromagnetic thin films by Corciovei and coworkers.[215–220] We will first work out the complete partition function and stress the assumptions which are involved in this theory when applied to ferromagnetic thin films. We assume that $[\mathscr{H}, S^z] = 0$, where $S^z = \sum_{p=1}^{q} \sum_{f\in p} S_{fp}^z = \sum_p S_p{}^z$ but $[\mathscr{H}, S_p{}^z]$ is not necessarily zero (in fact these two operators do not generally commute for many actual systems). Let $\{|m, \rho\rangle\}$ be a complete set of eigenfunctions of S^z which are not necessarily eigenfunctions of $\mathscr{H}$ such that

$$g\mu_B S^z|m, \rho\rangle = m|m, \rho\rangle. \tag{7.59}$$

Then we may write

$$Z = \sum_m \sum_\rho \langle m\rho|e^{-\beta\mathscr{H}}|m\rho\rangle \tag{7.60}$$

and let Z_m be

$$Z_m = \sum_\rho \langle m\rho|e^{-\beta\mathscr{H}}|m\rho\rangle, \tag{7.61}$$

[215] A. Corciovei, *Czech. J. Phys.* **10**, 568 (1960).
[216] A. Corciovei, *Czech. J. Phys.* **10**, 917 (1960).
[217] A. Corciovei and G. Costache, *Phys. Status Solidi* **16**, 329 (1966).
[218] A. Corciovei and G. Costache, *Rev. Roum. Phys.* **11**, 885 (1966).
[219] A. Corciovei and G. Ghika, *Czech. J. Phys.* **22**, 278 (1962).
[220] A. Corciovei and D. Vamanu, *J. Appl. Phys.* **39**, 1381 (1968).

so that the statistical sum may be finally presented as

$$Z = \sum_m Z_m \tag{7.61'}$$

As $[S^z, S_p{}^z] = 0, p = 1, 2,..., q$ there is a basis on which the operators S^z and $S_p{}^z$ can be simultaneously diagonalized. Let this basis be

$$|m, \rho\rangle = |m_1, \alpha\rangle \cdots |m_q, \alpha\rangle \equiv |m_1, m_2, ..., m_q; \alpha\rangle, \tag{7.62}$$

where $g\mu_B S_p{}^z|m_p, \alpha\rangle = m_p|m_p, \alpha\rangle$ and $\sum m_p = m$. If we denote $Z_{\{m_p\}}$ the following expression

$$Z_{\{m_p\}} \equiv \sum_\alpha \langle m_1, ..., m_q; \alpha|e^{-\beta\mathscr{H}}|m_1, ..., m_q; \alpha\rangle \tag{7.63}$$

then

$$Z_m = \sum_{\{m_p\}(\Sigma m_p = m)} Z_{\{m_p\}} \tag{7.64}$$

It may be proved to a very good approximation in the study of the complete partition function for a bulk body that it suffices to consider only the partition function with given m, Z_m.[221] The value of m for which

$$Z \simeq Z_m \tag{7.65}$$

is determined from the equation

$$\partial \ln Z_m/\partial m = 0. \tag{7.66}$$

On the other hand, since the molecular field theory has led to the conclusion that the quantities σ_p are thermodynamical quantities, it is reasonable to assume that for thin films

$$Z_m \simeq Z_{\{m_p\}} \tag{7.67}$$

with $Z_{\{m_p\}}$ determined from the condition

$$\partial \ln Z_{\{m_p\}}/\partial m_p = 0. \tag{7.68}$$

That is we have assumed that for all the configurations $\{m_p\}$ for which $\sum m_p = m$, only that satisfying Eq. (7.68) is realized.

Thus we have to evaluate the partition function

$$Z \simeq Z_{\{m_p\}} \tag{7.69}$$

[221] This means that from all the terms Z_m of Eq. (7.61′) only one of them gives the most important contribution to the partition function.

where $m_1, \ldots, m_q$ are fixed and then determine these quantities from Eq. (7.68).

Now, writing

$$\begin{aligned}\ln Z &\simeq \ln(\mathrm{tr}_{\{m_p\}} e^{-\beta\mathscr{H}})/(\mathrm{tr}_{\{m_p\}} 1) + \ln(\mathrm{tr}_{\{m_p\}} 1) \\ &= \ln\langle e^{-\beta\mathscr{H}}\rangle_{\{m_p\}} + \ln w(\{m_p\}),\end{aligned} \tag{7.70}$$

where

$$\langle 0\rangle_{\{m_p\}} \equiv \mathrm{tr}_{\{m_p\}}\, 0/\mathrm{tr}_{\{m_p\}}\, 1 \tag{7.71}$$

and

$$w(\{m_p\}) \equiv \prod_p \binom{N}{(N/2) - m_p}$$

we may use a semi-invariant technique, i.e., we shall write

$$\ln\langle e^{-\beta\mathscr{H}}\rangle_{\{m_p\}} = \sum_{n=1}^{\infty} [(-\beta)^n/(n!)]\, M_n \tag{7.72}$$

The first semi-invariants M_n are

$$\begin{aligned} M_1 &= \langle\mathscr{H}\rangle_{\{m_p\}}, \qquad M_2 = \langle\mathscr{H}^2\rangle_{\{m_p\}} - \langle\mathscr{H}\rangle^2_{\{m_p\}}, \\ M_3 &= \langle\mathscr{H}^3\rangle_{\{m_p\}} - 3\langle\mathscr{H}^2\rangle_{\{m_p\}}\langle\mathscr{H}\rangle_{\{m_p\}} + 2\langle\mathscr{H}\rangle^3_{\{m_p\}}\end{aligned} \tag{7.73}$$

In the calculation of these semi-invariants a diagramatic representation is quite useful.

Let us suppose that

$$\mathscr{H} = -\sum A_{\mathbf{f}p,\mathbf{g}p'}\, S_{\mathbf{f}p}\, S_{\mathbf{g}p'}$$

If we attach a vertex to every spin and a solid bond to every pair of spins interacting by the potential, then the diagrams shown in Fig. 10, would appear. Thus we may see that our expansion of the partition function is an expansion in *clusters* of linkages (or spin pairs). In the cluster

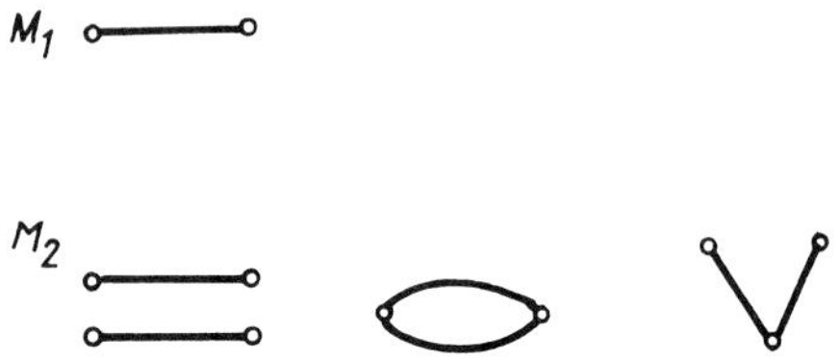

FIG. 10. The types of graphs appearing in the calculation of the first two semi-invariants.

theories the terms of this series are regrouped in such a way that they refer to clusters of spins. Therefore, it may be supposed that in the neighborhood of the Curie temperature the comparison between the results of this theory and those presented in Section 7a–c is meaningful (Fig. 11). As expected, the first approximation produces results identical

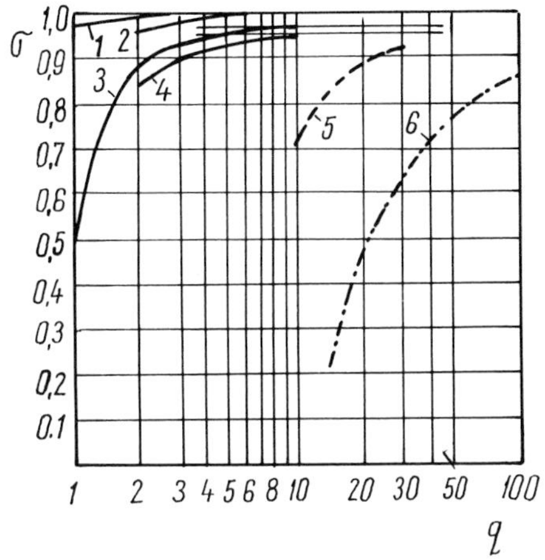

FIG. 11. Plot of $\sigma = \sigma(q)$. Theoretical results: (1) Valenta (Co) [L. Valenta, *Izv. Akad. Nauk SSSR, Ser. Fiz.* **21**, 879 (1957) (Russian)]; (2) Corciovei (Co) [A. Corciovei, *Czech. J. Phys.* **B10**, 568 (1960)]; (3) Valenta (Ni); (4) Corciovei (Ni). Experimental results: (5) Drigo (Co) [A. Drigo, *Nuovo Cimento* **VIII**, 498 (1951)]; (6) Crittenden and Hoffman (Ni) [E. C. Crittenden and R. W. Hoffman, *Rev. Mod. Phys.* **25**, 310 (1953)].

to those of the molecular field theory,[222] although the approaches are quite different.

As the calculations of the semi-invariants are rather lengthy and tedious we shall not be concerned with them here. It suffices to note that if the exchange semi-invariants are calculated (the first two moments were written down in this case by Corciovei[215]) then it is quite easy to obtain[217] the semi-invariants for the following Hamiltonian

$$\mathscr{H} = -\sum A_{\mathbf{f}p,\mathbf{g}p'}\mathbf{S}_{\mathbf{f}p}\cdot\mathbf{S}_{\mathbf{g}p'} - \sum K_{\parallel\,\mathbf{f}p,\mathbf{g}p'}S^z_{\mathbf{f}p}S^z_{\mathbf{g}p'} \qquad (7.74)$$

it being supposed that the thin film has a *uniaxial* anisotropy with the easy axis in the plane of the film (K is here a *microscopic* constant of anisotropy).

[222] Each method casts its own light on the nature of the approximation involved in the evaluation of the partition function.

Alternatively systems with perpendicular anisotropy or with parallel and perpendicular anisotropy have been studied[216,218]. When it was assumed that the magnetization vector was directed along one of the anisotropy axes it was found that the Curie temperature of the system was increasing and could equal the bulk Curie temperature even for films with few layers. This conclusion is related to the fact that, if the spontaneous magnetization is along the easy axis of anisotropy then, a gap appears, in the spin wave spectrum of the system, the magnitude of which increases with increasing values of the "constant" of anisotropy. Hence the energy required for an excitation of a spin wave has to increase, i.e., the energy involved in diminishing the value of the spontaneous magnetization increases. If two nearest neighbors anisotropies are introduced in the Hamiltonian of the system[218] one in the plane of the film described by $K_{\parallel}$ (along the axis of which the magnetization is supposed to be directed) and the other perpendicular to the film surface, $K_{\perp}$, then the perpendicular anisotropy tends to decrease the gap in the spin wave spectrum and hence to reduce the energy required for the excitation of spin waves. Therefore, it is expected that the Curie temperature depends on the difference $K_{\parallel} - K_{\perp}$, and becomes smaller and smaller (with smaller values of the difference $K_{\parallel} - K_{\perp}$) as compared with the bulk value of the Curie temperature. Similar conclusions were found in the Green's function theory of ferromagnetic thin films (see Corciovei *et al.*[223]). The above statements can be more easily understood if the Curie temperature of such systems is analytically expressed as

$$A\beta_c(q) = [(1 + \eta_{\parallel}/2)(2 + \alpha)\, h(x)]^{-1} \times\{1 - [1 - (8/\Gamma_0^0)(1 + \alpha/2)\, h(x)[1 + 2(\Gamma_1^0/\Gamma_0^0)x]^{-1}]^{1/2}\}, \tag{7.75}$$

where

$$\beta_c(q) = (k_B T_c)^{-1}, \qquad \alpha = (\eta_{\perp} - \eta_{\parallel} - \eta_{\parallel}^2/4)/(1 + \eta_{\parallel}/2)^2,$$
$$\eta_{\perp} = 2K_{\perp}/A, \qquad \eta_{\parallel} = 2K_{\parallel}/A, \qquad x = \cos(\pi/(q + 1)); \tag{7.76}$$
$$h(x) = 1 + (\Gamma_1^0/\Gamma_0^0)(1 - x)(1 + \alpha/2)^{-1}[1 + 2(\Gamma_1^0/\Gamma_0^0)x]^{-1}$$

The thickness dependence of the Curie temperature (spontaneous magnetization) is one of the most controversial subjects of magnetic film physics. From the tremendous number of experimental investigations

[223] A. Corciovei, G. Costache, W. Haubenreisser, and W. Brodkorb, *Rev. Roum. Phys.* **14**, 447 (1969).

dealing with this subject, one may conclude that they support either the idea of a "thin film effect" (thickness dependence) or the idea that the magnetic properties of films and bulk are identical at least for films thicker than about 20 Å.

As a typical alternative we mention here Neugebauer's experiments. Polycrystalline Ni films thinner than 100 Å prepared in high vacuum[224] exhibited saturation magnetization equal to that of the bulk. When air contamination was allowed[225] it was found that the saturation magnetization decreased. This decrease was attributed partly to the oxidation which caused an increase of the perpendicular anisotropy.

We stress that the thin film model, as discussed in this section is unable to account for the full variety of experimental results. It is generally thought that a more physical approach to the problem of the "thin film effect" should give much more consideration to the effects of the conditions.

8. Spin Waves

Thus far we have studied various expansions of the free energy which are valid at high temperatures. An expansion of the free energy in powers of T/T_C valid at low temperatures was first investigated by Bloch.[226] It has been predicted that a two-dimensional lattice would be nonferromagnetic for $T > 0°K$, while the bulk material would be ferromagnetic and follow the $T^{3/2}$ law. The prediction concerning the two-dimensional lattices was of questionable validity since it stemmed from a supposed breakdown of the spin wave theory. It was pointed out how such a difficulty could be overcome, that is how one may keep the picture of the simple spin wave theory. A modification of the Heisenberg isotropic Hamiltonian was suggested. It was asserted that if this Hamiltonian is supplemented with some anisotropy terms such that the system becomes weakly anisotropic, then the bound magnon states are not significant and the approximation of simple spin waves works.

The transition between the two-dimensional and three-dimensional lattices was first examined by Klein and Smith.[227] They calculated the spontaneous magnetization of the film and found that the temperature dependence of the magnetization, considered also a function of the film thickness, exhibits strong deviations from the $T^{3/2}$ law. These deviations become appreciable for thin enough films, thinner than about 100 Å.

[224] C. A. Neugebauer, *Phys. Rev.* **116**, 1441 (1959).
[225] C. A. Neugebauer, General Electric Sci. Rep., No. 2 (June 1961).
[226] F. Bloch, *Z. Phys.* **74**, 295 (1932).
[227] M. Klein and R. Smith, *Phys. Rev.* **81**, 378 (1951).

Following Keffer and Davis,[228,229] we can understand at least qualitatively why the magnetization of the film as calculated by Klein and Smith falls off with temperature with a power smaller than $\frac{3}{2}$. For a two-dimensional lattice, the piling-up of the density of states in the neighborhood of the $k_x = 0$ mode would produce a rapid decrease of the spontaneous magnetization with increasing temperature and thus a pronounced difference between the two-dimensional and three-dimensional body. Now, in a thin film for $k_x = 0$, if periodic boundary conditions in all dimensions are used, the number of spin waves with the wave number less than k_0 is $\nu_f = (4\pi)^{-1}Na^2k_0^2$, where a is the lattice spacing and N is the number of atoms in a layer. Likewise the number of spin waves with a wave number less than k_0 in a cube sample with Nq atoms is $\nu_b = (6\pi^2)^{-1}Nqa^3k_0^3$ and the ratio $\nu_f/\nu_b = (3\pi/2)(1/(qa))\, k_0^{-1}$ is very large for k_0 smaller than $\pi/(qa)$. It may be concluded that when periodic boundary conditions are applied at the film boundaries a piling up of low energy modes with $k_x = 0$ will appear, with the result that the magnetization decreases faster than $T^{3/2}$, i.e., with a smaller power of T.

Let us now remember that Klein and Smith[227] discarded the $\mathbf{k} = 0$ mode on the ground that this state brings an infinite contribution to the sample magnetization, which is an impossible result. Hence they let the minimum value of k_0 be $2\pi/(\sqrt{N}a)$, which is much lower than the value $\pi/(qa)$. Thus the elimination of the $\mathbf{k} = 0$ mode is insufficient to reduce the high density of states below $k_0 = \pi/(qa)$. This result is related to the fact that it is no longer possible to use periodic boundary conditions in all directions of the film.

In fact the thin film is a many body system finite in only one direction. The peculiar feature of such a system is its surface defect. There must be many concurrent aspects of the influence of the surface defect on the spectrum of elementary excitations like spin waves (magnons). All of these aspects are expected to contribute to the surface anisotropy of the film. The best defined and most relevant of them may be summarized as (i) the *natural* surface defect and (ii) the surface *contamination* and *substrate* influence.

(i) The *natural* surface defect results from the lack of neighbors of the surface atoms, in directions perpendicular to the surfaces. From the viewpoint of the molecular field picture, such a lack results in a difference in the fields acting on the surface and inner spins. From the point of view

[228] F. Keffer, *in* "Handbuch der Physik" (S. Flügge, ed.), Vol. Band XVIII/2, p. 1. Springer Verlag, Berlin and New York, 1966.

[229] J. Davis and F. Keffer, *J. Appl. Phys.* **34**, 1135 (1963).

of the spin wave picture this kind of defect leads to a lack of periodicity and translational symmetry in the direction perpendicular to the film, in contrast with the perfect periodicity and translational invariance along in-plane directions. Restricting the validity of the Bloch's theorem only to the in-plane directions, this circumstance determines the peculiar character of the spin waves in films,[230–233] namely that they are *plane waves* running in the plane of the film, but *standing waves* with respect to the perpendicular direction. Furthermore, while the quantization of the wave vector in the plane of the film has to obey the usual Born–Kármán cyclic condition there is no reason to assume the same rule for the perpendicular direction: one has to label the spin waves with respect to the perpendicular direction by a linear momentum whose values are determined by the actual *boundary conditions* on the film surfaces. Complex values of this "perpendicular momentum" are possible, so that surface localized modes may be expected (see, e.g., Jelitto[200]). Since it may be proved that at least one of the surface modes requires the lowest energy of excitation of the spin wave spectrum, one can conclude that the surface modes make a very important contribution in the statistical calculation of the saturation magnetization.

(ii) The aspects discussed are always involved to a certain degree in any reasonable approach. However they do not exhaust the question of the surface anisotropy. With a view to improve the model so as to make it more realistic for actual thin films, the *surface contamination* and *substrate influence* should be taken into account. It is now well known that the presence of gases (residual or used in the technology) like O_2, H_2, N_2, H_2O (see, e.g., Meiklejohn and Bean,[16] Kooi, Holmquist *et al.*,[234] Soohoo[235]) as well as additional layers of materials of technological interest like Ni, Fe, Mn[236,237] lead generally to important changes in the magnetic properties of thin specimens, such as spin wave resonance spectra, saturation magnetization, anisotropy, hysteresis loops, coercive force. The influence of the substrate also is very important and it is somewhat analogous to the contamination aspects. By protecting one

[230] W. Döring, *Z. Naturforsch.* **16a**, 1008 (1961).
[231] R. Abbel, *Z. Naturforsch.* **18a**, 371 (1963).
[232] A. Corciovei, *Phys. Rev.* **130**, 2223 (1963).
[233] L. Valenta and L. Wojtczak, *Proc. Int. Conf. Magn. Nottingham*, p. 830 (1964).
[234] S. F. Kooi, W. R. Holmquist, P. E. Wigen, and J. T. Doherty, *J. Phys. Soc. Japan, Suppl. BI* **17**, 599 (1962).
[235] R. F. Soohoo, *in* "Physica Magnitnyh Plenok," p. 378. Irkutsk, 1968.
[236] A. Stankoff, *in* "Physica Magnitnyh Plenok," p. 422. Irkutsk, 1968.
[237] B. Waksmann, O. Massenet, and S. F. Kooi, *in* "Physica Magnitnyh Plenok," p. 308. Irkutsk, 1968.

surface against the factors which act on the other one, introducing additional stresses, determining some features of the deposition and crystallization processes, the substrate makes conditions on its boundary with the specimen quite different from those at the opposite boundary, so that the thin film may be thought of generally as an *asymmetric* specimen.

Faced with these complications, an improved model should at least give the right number of *normal* spin wave modes, as prescribed by its boundary conditions. The central idea is that the surface anisotropy tends mainly to keep the surface spins fixed, thereby reducing the number of normal modes and modifying their sequence. The phenomenon is called *surface spin pinning*. The pinning mechanism has been discussed by Kittel[238] and Pincus[239] for linear chains of spins, and used by Soohoo,[240] Ayukawa,[241] Portis,[242] Davis,[243] Wojtczak,[244] Puszkarski,[245] Stankoff,[236] Corciovei and Vamanu,[246] to account for thin film properties. In the present paper we shall discuss a simple model of thin films with *pinned surfaces*.

A natural way of including the pinning effect in the model is to replace the usual *volume* anisotropic term of the Hamiltonian (see Eq. (7.76)) by an appropriate one, which should exhibit special anisotropic constants for the surfaces, and stressing also the asymmetry of the surfaces. We deal in the following only with the quasi-saturation approximation which corresponds to weakly excited spin waves, at low temperatures.[247]

As is well known, in this approximation the mean value of the spin displacements $S_{\mathbf{f}p} - S_{\mathbf{f}p}^{z'}$ come close to zero, so that the right-hand side of the commutation rule of the spin operators $S_{\mathbf{f}p}^{+'}$ and $S_{\mathbf{f}p}^{-'}$ can be replaced approximately by a c-number:

$$[S_{\mathbf{f}p}^{+'}, S_{\mathbf{g}p'}^{-'}] = 2S_{\mathbf{f}p}^{z'}\delta_{\mathbf{f}p,\mathbf{g}p'} \simeq 2S_{\mathbf{f}p}\delta_{\mathbf{f}p,\mathbf{g}p'}\,. \tag{8.1}$$

[238] C. Kittel, *Phys. Rev.* **110**, 1295 (1958).
[239] P. Pincus, *Phys. Rev.* **118**, 658 (1960).
[240] R. F. Soohoo, *Phys. Rev.* **131**, 594 (1963).
[241] T. Ayukawa, *J. Phys. Soc. Japan* **18**, 970 (1963).
[242] A. M. Portis, *Appl. Phys. Lett.* **2**, 69 (1963).
[243] I. Davis, *J. Appl. Phys.* **36**, 3520 (1965).
[244] L. Wojtczak, *Rev. Roum. Phys.* **12**, 577 (1967).
[245] H. Puszkarski, *Phys. Status Solidi* **22**, 355 (1967).
[246] A. Corciovei and D. Vamanu, *IEEE Trans. Magn.* **MAG-5**, 180 (1969).
[247] This approximation corresponds to the first step of the Holstein–Primakoff expansion. Here the spin wave picture in the second quantization formulation is used.[93]

In this way the spin variables acquire a bosonic character since the obvious normalization

$$S_{\mathbf{f}p}^{+'} = \sqrt{2S_{\mathbf{f}p}}\, b_{\mathbf{f}p}\,, \qquad S_{\mathbf{f}p}^{-'} = \sqrt{2S_{\mathbf{f}p}}\, b_{\mathbf{f}p}^{\dagger} \tag{8.2}$$

leads to the bosoniclike operators $b_{\mathbf{f}p}$, obeying the rules

$$[b_{\mathbf{f}p}\,, b_{\mathbf{g}p'}^{\dagger}] = \delta_{\mathbf{f}p,\mathbf{g}p'}\,, \qquad [b_{\mathbf{f}p}\,, b_{\mathbf{g}p'}] = 0. \tag{8.3}$$

The spin displacement operator becomes accordingly:

$$S_{\mathbf{f}p} - S_{\mathbf{f}p}^{z'} = b_{\mathbf{f}p}^{\dagger} b_{\mathbf{f}p} = n_{\mathbf{f}p}\,. \tag{8.4}$$

Now let us recall the standard expansion (3.22) of the Hamiltonian relative to the spin deviations. In this case it is convenient to separate it in three parts as follows

$$\mathscr{H}(S_{\mathbf{f}p}^{\alpha}) \simeq \mathscr{E}_0 + \mathscr{H}_{\mathrm{SW}} + h.$$

Here $\mathscr{E}_0$ denotes, as usual the ground state energy. It is only a function of the magnetic configuration as prescribed by the Eqs. (3.21) for the direction cosines $\gamma_{\mathbf{f}p}^{\alpha}$. The part h includes all the terms involving more than two bosonic operators $b_{\mathbf{f}p}$, $b_{\mathbf{f}p}^{\dagger}$. In the quasi-saturation approximation this part will be neglected.

Thus the spin wave analysis of the saturation magnetization at low temperatures will focus on the part $\mathscr{H}_{\mathrm{SW}}$ which has a quadratic form in the bosonic variables. Generally it has the following form

$$\mathscr{H}_{\mathrm{SW}} = -2 \sum_{\mathbf{f}p} \lambda_{\mathbf{f}p} n_{\mathbf{f}p} + \mathscr{H}_2\,. \tag{8.5}$$

The quantities $\lambda_{\mathbf{f}p}$ are the Lagrange factors involved in Eq. (3.21). They may be expressed as

$$2\lambda_{\mathbf{f}p} = \sum_{\alpha} (\partial\mathscr{E}_0/\partial\gamma_{\mathbf{f}p}^{\alpha})\, S_{\mathbf{f}p}^{-1} \gamma_{\mathbf{f}p}^{\alpha}\,. \tag{8.6}$$

The part $\mathscr{H}_2$ involves a bilinear mixture of the bosonic operators b and $b^{\dagger}$.

The whole spin wave Hamiltonian $\mathscr{H}_{\mathrm{SW}}$ depends on the magnetic configuration through the direction parameters $\gamma_{\mathbf{f}p}^{\alpha}$ and the related $\mathscr{A}_{\mathbf{f}p}^{\alpha}$. Assuming a single domain structure with the magnetization in the $x0z$ plane ($0x$ is perpendicular to the film; see Fig. 1) we have

$$A^x = \gamma^z/2, \qquad A^y = -\, i/2, \qquad A^z = -\,\gamma^x/2 \tag{8.7}$$

Now, because of the perfect periodicity in the film plane one assumes for the bosonic operators $b_{\mathbf{f}p}$ and $b_{\mathbf{f}p}^{\dagger}$ a plane wavelike character, using their Fourier expansions only in the plane of sublattices:

$$\begin{aligned} b_{\mathbf{f}p} &= (1/N^{1/2}) \sum_{\mathbf{v}} b_{p\mathbf{v}} \exp[i\mathbf{f} \cdot \mathbf{v}], \\ b_{\mathbf{f}p}^{\dagger} &= (1/N^{1/2}) \sum_{\mathbf{v}} b_{p\mathbf{v}}^{\dagger} \exp[-i\mathbf{f} \cdot \mathbf{v}]. \end{aligned} \tag{8.8}$$

In this way, the effective spin wave Hamiltonian $\mathscr{H}_{\mathrm{SW}}$ becomes a quadratic combination of bosonic amplitudes $b_{p\mathbf{v}}$ and $b_{p\mathbf{v}}^{\dagger}$ of the form

$$\begin{aligned} \mathscr{H}_{\mathrm{SW}} = \sum_{\mathbf{v}} \sum_{p,p'} &((1/2)\, \mathscr{R}_{pp'}^{(*)\mathbf{v}} b_{p\mathbf{v}}^{\dagger} b_{p',-\mathbf{v}}^{\dagger} + \mathscr{S}_{pp'}^{\mathbf{v}} b_{p\mathbf{v}}^{\dagger} b_{p'\mathbf{v}} \\ &+ (1/2)\, \mathscr{R}_{pp'}^{\mathbf{v}}\, b_{p,-\mathbf{v}} b_{p'\mathbf{v}}) \end{aligned} \tag{8.9}$$

The coefficients $\mathscr{R}_{pp'}^{\mathbf{v}}$, $\mathscr{R}_{pp'}^{(*)\mathbf{v}}$, $\mathscr{S}_{pp'}^{\mathbf{v}}$ follow from the composition of the model Hamiltonian. They include plane Fourier transforms of the coupling parameter, like the exchange integral $A_{\mathbf{f}p,\mathbf{g}p'}$, the uniaxial anisotropy integrals $K_{\mathbf{f}p,\mathbf{g}p'}$, and the dipolar coupling parameter $D_{\mathbf{f}p,\mathbf{g}p'}$ with the general form:

$$Q_{pp'}^{\mathbf{v}} = \sum_{\mathbf{g}\in p'} Q_{\mathbf{f}p,\mathbf{g}p'} \exp[-i(\mathbf{f} - \mathbf{g}) \cdot \mathbf{v}]. \tag{8.10}$$

The following relations are obeyed:

$$\begin{aligned} \mathscr{R}_{pp'}^{\mathbf{v}} &= \mathscr{R}_{p'p}^{-\mathbf{v}}, \ \mathscr{R}_{pp'}^{(*)\mathbf{v}} \equiv (\mathscr{R}_{pp'}^{-\mathbf{v}})^{*}; \\ \mathscr{S}_{pp'}^{\mathbf{v}} &= (\mathscr{S}_{p'p}^{\mathbf{v}})^{*}, \ \mathscr{S}_{pp'}^{(*)\mathbf{v}} \equiv (\mathscr{S}_{pp'}^{-\mathbf{v}})^{*}. \end{aligned} \tag{8.11}$$

It is obvious that they are consistent with the hermiticity of the spin wave Hamiltonian.

The diagonalization of quadratic forms like (8.9) was widely studied by Bogoljubov and Valatin for fermions and by Bogoljubov and Tyablikov[248] for bosons. The procedure consists in performing a canonical transformation of the form

$$b_{p\mathbf{v}} = \sum_{\tau} (u_{p\tau}^{\mathbf{v}} \xi_{\tau\mathbf{v}} + v_{p\tau}^{-\mathbf{v}*} \xi_{\tau,-\mathbf{v}}^{\dagger}), \tag{8.12}$$

where $\xi_{\tau\mathbf{v}}$ and $\xi_{\tau\mathbf{v}}^{\dagger}$ are also bosonic variables obeying the rules

$$[\xi_{\tau\mathbf{v}}, \xi_{\tau'\mathbf{v}'}^{\dagger}] = \delta_{\tau\tau'} \delta_{\mathbf{v}\mathbf{v}'}, \qquad [\xi_{\tau\mathbf{v}}, \xi_{\tau'\mathbf{v}'}] = 0. \tag{8.13}$$

[248] N. N. Bogoljubov and S. V. Tyablikov, *JETP* **19**, 256 (1949).

Assuming that the new variables satisfy the dynamical equations

$$\begin{aligned} i\, d\xi_{\tau\nu}/dt &= E_{\tau\nu}\xi_{\tau\nu}\,, \\ -i\, d\xi^{\dagger}_{\tau\nu}/dt &= E_{\tau\nu}\xi^{\dagger}_{\tau\nu}\,, \end{aligned} \tag{8.14}$$

the following system of equations involving the amplitudes u, v is found:

$$\begin{aligned} E_{\tau\nu}u^{\nu}_{p\tau} &= \sum_{p'} \mathscr{S}^{\nu}_{pp'}u^{\nu}_{p'\tau} + \sum_{p'} \mathscr{R}^{(*)\nu}_{pp'}v^{\nu}_{p'\tau}\,, \\ -E_{\tau\nu}v^{\nu}_{p\tau} &= \sum_{p'} \mathscr{R}^{\nu}_{pp'}\, u^{\nu}_{p'\tau} + \sum_{p'} \mathscr{S}^{(*)\nu}_{pp'}v^{\nu}_{p'\tau}\,. \end{aligned} \tag{8.15}$$

Then one finds that the amplitudes $u^{\nu}_{p\tau}$ and $v^{\nu}_{p\tau}$ obey also the orthogonality rules

$$\begin{aligned} &\sum_{p} (u^{\nu *}_{p\tau}u^{\nu}_{p\tau'} - v^{\nu *}_{p\tau}v^{\nu}_{p\tau'}) = \delta_{\tau\tau'}\,, \\ &\sum_{p} (u^{\nu}_{p\tau}v^{-\nu}_{p\tau'} - v^{\nu}_{p\tau}u^{-\nu}_{p\tau'}) = 0. \end{aligned} \tag{8.16}$$

In this way the effective spin wave Hamiltonian takes the following form

$$\mathscr{H}_{\mathrm{SW}} = \sum_{\tau}\sum_{\nu} E_{\tau\nu}\xi^{\dagger}_{\tau\nu}\xi_{\tau\nu} + \Delta\mathscr{E}_0\,. \tag{8.17}$$

Thus the quantity $E_{\tau\nu}$ is revealed as the energy of an elementary excitation (spin wave or magnon). The quantity

$$\Delta\mathscr{E}_0 = -\sum_{\tau}\sum_{\nu} E_{\tau\nu} \sum_{p} |v^{\nu}_{p\tau}|^2 \tag{8.18}$$

is related to the zero point energy of the system of vibrating spins. The amplitudes $v^{\nu}_{p\tau}$ disappear if the effective spin wave Hamiltonian (8.9) does not contain the terms $b^{\dagger}_{p\nu}b^{\dagger}_{p',-\nu}$ and $b_{p,-\nu}b_{p'\nu}$. The corresponding coefficients $\mathscr{R}^{\nu}_{pp'}$, $\mathscr{R}^{(*)\nu}_{pp'}$, vanish only if the thin film Hamiltonian *does not contain* quadratic anisotropic terms for axes perpendicular to the magnetization. As was discussed by Corciovei and Vamanu,[249] such terms may be of the uniaxial form $K_{\perp}S^{x}_{\mathrm{f}p}S^{x}_{\mathrm{g}p'}$; when the magnetization is along the z axis, of the dipolar form, etc.

For vanishing $v^{\nu}_{p\tau}$ the canonical transform (8.12) takes a simpler form, namely,

$$b_{p\nu} = \sum_{\tau} u^{\nu}_{p\tau}\xi_{\tau\nu}\,. \tag{8.12$'$}$$

[249] A. Corciovei and D. Vamanu, *in* "Physica Magnitnyh Plenok," p. 279. Irkutsk, 1968.

The spin wave amplitudes are now solutions of the eigenequations

$$E_{\tau\nu}\, u^{\nu}_{\tau p} = \sum_{p'=1}^{q} \mathscr{S}_{pp'} u^{\nu}_{p'\tau}\,, \tag{8.15$'$}$$

and the orthogonality rules (8.16) becomes

$$\sum_{p=1}^{q} u^{\nu *}_{p\tau}\, u^{\nu}_{p\tau'} = \delta_{\tau\tau'}\,. \tag{8.16$'$}$$

Accordingly, the diagonal in the magnon occupation number spin wave Hamiltonian is

$$\mathscr{H}_{\mathrm{SW}} = \sum_{\tau}\sum_{\nu} E_{\tau\nu}\xi^{\dagger}_{\tau\nu}\xi_{\tau\nu}\,. \tag{8.17$'$}$$

Now we calculate the *monolayer relative magnetization* as well as the total relative magnetization of the film. The distribution of the magnetization across the film is given by the average of the sum over the local magnetic moments lying in a monolayer, which is proportional to

$$\langle S_p{}^z\rangle = \sum_{\mathbf{f}\in p} \langle S^z_{\mathbf{f}p}\rangle. \tag{8.19}$$

In terms of bosonic variables we have

$$\langle S_p{}^z\rangle = NS - \sum_{\mathbf{f}\in p} \langle b^{\dagger}_{\mathbf{f}p} b_{\mathbf{f}p}\rangle. \tag{8.20}$$

Here the angular brackets denote statistical averages performed using $\exp\{-\beta(\mathscr{E}_0 + \mathscr{H}_{\mathrm{SW}})\}$. When the standard canonical transformations (8.8) and (8.12) are used, the *relative monolayer magnetization* appears as a sum over all spin wave modes allowed by the surface conditions

$$\langle S_p{}^z\rangle/S_p{}^0 = 1 - (1/S_p{}^0)\sum_{\tau}\sum_{\nu}\{(|u^{\nu}_{p\tau}|^2 + |v^{\nu}_{p\tau}|^2)\,\mathscr{N}_{\tau\nu}\}$$

$$-(1/S_p{}^0)\sum_{\tau}\sum_{\nu}|v^{\nu}_{p\tau}|^2, \tag{8.21}$$

where $S_p{}^0 = NS$ and $\mathscr{N}_{\tau\nu}$ denotes the usual Bose distribution

$$\mathscr{N}_{\tau\nu} = [\exp(\beta E_{\tau\nu}) - 1]^{-1}. \tag{8.22}$$

For $v^{\nu}_{p\tau} = 0$ the magnetization distribution takes a simpler form

$$\sigma_p = \langle S_p{}^z\rangle/S_p{}^0 = 1 - (1/S_p{}^0)\sum_{\tau}\sum_{\nu}|u^{\nu}_{p\tau}|^2/(\exp(\beta E_{\tau\nu}) - 1). \tag{8.21$'$}$$

Taking into account the orthogonality rule (8.16′), the relative magnetization of the whole film becomes

$$\begin{aligned}\sigma &= \langle S^z \rangle / S^0 \\ &= 1 - (1/S^0) \sum_{\tau} \sum_{\nu} (\exp(\beta E_{\tau\nu}) - 1)^{-1},\end{aligned} \tag{8.23}$$

where $S^0 = qS^0{}_p$.

The canonical transformations (8.12) and (8.12′) are very characteristic. They indicate that the spin wave modes in thin films depend on more than the plane wave vector **ν**. The diagonalization of the spin wave Hamiltonian needs a *new index* τ, which is taken to label the perpendicular standing component of the spin wave. The corresponding spectrum of the energy can not be determined by the cyclic condition, which does not work in the perpendicular direction, but by the *actual surface boundary conditions*, following from the model.

The treatment of the boundary conditions in thin films as a modification of the cyclic conditions has been performed by Corciovei.[232,250] It was pointed out that although problems like thermal or magnetic vibrations in finite chains or thin films are not perfectly periodical, the correct solution of the secular problem of a finite chain (film) of a given length (thickness) is equivalent to the solution of the secular problem of a *cyclic chain* (film) of *double length* (thickness). On this ground a *canonical transform* of the (8.12′) type was proposed, with a view to writing the Hamiltonian as a sum of harmonic oscillator Hamiltonians. Corciovei's canonical transform has the form[232]

$$Q_{p\nu} = \sum_{\tau} T_{p\tau}^{(\nu)} \bar{q}_{\tau\nu}, \qquad P_{p\nu} = \sum_{\tau} T_{p\tau}^{(\nu)} \bar{q}_{\tau\nu}, \tag{8.24}$$

where $Q_{p\nu}$, $P_{p\nu}$ are hermitian canonical conjugated operators, related to the bosonic amplitudes $b_{p\nu}$ by the well-known equation

$$b_{p\nu} = (\tfrac{1}{2})^{1/2} (Q_{p\nu} + iP_{p\nu}).$$

In this way the case of simple cubic thin films with (100) surfaces was fully treated.

Corciovei's method was generalized by Wojtczak[251] for the case of other types of lattices and surface orientations. Special attention was given to the case of thin Co films.[252] Since near room temperature Co has a hexagonal closed packed structure with two atoms per elementary

[250] A. Corciovei, *Rev. Roum. Phys.* **10**, 3 (1965).
[251] L. Wojtczak, *Acta Phys. Pol.* **28**, 25 (1965).
[252] L. Wojtczak, *Phys. Status Solidi* **13**, 245 (1966).

cell it becomes necessary to treat the thin film as a superposition of two kinds of Valenta's plane monolayers. Accordingly, we introduce two sets of bosonic spin wave operators and corresponding canonical transforms of the form

$$a_{\mathbf{f}p}^{\pm} = (2\sqrt{N})^{-1} \sum_{\tau} \sum_{\mathbf{v}} \exp[\pm i\mathbf{f} \cdot \mathbf{v}]$$

$$\times [\mathscr{I}_1^{\pm} T_{p\tau_A}^{\mathbf{v}_A} A_{\tau\mathbf{v}}^{\pm} + \mathscr{I}_1^{\pm} T_{p\tau_B}^{\mathbf{v}_B} B_{\tau\mathbf{v}}^{\pm}],$$

$$b_{\mathbf{f}p}^{\pm} = (2\sqrt{N})^{-1} \sum_{\tau} \sum_{\mathbf{v}} \exp[\pm i\mathbf{f} \cdot \mathbf{v}]$$

$$\times [\mathscr{I}_2^{\pm} T_{p\tau_A}^{\mathbf{v}_A} A_{\tau\mathbf{v}}^{\pm} + \mathscr{I}_2^{\pm} T_{p\tau_B}^{\mathbf{v}_B} B_{\tau\mathbf{v}}^{\pm}], \tag{8.25}$$

where $\mathbf{f}$, p, $\mathbf{v}$, τ, N are defined previously, $T_{p\tau}^{\mathbf{v}}$ are spin wave amplitudes of the type involved in (8.24) and (8.12′) and $\mathscr{I}^{\pm}$ are some quantities related to the lattice parameters and obeying the rule

$$\mathscr{I}_1^{-}\mathscr{I}_1^{+} = \mathscr{I}_2^{-}\mathscr{I}_2^{+} = 1.$$

The procedure leads to a dispersion relation for the spin wave energy showing an acoustical and an optical branch, which is different from the case of the cubic lattices, with one atom per elementary cell. A full discussion has been given by Valenta and Wojtczak,[253] on the basis of the generalized Corciovei's transformation (8.24).

For simplicity, the following development will be restricted to the case of thin films of cubic structure. One uses immediately the results (8.12′), (8.15′), (8.16′), and (8.17′) which lead to a relevant formulation of the spin wave problem in thin films in terms of surface boundary conditions. We assume for the model Hamiltonian:

$$\mathscr{H} = - \sum_{\mathbf{f}p,\mathbf{g}p'} A_{\mathbf{f}p,\mathbf{g}p'} \mathbf{S}_{\mathbf{f}p} \cdot \mathbf{S}_{\mathbf{g}p'} + \mathscr{H}_{\text{anis}}. \tag{8.26}$$

The exchange isotropic part is perturbed by an anisotropic contribution which can be of a uniaxial or unidirectional type. The *uniaxial* anisotropic term has the well-known form

$$- \sum_{\mathbf{f}p,\mathbf{g}p'} K_{\mathbf{f}p,\mathbf{g}p'} S_{\mathbf{f}p}^{z} S_{\mathbf{g}p'}^{z} \tag{8.27}$$

but this time the anisotropic coupling parameter $K_{\mathbf{f}p,\mathbf{g}p'}$ will take into

[253] L. Valenta and L. Wojtczak, *Z. Naturforsch.* **22a**, 620 (1967).

account the *surface anisotropy* as well as the asymmetry of the surfaces.[254] For the nearest neighbors we have[255]

$$K_{\mathbf{f}p,\mathbf{g}p'} = \begin{cases} K_0 & \text{as} \quad p = p' = 1, \\ \bar{K}_0 & \text{as} \quad p = p' = q, \\ K & \text{otherwise.} \end{cases} \tag{8.28}$$

Assuming that the magnetization is along the z axis, the spin wave Hamiltonian (8.9) is reduced to

$$\mathscr{H}_{\mathrm{SW}} = \sum_{\boldsymbol{\nu}} \sum_{p,p'} \mathscr{S}^{\boldsymbol{\nu}}_{pp'} b^{\dagger}_{p\boldsymbol{\nu}} b_{p'\boldsymbol{\nu}} , \tag{8.29}$$

where $\mathscr{S}^{\boldsymbol{\nu}}_{pp'}$ has the form

$$\mathscr{S}^{\boldsymbol{\nu}}_{pp'} = 2S\delta_{pp'} \sum_{p''} (A^0_{pp''} + K^0_{pp''}) - 2SA^{\boldsymbol{\nu}}_{pp'} . \tag{8.30}$$

Restricting the calculation to nearest neighbors only and taking into account the definition (8.28), the plane Fourier transform in the sense of Eq. (8.10) for the coupling parameters, may be written as

$$\begin{aligned} A^{\boldsymbol{\nu}}_{pp'} &= A\Gamma^{\boldsymbol{\nu}}_{p'-p} \\ K^{\boldsymbol{\nu}}_{pp'} &= K_0\Gamma_0{}^{\boldsymbol{\nu}}\delta_{1p}\delta_{1p'} + \bar{K}_0\Gamma_0{}^{\boldsymbol{\nu}}\delta_{qp}\delta_{qp'} \\ &\quad + K\Gamma^{\boldsymbol{\nu}}_{p'-p}(1 - \delta_{1p}\delta_{1p'} - \delta_{qp}\delta_{qp'}). \end{aligned} \tag{8.31}$$

where $\Gamma^{\boldsymbol{\nu}}_{p'-p}$ is the structure factor (7.27) with values depending on the lattice type and surface orientation (see Table I); $\delta_{pp'}$ is the Kronecker symbol.

The canonical transformation (8.12) uses a single set of spin wave amplitudes $u^{\boldsymbol{\nu}}_{p\tau}$ which occurs as solutions of the eigenequations (8.15′) restricted by the orthogonality rule (8.16′). In order to solve the eigenequations (8.15′), the boundary condition method proposed by Jelitto[200]

[254] This way of treating the surface pinning looks exceedingly simple. It does not assume explicitly any mechanism to account for the differences between the value of the anisotropy constants inside the film and on the surfaces. It may be conceived as an *effective* pinning, which affects only the coupling of the surface spins and may be caused by various mechanisms, as reported in the literature.[238–243]

[255] A surface *unidirectional* anisotropy may be considered alternatively, using a Hamiltonian term $-g\mu_{\mathrm{B}} \sum_{\mathbf{f}p} H_p S^z_{\mathbf{f}p}$, with the effective field distributed as follows:

$$H_p = \begin{cases} H + H_0 & \text{as} \quad p = 1, \\ H + \bar{H}_0 & \text{as} \quad p = q, \\ H & \text{otherwise.} \end{cases}$$

TABLE I. SYSTEMATICS OF STRUCTURE CONSTANTS IN CUBIC THIN FILMS WITH VARIOUS ORIENTATIONS OF THE SURFACES[a,b]

Bulk layer	Surface orientation	Surface layer	D	Γ_0^ν	$\lvert \Gamma_1^\nu \rvert$	Γ_2^ν	Γ_3^ν
sc	{100}	Quadratic	1	$2(\cos \nu_1 + \cos \nu_2)$	1	—	—
$\Gamma = 6$	{110}	Rectangular	1	$2 \cos \nu_1$	$2 \cos(\nu_2/2)$	—	—
	{111}	Hexagonal	1	0	$\{3 + 2[\cos \nu_1 + \cos \nu_2 + \cos(\nu_1 + \nu_2)]\}^{1/2}$	—	—
fcc	{100}	Quadratic	1	$2(\cos \nu_1 + \cos \nu_2)$	$4 \cos(\nu_1/2) \cos(\nu_2/2)$	—	—
$\Gamma = 12$	{111}	Hexagonal	1	$2[\cos \nu_1 + \cos \nu_2 + \cos(\nu_1 + \nu_2)]$	$\{3 + 2[\cos \nu_1 + \cos \nu_2 + \cos(\nu_1 + \nu_2)]\}^{1/2}$	—	—
	{110}	Rectangular	2	$2 \cos \nu_1$	$4 \cos(\nu_1/2) \cos(\nu_2/2)$	1	—
bcc	{100}	Quadratic	1	0	$4 \cos(\nu_1/2) \cos(\nu_2/2)$	—	—
$\Gamma = 8$	{110}	Rhomboedral	1	$2(\cos \nu_1 + \cos \nu_2)$	$2 \lvert \cos(\frac{1}{2}(\nu_1 - \nu_2))\rvert$	—	—
	{111}	Hexagonal	3	0	$\{3 + 2[\cos \nu_1 + \cos \nu_2 + \cos(\nu_1 + \nu_2)]\}^{1/2}$	0	1

[a] ν_1, ν_2 are components of the plane spin wave vectors.
[b] After R. J. Jelitto, *Z. Naturforschg.* **19a**, 1567 (1964).

and summarized in Section 7 may be used. Making the system (8.15′) uniform the problem becomes the solution of the system[256]

$$-\epsilon_\nu u_p{}^\nu = \Gamma_1{}^\nu u_{p+1}^\nu + \Gamma_{-1}^\nu u_{p-1}^\nu, \qquad p = 1, 2, \ldots, q \tag{8.32}$$

with the *boundary conditions*

$$\mathscr{P}\Gamma_1{}^0 u_1 - \Gamma_{-1}^\nu u_0 = 0, \qquad \bar{\mathscr{P}}\Gamma_1{}^0 u_q - \Gamma_1{}^\nu u_{q+1} = 0. \tag{8.33}$$

The eigenvalue $\epsilon_{\tau\nu}$ is related to the spin wave energy $E_{\tau\nu}$ by the equation

$$E_{\tau\nu} = 2AS(\Gamma - \Gamma_0{}^\nu + \kappa + \epsilon_{\tau\nu}), \tag{8.34}$$

where $\Gamma = \Gamma_0{}^0 + 2\Gamma_1{}^0$ denotes the total number of the nearest neighbors of an inner atom and κ carries the contribution of the volume anisotropy: $\kappa = \Gamma K/A$.

The quantities $\mathscr{P}$ and $\bar{\mathscr{P}}$ involved in the boundary conditions may be called *pinning parameters*. They are

$$\begin{aligned} \mathscr{P} &= 1 + (K/A)(1 + \Gamma_0{}^0/\Gamma_1{}^0) - (K_0/A)(\Gamma_0{}^0/\Gamma_1{}^0),, \\ \bar{\mathscr{P}} &= 1 + (K/A)(1 + \Gamma_0{}^0/\Gamma_1{}^0) - (\bar{K}_0/A)(\Gamma_0{}^0/\Gamma_1{}^0). \end{aligned} \tag{8.35}$$

The pinning parameters separate out in a striking manner the *surface effects*. In fact K_0/A and $\bar{K}_0/A$ are surface anisotropy constants, whose difference take account of the film asymmetry. The natural surface defect is involved in the pinning parameters by the ratio $\Gamma_0{}^0/\Gamma_1{}^0$. This quantity depends on the lattice type on the orientation of the film surface (see Tables I and II). It is intuitively clear that if the pinning parameters were infinite the existence of any spin wave on the surfaces would be impossible. This is the case of *completely pinned surfaces*. The value 1

TABLE II. THE STRUCTURE CONSTANT $\Gamma_1{}^0$

Structure and orientation[a]	sc			fcc		bcc	
	{100}	{110}	{111}	{100}	{111}	{100}	{110}
$\Gamma_1{}^0$	1	2	3	4	3	4	2

[a] $\Gamma_1{}^0$ denotes the number of nearest neighbors lying in the first neighboring monolayer.

[256] Since all the amplitudes refer to the same index τ it is no longer mentioned in the following.

of the pinning parameters is related to a certain compensation of the effects of the surface and volume anisotropy:

$$K_0/A = \bar{K}_0/A = (K/A)(1 + \Gamma_1{}^0/\Gamma_0{}^0).$$

Since the pinning parameters no longer include anisotropic effects, this case may be called the *free surface* case.

As is customary we look for solutions of the uniform system (8.32) in the form of a *superposition* of two plane waves

$$u_p{}^{\mathbf{v}} = c_1 e^{ik_1 p} + c_2 e^{ik_2 p}. \tag{8.36}$$

Noting that generally the structure factor $\Gamma_1{}^{\mathbf{v}}$ is a complex quantity, $\Gamma_1{}^{\mathbf{v}} = |\Gamma_1{}^{\mathbf{v}}| e^{i\varphi_{\mathbf{v}}}$, and assuming also complex values of the perpendicular linear momentum, it is easy to see[200] that real eigenvalues $\epsilon_{\tau\mathbf{v}}$, as required by the hermiticity of the Hamiltonian, are allowed by only two choices of k_1 and k_2, namely

$$\begin{aligned} &(1)\ \text{real momenta:} && k_1 = s - \varphi_{\mathbf{v}}, && k_2 = -s - \varphi_{\mathbf{v}}, \\ &(2)\ \text{complex momenta:} && k_1 = (n\pi - \varphi_{\mathbf{v}}) + i\sigma, && k_2 = (n\pi - \varphi_{\mathbf{v}}) - i\sigma, \end{aligned} \tag{8.37}$$

where n is an arbitrary integer. The quantities s and σ labels the *perpendicular component* of the spin wave. The corresponding eigenvalues are respectively

$$\epsilon_{s\mathbf{v}} = -2|\Gamma_1{}^{\mathbf{v}}| \cos s, \tag{8.38}$$

$$\epsilon_{\sigma\mathbf{v}} = -(-1)^n\, 2|\Gamma_1{}^{\mathbf{v}}| \cosh \sigma. \tag{8.38$'$}$$

Now, fitting the general solution (8.36) of the uniform system (8.32) so as to comply with the actual boundary conditions (8.33), one finds the actual spin wave amplitudes. For *real* perpendicular linear momenta they show a trigonometric dependence on the perpendicular coordinate:

$$u_{ps}^{\mathbf{v}} = \text{const} \exp[-i\varphi_{\mathbf{v}} p]\{\mathscr{P}\Gamma_1{}^0 \sin((p-1)s) - |\Gamma_1{}^{\mathbf{v}}| \sin(ps)\}. \tag{8.39}$$

The spectrum of s is found from the equation

$$\begin{aligned} \mathscr{P}\bar{\mathscr{P}}(\Gamma_1{}^0)^2 \sin((q-1)s) - (\mathscr{P} + \bar{\mathscr{P}})\, \Gamma_1{}^0 |\Gamma_1{}^{\mathbf{v}}| \sin(qs) \\ + |\Gamma_1{}^{\mathbf{v}}|^2 \sin((q+1)s) = 0. \end{aligned} \tag{8.40}$$

Complex perpendicular linear momenta lead to a hyperbolic dependence of the amplitudes on the perpendicular coordinate:

$$\begin{aligned} u_{p\sigma}^{\mathbf{v}} = \text{const}(-1)^{np} \exp[i\varphi_{\mathbf{v}} p]\{(-1)^n\, \mathscr{P}\Gamma_1{}^0 \sinh((p-1)\sigma) \\ -|\Gamma_1{}^{\mathbf{v}}| \sinh(p\sigma)\}, \end{aligned} \tag{8.39$'$}$$

where σ follows from the equation

$$\mathscr{P}\bar{\mathscr{P}}(\Gamma_1{}^0)^2 \sinh((q-1)\sigma) - (-1)^n(\mathscr{P}+\bar{\mathscr{P}})\, \Gamma_1{}^0|\Gamma_1{}^\nu| \sinh(q\sigma)$$
$$+ |\Gamma_1{}^\nu|^2 \sinh((q+1)\sigma) = 0 \tag{8.40'}$$

Thus one sees that in thin films the spin waves exhibit the expected *standing* character. Their symmetry relative to the middle of the film definitely determines the magnetization distribution. The results (8.38) and (8.38′) for the dispersion law and (8.39), (8.39′) for the amplitudes enable us to perform the spin wave analysis of the saturation magnetization for models with various degrees of symmetric or asymmetric pinning of spins on the surfaces. We review in the following some relevant, particular cases.

(*i*) *Symmetric films*, $\mathscr{P} = \bar{\mathscr{P}}$. In this case the specimens are equally pinned on both surfaces. The spectrum of the perpendicular momentum is given by the equations

$$\mathscr{P} \begin{Bmatrix}\cos\\ \sin\end{Bmatrix} ((q-1)\, s/2) = \alpha_\nu \begin{Bmatrix}\cos\\ \sin\end{Bmatrix} ((q+1)\, s/2), \tag{8.41}$$

$$(-1)^n\, \mathscr{P} \begin{Bmatrix}\cosh\\ \sinh\end{Bmatrix} ((q-1)\, \sigma/2) = \alpha_\nu \begin{Bmatrix}\cosh\\ \sinh\end{Bmatrix} ((q+1)\, \sigma/2), \tag{8.41'}$$

where $\alpha_\nu = |\Gamma_1{}^\nu|/\Gamma_1{}^0$, for the trigonometric and hyperbolic modes respectively. These equations result from the general ones (8.40), (8.41′). The spin wave amplitudes become

$$u^\nu_{ps} = \exp[-i\varphi_\nu p](2/q)^{1/2}\, [1 \pm (\sin qs)/(q \sin s)]^{-1/2}$$
$$\times \begin{Bmatrix}\cos\\ \sin\end{Bmatrix} [(q+1)/2 - p)s], \tag{8.42}$$

$$u^\nu_{p\sigma} = (-1)^{np} \exp[-i\varphi_\nu p](2/q)^{1/2}\, [(\sinh q\sigma)/(q \sinh \sigma) \pm 1]^{-1/2}$$
$$\times \begin{Bmatrix}\cosh\\ \sinh\end{Bmatrix} [((q+1)/2 - p)\sigma]. \tag{8.42'}$$

They are normalized according to the orthogonality rule (8.16′). The dependence of the spin wave amplitudes on the perpendicular coordinate p denotes standing waves which are *symmetric* (cos, cosh) or *antisymmetric* (sin, sinh) relative to the middle plane of the film.

It follows from the general formalism that the probability of exciting a wave of the mode τ, ν in the pth monolayer of the film is proportional to $|u^\nu_{p\tau}|^2$, so that the hyperbolic modes occur excited *mainly on the surfaces*.

TABLE III. SURFACE AND SPACE MODES IN THIN FILMS; SYMMETRIC PINNING

$a = \alpha_{\mathbf{v}}/\mathscr{P}$	cosh		sinh		sin		cos		Number of modes
	q odd	q even	q odd	q even	q odd	q even	q odd	q even	
0	0	0	0	0	$(q-3)/2$	$q/2-1$	$(q-1)/2$	$q/2-1$	$q-2$
$\lvert a \rvert < (q-1)/(q+1)$	1	1	1	1	$(q-3)/2$	$q/2-1$	$(q-1)/2$	$q/2-1$	q
$(q-1)/(q+1) < \lvert a \rvert \leqslant 1$	1	1	0	0	$(q-1)/2$	$q/2$	$(q-1)/2$	$q/2-1$	q
$1 < \lvert a \rvert$	0	0	0	0	$(q-1)/2$	$q/2$	$(q+1)/2$	$q/2$	q

For this reason, the hyperbolic modes are called *surface localized modes*, while the trigonometric ones are known as *space modes*.

Solving graphically the equations of the perpendicular momenta (8.41) and (8.41′), a complete analysis can be carried out, as shown by Jelitto,[200] concerning the number of symmetric and antisymmetric surface and space modes. The conclusions are summarized in Table III. The first row describes the limiting case in which the pinning parameters of the surfaces tend to infinity. It corresponds to a *complete pinning* of the surface spins by the surface anisotropy, as discussed above. The spectrum is

$$s_\tau = \tau\pi/(q-1), \qquad \tau = 1, 2, \ldots, q-2. \tag{8.43}$$

and the spin wave amplitudes are

$$u^{\mathbf{v}}_{p\tau} \approx \exp[-i\varphi_{\mathbf{v}} p](2/(q-1))^{1/2} \sin((p-1)\,\tau\pi/(q-1)), \qquad \tau = 1, 2, \ldots, q-2. \tag{8.44}$$

They denote *standing space waves* with nodes on both surfaces, meaning that no spin wave is excited in the end "frozen" monolayers, as expected. Thus the complete pinning of the spins on the surfaces reduces by $2N$ the total number qN of the degrees of freedom.

The other three cases stated in the Table IV correspond to finite decreased values of the pinning parameters. One sees that when the

TABLE IV. SURFACE AND SPACE MODES IN THIN FILMS; ASYMMETRIC PINNING[a]

$a = \alpha_{\mathbf{v}}/\mathscr{P}$	sinh	sin	Number of modes
0	0	$q-2$	$q-2$
$\lvert a \rvert < (q-1)/q$	1	$q-2$	$q-1$
$(q-1)/q < \lvert a \rvert$	0	$q-1$	$q-1$

[a] The qth monolayer is completely pinned.

ratio $|a(\mathbf{v}, K_0)|$ between the structure factor $|\alpha_{\mathbf{v}}|$ and the pinning parameter $|\mathscr{P}|$ remains under certain limit, depending on the film thickness, both symmetric (cosh) and antisymmetric (sinh) *surface* modes are excited. The number of space modes is then decreased correspondingly, so that the total number of modes is just equal to the number of the degrees of freedom.

As the ratio $|a(\mathbf{v}, K_0)|$ increases the antisymmetric surface mode sinh disappears while an additional space mode sin becomes possible.

Finally as the ratio $|a(\nu, K_0)|$ exceeds unity surface modes are no longer excited and all qN modes are *spacelike*. In particular, as the total ground energy per spin is uniformly distributed across the film ($\mathscr{P} = 0$) the spectrum of the perpendicular momentum includes q values

$$s_\tau = \tau\pi/(q + 1), \qquad \tau = 1, 2,..., q. \tag{8.45}$$

The corresponding amplitudes are

$$u^\nu_{p\tau} = \exp[-i\varphi_\nu p](2/(q + 1))^{1/2} \sin(p\pi\tau/(q + 1)), \qquad \tau = 1, 2,..., q. \tag{8.46}$$

Let us consider now a case of special interest, namely that of *free surfaces* ($\mathscr{P} = \bar{\mathscr{P}} = 1$). In this case, the number and the symmetry of the spin wave modes are determined only by the structure factor α_ν. In the long wavelength range the structure factor α_ν approaches unity so that the following approximation[257] may be used

$$(q - 1)/(q + 1) < \alpha_\nu < 1. \tag{8.47}$$

According to Table IV, in this case there is only one symmetric surface mode, the remaining modes being spacelike. For calculations of magnetization distribution, the spin wave amplitudes can be taken as

$$u^\nu_{p\tau} \approx u^0_{p\tau} = ((2 - \delta_{0\tau})/q)^{1/2} \cos((p - 1/2)\, \pi\tau/q). \tag{8.48}$$

They correspond to a sequence of approximated energy levels of the form[258]

$$s_\tau = \tau\pi/q, \qquad \tau = 0, 1, 2,..., q - 1. \tag{8.49}$$

Note that this is an exact result for sc thin films with {100} surfaces, as pointed by Corciovei.[232] In this case, which is commonly used in thin film approaches, the symmetric surface mode transforms into a "mode" which is uniformly excited across the film. Accordingly, many authors state that simple cubic thin films with {100} surfaces do not exhibit surface modes.

One sees (Eqs.(8.38) and (8.38′)) that the sequence of energy levels of the space modes is bounded by *surface levels*, either at the lowest end (n even), or at the highest end (n odd). The excitation of the lowest

[257] R. J. Jelitto, *Z. Naturforsch.* **19a**, 1580 (1964).

[258] Following Jelitto,[257] an improved sequence of momentum levels may be used for the dispersion law, found by series expansions of the momentum equations (8.41) and (8.41′) near the levels of the approximate sequence (8.49).

surface branch is related to positive values of the pinning parameter, $0 < \mathscr{P} < 1$, while negative values $-1 < \mathscr{P} < 0$ of the pinning parameter require the excitation of the highest surface branch.

(*ii*) *Asymmetric films*, $\mathscr{P} \neq \bar{\mathscr{P}}$. In the case of films with asymmetrically pinned surfaces, relevant results in compact form are found by assuming that one of the surfaces is completely pinned. Let us take, for example, $\bar{\mathscr{P}} \to \infty$, on the q end monolayer. The perpendicular momentum equations (8.40), (8.40′) becomes

$$\mathscr{P} \sin((q-1)s) = \alpha_{\mathrm{v}} \sin(qs), \tag{8.50}$$

$$(-1)^n \mathscr{P} \sinh((q-1)\sigma) = \alpha_{\mathrm{v}} \sinh(q\sigma), \tag{8.50′}$$

corresponding to the normalized amplitudes

$$u_{ps}^{\mathrm{v}} = \exp[-i\varphi_{\mathrm{v}} p] 2/(2q-1)^{1/2} \times [1 - (\sin(2q-1)s)/((2q-1)\sin s)]^{-1/2} \sin(q-p)s, \tag{8.51}$$

$$u_{p\sigma}^{\mathrm{v}} = (-1)^{np} \exp[-i\varphi_{\mathrm{v}} p] 2/(2q-1)^{1/2} \times [(\sinh(2q-1)\sigma)/((2q-1)\sinh\sigma) - 1]^{-1/2} \sinh(q-p)\sigma, \tag{8.51′}$$

for the *space* and *surface* modes respectively. These amplitudes denote standing waves with a node on the pinned surface, $p = q$. Due to this "frozen" monolayer the number of perpendicular modes can not exceed $q - 1$ (see Table IV). Let us assume that the surface $p = 1$ is free ($\mathscr{P} = 1$). In the long wavelength range the structure factor α_{v} approaches unity so that the inequality $(q-1)/q < \alpha_{\mathrm{v}}$ can be assumed. Then there are only $q - 1$ space modes, with momenta approaching the sequence

$$s_\tau = (2\tau - 1)\pi/(2q-1), \qquad \tau = 1, 2, \ldots, q-1. \tag{8.52}$$

In this way, the spin wave amplitudes become

$$u_{p\tau}^{\mathrm{v}} \approx u_{p\tau}^{0} = 2/(2q-1)^{1/2} \times \cos((p - 1/2)(2\tau - 1)\pi/(2q-1)), \qquad \tau = 1, 2, \ldots, q-1. \tag{8.53}$$

The *above results* allow calculations of the *magnetization distribution* (8.21′) and of the total relative magnetization of the film (8.23) in three striking cases, *free surfaces*, *completely pinned surfaces*, and *one pinned–one free surfaces*. Thanks to the great density of the plane modes the sums over the wave vector ν can be replaced by integrals, so that finally the sublattice magnetization and the total relative magnetization are given by sums only over the perpendicular momenta.

In the case of free surfaces, the contribution of the symmetric surface mode to the mean magnon number is proportional to $\ln(1 - e^{-A\kappa\beta})^{-1}$. It has a singularity for vanishing volume anisotropy $\kappa = 0$. This behavior reveals that the absence of anisotropy terms in the Hamiltonian leads to an excessive contribution of the lowest spin wave modes, which are usually *surface* modes, to the mean magnon number, so that the stability of the ferromagnetic phase is destroyed. Note that such a difficulty does not appear either in the theory of the bulk ferromagnetic body, thanks to an appropriate density of states in three dimensional specimens, or in the case of thin films with pinned surfaces, due to the natural elimination of the $k = 0$ mode.

Numerical calculations of the monolayer relative magnetization and of the total relative magnetization of the film were performed by Corciovei and Vamanu[246] for films of cubic symmetry. Some selected results are plotted (Figs. 12–14, 16, 17). These results indicate mainly that the surface conditions determine in a decisive manner the magnetization of the whole film as well as the *distribution* of the magnetization across the film. The case of free surfaces corresponds to a collapse of the total magnetization, but especially of the magnetization of the end sublattices, while complete pinning tends to increase these quantities. As the surface defect is more stressed (larger values of the ratio $\Gamma_1{}^0/\Gamma_0{}^0$, for a given structure but different surface orientations) the effect of the surfaces is increased.

Attention must be given to the asymmetric case of pinning on one of the surfaces only (Fig. 14) which may be a simple model of a thin ferromagnetic film lying on a substrate. This case appears as a mixture of the two other cases, which indicates opposing effects of the free and completely pinned surfaces. As can be seen from the figures, in this situation the pinned surface has the greater effect, increasing the magnetization of all monolayers, even in the neighborhood of the free surface.

Concerning the effect of the temperature variations on the magnetization distribution, it may be shown that *increasing temperature* causes a continuous decrease of the monolayer magnetization, as expected, but also the magnetization "collapses" near the surfaces. Anticipating here a result of the Green's function approach we must point out that this behavior is rightly described only at low temperatures. The Green's function method, which also works at high temperatures, shows[210] that the progressive collapse of the magnetization near the surfaces is

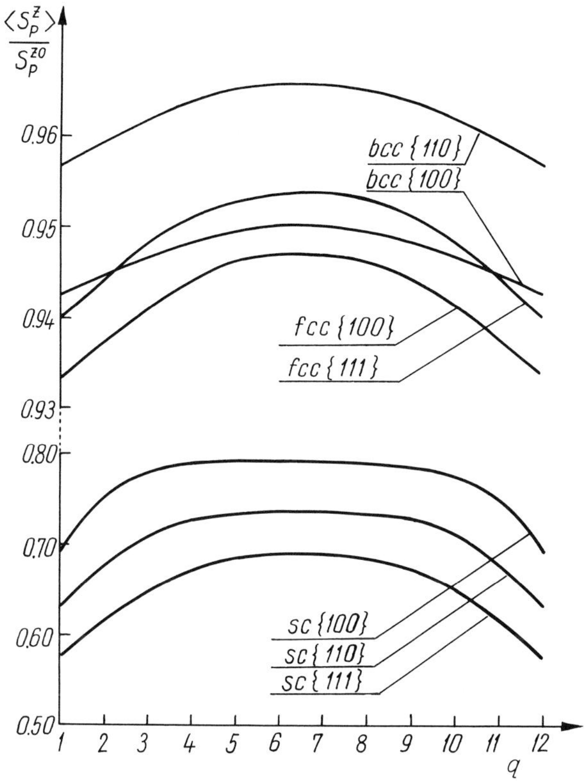

FIG. 12. Spatial distribution of the magnetization through the film thickness (free surfaces); $I = 80°K$, $\kappa = 5 \times 10^{-5}$ [after A. Corciovei and D. Vamanu, *Rev. Roum. Phys.* **15**, 473 (1970)].

replaced at high temperatures by a progressive smoothing of the magnetization distribution, so that at the Curie temperature the ferromagnetism of all monolayers disappears simultaneously (Fig. 15). One sees that the films with symmetrically and asymmetrically pinned surfaces are less sensitive to the variations of the temperature as compared with films having free surfaces.

Similar conclusions apply concerning the dependence of the monolayer magnetization on the volume anisotropy. Note also that a strong surface anisotropy obscures the effects of the variations of the volume anisotropy. Figure 16 shows the distribution of the magnetization for films of different thicknesses. A very significant result is that the magnetization of the films with free surfaces lies always below the bulk magnetization, while the magnetization of the films with completely pinned surfaces is

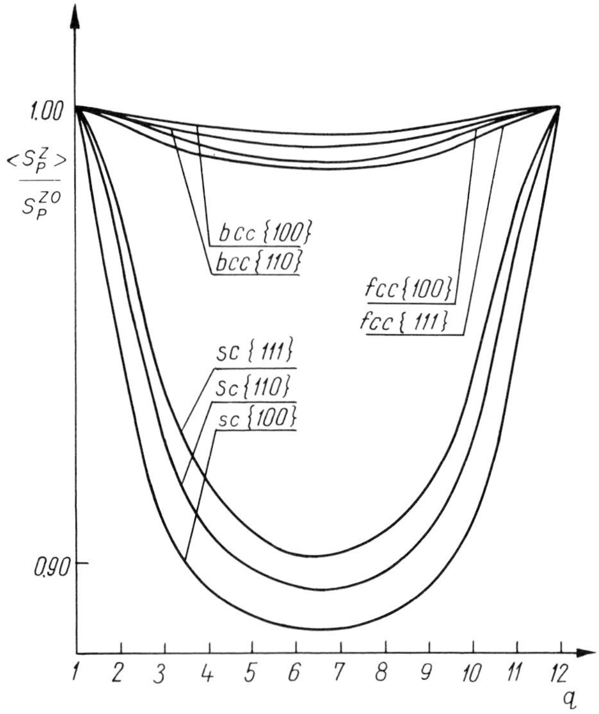

FIG. 13. Spatial distribution of the magnetization through the film thickness (complete symmetric pinning); $T = 80°K$, $\kappa = 5 \times 10^{-5}$ [after A. Corciovei and D. Vamanu, *Rev. Roum. Phys.* **15**, 473 (1970)].

above the bulk limit. The asymmetric case of films with one free and one pinned surfaces exhibits in this respect an intermediate position: their magnetization approaches closely the bulk limit, even for small thicknesses. In our case (Fig. 17), the influence of the pinned surface seems to be predominant, but a situation can be imagined in which a particular balance of anisotropy on the surfaces would lead to the result that the magnetization essentially would not depend on the film thickness. Thus the simple model discussed above seems explain quite satisfactorily (see, e.g., Davis[243]) a large variety of experimental situations in terms of the surface conditions of the specimens.

Spin Wave Resonance

The spin wave theory treats the ferromagnetic body like a system containing an ideal gas of quasi particles—the magnons. The state of magnetization which corresponds to the thermodynamical equilibrium of the magnonic gas can be analyzed in terms of normal modes of spin

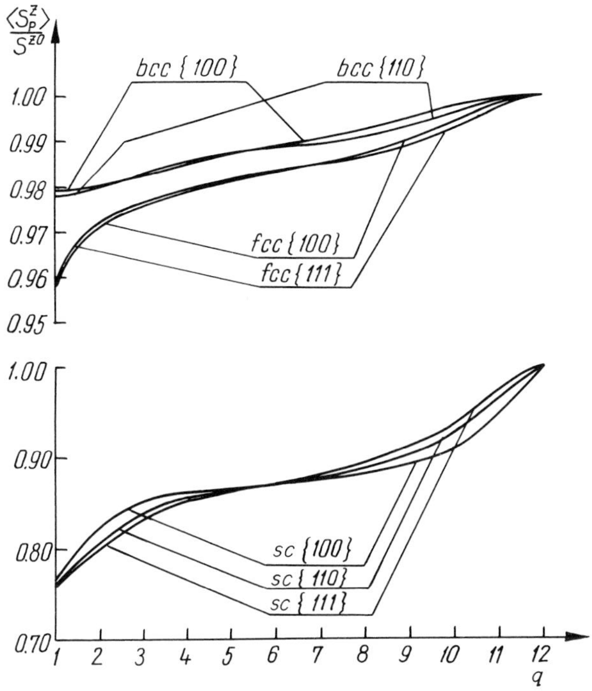

FIG. 14. Spatial distribution of the magnetization through the film thickness (asymmetric pinning on one free-one pinned surface); $T = 80°K$, $\kappa = 5 \times 10^{-5}$ [after A. Corciovei and D. Vamanu, *Rev. Roum. Phys.* **15**, 473 (1970)].

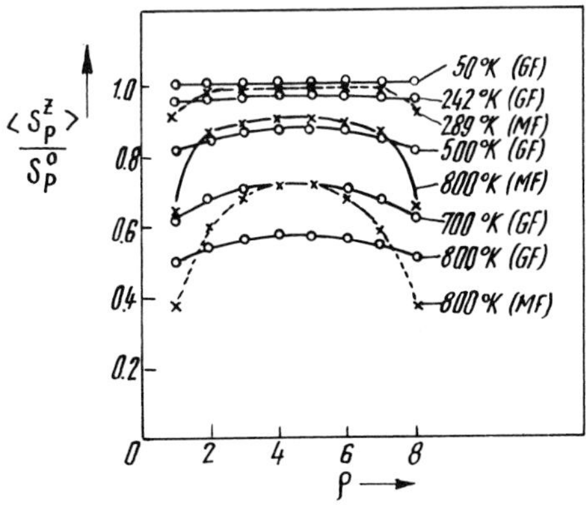

FIG. 15. Temperature dependence of the spatial distribution of the magnetization through the film thickness for $q = 8$; Fe = (100); $S = 1$; (solid line) $A = 3.76 \times 10^{-16}$ erg; (dashed line) $A = 27 \times 10^{-15}$ erg [after L. Valenta, W. Haubenreisser, and H. Brodkorb, *Phys. Status Solidi* **26**, 191 (1968)].

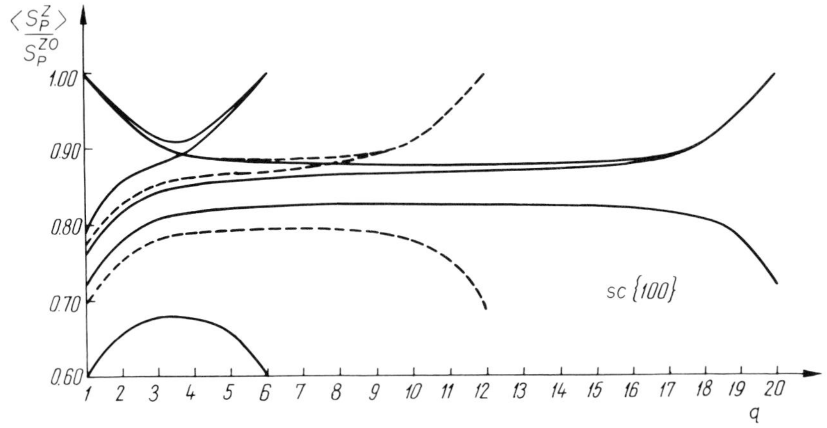

FIG. 16. Spatial distribution of the magnetization as a function of the film thickness [after A. Corciovei and D. Vamanu, *Rev. Roum. Phys.* **15**, 473 (1970)]. Typical cases of free, symmetrically pinned, and asymmetrically pinned surfaces plotted for $q = 6$, 12, and 20 monolayers; $T = 8°\text{K}$, $\kappa = 5 \times 10^{-5}$.

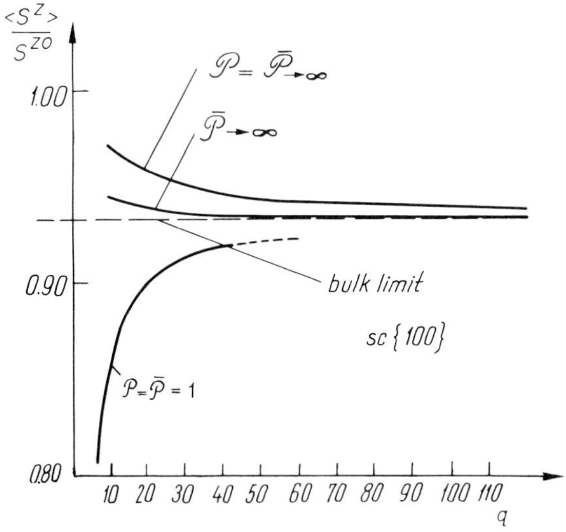

FIG. 17. Thin film magnetization versus thickness [after A. Corciovei and D. Vamanu, *Rev. Roum. Phys.* **15**, 473 (1970)]. Typical positions of free ($\mathcal{P} = \bar{\mathcal{P}} = 1$), symmetrically pinned ($\mathcal{P} = \bar{\mathcal{P}} \to \infty$), and asymmetrically pinned ($\mathcal{P} \to \infty$, $\bar{\mathcal{P}} = 1$) surfaces are plotted; $T = 50°\text{K}$, $\kappa = 5 \times 10^{-5}$.

waves. The normal modes are classified according to their wave number. The spectrum of the wave momentum as well as the shape of the waves are essentially determined by the shape of the body, i.e., by the boundary conditions. To test the validity of the picture of *normal modes* of magnetic vibrations it seems desirable to excite spin waves by coupling the normal modes to an external magnetic field. The standard configuration of such an experiment involves a *constant magnetic field* which supports the equilibrium direction of the magnetization and a rf magnetic field which usually is taken as linearly (circularly) polarized in a plane perpendicular to the constant field.[259]

Due to its perfect translational symmetry involving the equivalence of all the atom sites, the infinite bulk ferromagnet is able to absorb rf power only in the uniform mode of zero wave vector. This is the *ferromagnetic resonance*. To carry out further details concerning the statistics of magnetics, higher spin wave modes have to be excited and detected. A natural way of reaching this aim is to reduce at least one of the dimensions of the specimen, so that the translational symmetry would be at least partly destroyed. The resonant absorption of power from rf magnetic fields in higher spin wave modes has been called *spin wave resonance*.

The interaction of the Heisenberg system of localized spins with the rf component of the magnetic field is described by the Hamiltonian term[93]

$$\mathscr{H}_{\mathrm{SWR}} = -g\mu_{\mathrm{B}} \sum_{\mathbf{f}p} \mathbf{H} \cdot \mathbf{S}_{\mathbf{f}p} \,. \tag{8.54}$$

The practically available values of the magnetic rf field $\mathbf{H}$, permits treating the coupling of the spins with the rf field as a small perturbation of the strong internal exchange coupling. Accordingly, one proceeds by treating the perturbation (8.54) in terms of the bosonic operators which diagonalize the usual thin film Hamiltonian. To make the frame of discussion more precise let us take the z axis as the equilibrium direction of the magnetization, i.e., $\gamma = (0, 0, 1)$. To assure the applicability of the quasi saturation approximation, one assumes a single $0z$ axis of uniaxial anisotropy, so that $v_{p\tau}^{\nu} = 0$.

Let us take the rf magnetic field to be circularly polarized in the $x0y$ plane:

$$\mathbf{H} = (H_0 e^{-i\omega t}, -iH_0 e^{-i\omega t}, 0). \tag{8.55}$$

In this way, the interaction Hamiltonian becomes

$$\mathscr{H}_{\mathrm{SWR}} = g\mu_{\mathrm{B}} H_0 (2SN)^{1/2} e^{-i\omega t} \sum_p \sum_\tau u_{p\tau}^{0*} \xi_{\tau 0}^{\dagger} \,. \tag{8.56}$$

[259] This type of energy transfer to the magnon gas of a ferromagnet is known as *perpendicular pumping*. We shall not deal here with the very interesting case of the *parallel pumping* (rf field parallel to the dc field), which is insufficiently elaborated and substantially supplements the theory, especially concerning magnon-phonon interaction (see H. Le Gall, *J. Phys.* **28**, Suppl. to No. 2, C-151 (1967)).

It includes the dynamical variables $\xi^\dagger$ which creates magnons in the τ *perpendicular* and $\nu = 0$ plane mode. This means that only modes which are uniformly excited in the directions obeying the translational invariance may be coupled with an external rf field. Now, the standard form of the time dependent perturbation theory easily gives the following probability per unit time of exciting one spin wave of the $(\tau, 0)$ mode in the state with $n_{\tau 0}$ magnons:

$$P = 2\pi\hbar^{-1}(g\mu_B H_0(2SN)^{1/2})^2 (n_{\tau 0} + 1)\left|\sum_{p=1}^{q} u_{p\tau}^{0*}\right|^2 \delta(\hbar\omega - E_{\tau 0}). \tag{8.57}$$

This result reveals new features of the phenomenon. Thus one sees that the absorption of power from a magnetic rf field exhibits indeed a *resonant character*, denoted by the delta function: to excite one spin wave of $E_{\tau 0}$ energy one needs a circular frequency $\omega = E_{\tau 0}/\hbar$.[260]

But more striking features follow from the dependence of the transition probability (8.59) on the quantity

$$U_\tau \equiv \left|\sum_{p=1}^{q} u_{p\tau}^{0*}\right|^2 \tag{8.58}$$

which involves the spin wave amplitudes $u_{p\tau}^0$. It denotes the dependence of the resonant absorption on the film thickness and on the surface conditions.

Thus in the case of symmetric films $\mathscr{P} = \bar{\mathscr{P}} > 0$ one finds:

$$U_\tau(\sin) = 0, \qquad U_\tau(\sinh) = 0, \tag{8.59}$$

meaning that no power is absorbed in the antisymmetric modes. For an appropriate degree of pinning ($|\mathscr{P}| < 1$) the power absorbed in the lowest symmetric surface mode is proportional to

$$\begin{aligned} U_\sigma(\cosh) = {} & 2q^{-1}|[\sinh(q\sigma)/(q \sinh \sigma) + 1]^{-1}| \\ & \times |(\sinh(q\sigma)/2)/\sinh(\sigma/2)|^2. \end{aligned} \tag{8.60}$$

This proves that the absorption outside the sequence of space modes near the uniform mode, falls off following approximately a k^{-1} law as shown by Puszkarski.[245] It may be shown that the absorption in the highest surface mode, which is located outside the upper bound of the

[260] There are two types of SWR experiments, namely at constant field H, or at constant frequency ω. It is clear that, as the increasing frequency of the rf field produces SW modes of increasing ordering number, the increasing dc magnetic field excites successively SW modes of decreasing ordering number.

space modes sequence is approximately q times weaker than the absorption in the lowest surface mode. In the ideal case of the free surfaces there is no absorption in the nonuniform modes, since, according to Eq. (8.48), as $\tau \neq 0$, then $\sum_p u_{p\tau}^{0*} = 0$. Thus the thin films with free surfaces exhibit only ordinary ferromagnetic resonance, corresponding to the uniform mode, for which $U_0 = q$. For the cosinelike space modes one finds

$$U_s(\cos) = 2q^{-1}|[1 + (\sin qs)/(q \sin s)]^{-1}| \times |(\sin(qs)/2)/\sin(s/2)|^2, \tag{8.61}$$

which proves that in the long wavelength range (small k) the peaks of absorption in the space modes falls off approaching a k^{-2} law.

The momentum dependence of the quantity U_τ indicates the essentials of a typical absorption curve in spin wave resonance. The main peak of the uniform mode has to be bounded by the surface modes contribution. The excited space modes follow each other as a consequence. Only *odd modes* are excited (in spacing units of π/q). The envelope of the spacelike peaks approaches the k^{-2} law while the envelope of the surface modes follows rather a k^{-1} law. Plotting the absorption curve on the spin wave energy scale the peak spacing increases following a k^2 law, which is a long wave approach of the cosinelike dependence (8.38) of the spin wave energy. The increase of the thickness obscures the higher-order spin wave resonance, since the higher mode tends to accumulate near the peak of the uniform mode.

In the case of thin films with *symmetrically pinned* surfaces, the absorption curve shows a marked tendency toward the k^{-2} law in the limit of small momenta. According to Eq. (8.44), one finds

$$U_\tau = \begin{cases} 2/(q-1)\cot^2(\tau\pi/(2(q-1)) & \text{as } \tau - \textit{odd}, \\ 0 & \text{as } \tau - \textit{even}. \end{cases} \tag{8.62}$$

A similar tendency is indicated for asymmetric thin films with one pinned and one free surface:

$$U_\tau = (2q-1)^{-1}\cot^2(2\tau - 1)\pi/(2(2q-1)), \qquad \tau = 1, 2, \ldots, q-1. \tag{8.63}$$

When counting the spin wave modes in specific unit-spacings, namely π/q for symmetric films with free surfaces, $\pi/(q-1)$ for complete pinning on both surfaces, $\pi/(q - \frac{1}{2})$ for one pinned and one free surfaces (see Eqs. (8.43), (8.49), and (8.52)), one can say that in symmetric films only the *odd modes* are involved in spin wave resonance, while in the asymmetric films both odd and even modes are excited. All the above remarks are indeed supported by experiment. A typical and accurate absorption curve is given by Tannenwald and Weber.[261] It seems to correspond to an asymmetric film, since a small admixture of even modes is recorded among the sharp peaks of the odd modes. The spin wave resonance studies on thin films were launched especially after Kittel[238] stated that spin wave modes can be excited not only by certain gradients of the magnetic field but also by an uniform rf field, provided appropriate *boundary conditions* are considered. Kittel's theory of spin wave excitation in linear chains, based on the gyromagnetic equation and on the pinning concept, includes some fundamental aspects of the phenomenon, such as the selection of the odd modes.

A comprehensive review of the works performed until 1962 is due to Frait.[262] It includes comments on some important theoretical contributions such as those due to Tannenwald and Seavey[263] (SWR by skin effect in metals), Fraitova[26] (effects of demagnetizing field, anisotropy and internal stresses on SWR), Pincus[239] (pinning model for SWR in films), Orbach and Pincus[264] (SWR in antiferromagnetics) as well as on some basic experimental investigations of Tannenwald,[265] Ondris and Frait,[266] Kooi *et al.*,[234] Tannenwald and Weber.[261] A quantum mechanical treatment of spin wave resonance in thin films, as followed in this paper, is due to Ferchmin.[267]

Attempts to improve the models were made in subsequent approaches, with a search for the best fit with the actual SWR features and parameters: the selection odd–even rule, the momentum dependence of the resonant absorbtion curve, the mode spacing, the line width, the line symmetry. The magnetic structure is in a wide measure responsible for the shifting, broadening, and distortions of the resonance lines. Hoffmann[268] studied the effect of the local fluctuations of the magnetization direction (ripple). The magnetization dispersion in polycrystalline films causes certain local demagnetizing fields which may considerably affect the dispersion law of the resonant frequency, especially for weak applied fields. Salanskii *et al.*[269] compared the influence of the ripple and "block"

[261] P. E. Tannenwald and R. Weber, *Phys. Rev.* **121**, 715 (1961).

[262] Z. Frait, *Phys. Status Solidi* **2**, 1417 (1962).

[263] P. E. Tannenwald and M. H. Seavey, Jr., *J. Phys. Radium* **20**, 323 (1959).

[264] R. Orbach and P. Pincus, *Phys. Rev.* **113**, 1213 (1959).

[265] P. E. Tannenwald, *Proc. Int. Conf. Structure Properties Thin Films, Bolton Landing, N.Y.*, p. 387 (1959).

[266] M. Ondris and Z. Frait, *Czech. J. Phys.* **B11**, 883 (1961).

[267] A. R. Ferchmin, *Phys. Lett.* **1**, 281 (1962).

[268] H. Hoffmann, *in* "Physica Magnitnyh Plenok," p. 389. Irkutsk, 1968.

[269] N. M. Salanskii, V. A. Ignatchenko, and B. V. Chroustalev, *in* "Physica Magnitnyh Plenok," p. 242. Irkutsk, 1968.

inhomogeneities on the SWR. It was shown that, as the ripple structure results in asymmetrizing and shifting the resonance peak, the "block" structure does not cause the shift and asymmetry of the resonance line, but only its broadening. Frait and Mitchel[270] investigated the way in which various relaxation mechanisms fit the experimental data.

Special attention is payed to the influence of surfaces on the SWR. Stankoff[236] investigated 80–20 permalloy films with additional surface layers of NiO, Ni, Fe_2O_3, and Fe. A selection rule for odd–even modes, as discussed above, was used for drawing conclusions about the degree of symmetry of the slabs. The existence of excited modes for fields greater than the main resonance field was used to establish the minus sign of the phenomenological anisotropy constant, following the theory of Pincus.[239] Some characteristic relations between the intensities of odd and even modes were used to estimate numerically the surface anisotropy value. Waksmann *et al.*[237] reported the influence of the surface layers of Mn on permalloy films, also taking into account the influence of the substrate. It was indicated that a diffusion mechanism caused the surface anisotropy. Soohoo[235] used an ultra-high vacuum system (3.5×10^{-11} Torr) and fine control devices to investigate the influence of residual gases like O_2, H_2O, H_2, N_2, on the SWR line width. He found that H_2, H_2O, and O_2 residuals cause a broadening of the SWR line as well as a decrease of the perpendicular anisotropy and of the magnetization, while N_2 has a relatively small influence on the SWR parameters. The oxidation and the inner stresses are taken to be responsible for the phenomena. Similar contaminating mechanisms have been investigated by Freedman[271] who found a decrease of the perpendicular anisotropy and an increase of the coercive force.

The spin wave resonance seems to be an appropriate and accurate method to investigate such involved matters as thin film magnetic structure and symmetry, dynamic processes, thickness, exchange, and anisotropy measurements using some simple and tractable models.

9. Green's Function Method

According to the familiar picture in which the magnetization of a system of localized spins is defined by the set of the local spin displacements $n_{fp} = S_{fp} - S_{fp}^{z'}$, the molecular field and the spin waves methods give complementary information by quite different approaches. Thus the molecular field method and its refinements (see Section 7) are well suited for high temperatures including those near the critical Curie

[270] Z. Frait and E. N. Mitchel, *in* "Physica Magnitnyh Plenok," p. 385. Irkutsk, 1968.
[271] J. F. Freedman, *J. Appl. Phys.* **36**, 964 (1965).

point, while the spin wave formalism refers to the low temperature range, including the vicinity of the ground state, of maximum magnetization.

To evaluate the statistical averages of the spin displacements over the entire temperature range the use of *double time and temperature dependent Green functions* was proposed.[272] In thin ferromagnetic film studies, the Green's functions method was first used by Corciovei and Ciobanu,[65] and Brodkorb and Haubenreisser.[66] The procedure is now sketched briefly.

The simplest model Hamiltonian,

$$\mathscr{H} = -\sum_{\mathbf{f}p,\mathbf{g}p'} A_{\mathbf{f}p,\mathbf{g}p'}\mathbf{S}_{\mathbf{f}p}\cdot\mathbf{S}_{\mathbf{g}p'} - g\mu_{\mathrm{B}}H_{\parallel}\sum_{\mathbf{f}p} S^z_{\mathbf{f}p}\,, \tag{9.1}$$

includes an exchange and a Zeeman-like term. The latter stresses the magnetization direction and provides the dispersion spin wave law with the gap, which assures the stability of the ferromagnetic phase (see Section 8). The standard expansion (3.22) of this Hamiltonian relative to the spin displacements has the following form

$$\begin{aligned}\mathscr{H} &= \mathscr{E}_0 + \sum_{\mathbf{f}p}\Big[2S\sum_{\mathbf{g}p'} A_{\mathbf{f}p,\mathbf{g}p'}\boldsymbol{\gamma}_{\mathbf{f}p}\cdot\boldsymbol{\gamma}_{\mathbf{g}p'} + g\mu_{\mathrm{B}}H_{\parallel}\gamma^z_{\mathbf{f}p}\Big]n_{\mathbf{f}p}\\ &\quad - \sum_{\mathbf{f}p,\mathbf{g}p'} A_{\mathbf{f}p,\mathbf{g}p'}T_{\mathbf{f}p}T_{\mathbf{g}p'} - \sum_{\mathbf{f}p,\mathbf{g}p'} A_{\mathbf{f}p,\mathbf{g}p'}\boldsymbol{\gamma}_{\mathbf{f}p}\cdot\boldsymbol{\gamma}_{\mathbf{g}p'}n_{\mathbf{f}p}n_{\mathbf{g}p'}\,.\end{aligned} \tag{9.2}$$

Assuming now that the thin film is a single domain with $\frac{1}{2}$ spins and parallel magnetization and eliminating the ground state energy, which is not essential here, the effective Hamiltonian becomes

$$\begin{aligned}\mathscr{H} &= \sum_{\mathbf{f}p}\Big[\sum_{p'} A^0_{pp'} + g\mu_{\mathrm{B}}H_{\parallel}\Big]n_{\mathbf{f}p}\\ &\quad - \sum_{\mathbf{f}p,\mathbf{g}p'} A_{\mathbf{f}p,\mathbf{g}p'}b^\dagger_{\mathbf{f}p}b_{\mathbf{g}p'} - \sum_{\mathbf{f}p,\mathbf{g}p'} A_{\mathbf{f}p,\mathbf{g}p'}n_{\mathbf{f}p}n_{\mathbf{g}p'}\,.\end{aligned} \tag{9.3}$$

This time it is convenient to express the spin displacements using Pauli operators, obeying the commutation rules

$$[b_{\mathbf{f}p}\,, b^\dagger_{\mathbf{g}p'}] = (1 - 2n_{\mathbf{f}p})\,\delta_{\mathbf{f}p,\mathbf{g}p'}\,, \qquad [b_{\mathbf{f}p}\,, b_{\mathbf{g}p'}] = 0, \tag{9.4}$$

by

$$\begin{aligned} S^x_{\mathbf{f}p} &= 2^{-1}(b^\dagger_{\mathbf{f}p} + b_{\mathbf{f}p}), \qquad S^y_{\mathbf{f}p} = (2i)^{-1}(b^\dagger_{\mathbf{f}p} - b_{\mathbf{f}p}),\\ S^z_{\mathbf{f}p} &= 2^{-1}(1 - 2b^\dagger_{\mathbf{f}p}b_{\mathbf{f}p}).\end{aligned} \tag{9.5}$$

[272] N. N. Bogoljubov and S. V. Tyablikov, *Dokl. Akad. Nauk SSSR* **126**, 53 (1959).

The relative magnetization per spin in the pth sublattice of the film is

$$\sigma_p = \langle 1 - 2n_{\mathbf{f}p} \rangle. \tag{9.6}$$

It is possible to find the average $\langle b^\dagger_{\mathbf{f}p} b_{\mathbf{f}p} \rangle$ as a particular value of the correlation function

$$\langle b^\dagger_{\mathbf{g}p'} b_{\mathbf{f}p} \rangle = \int_{-\infty}^{\infty} dE I_{b_{\mathbf{f}p} b^\dagger_{\mathbf{g}p'}}(E). \tag{9.7}$$

The spectral intensity $I_{b_{\mathbf{f}p} b^\dagger_{\mathbf{g}p'}}$ of the correlation function is related to the Fourier transforms of the double time and temperature dependent Green's function by the equation[273]

$$I_{b_{\mathbf{f}p} b^\dagger_{\mathbf{g}p'}}(E) = i(\langle\langle b_{\mathbf{f}p} | b^\dagger_{\mathbf{g}p'} \rangle\rangle_{E+i\eta} - \langle\langle b_{\mathbf{f}p} | b^\dagger_{\mathbf{g}p'} \rangle\rangle_{E-i\eta})/(e^{\beta E} - 1). \tag{9.8}$$

These Fourier transforms appearing in the right-hand side of Eq. (9.8) are found from the coupled equations

$$E\langle\langle b_{\mathbf{f}p} | b^\dagger_{\mathbf{g}p'} \rangle\rangle_E = \sigma_p (2\pi)^{-1} \delta_{\mathbf{f}p,\mathbf{g}p'} + \langle\langle [b_{\mathbf{f}p}, \mathscr{H}] | b^\dagger_{\mathbf{g}p'} \rangle\rangle_E . \tag{9.9}$$

With the expansion (9.3) of the effective Hamiltonian one finds

$$\begin{aligned} E\langle\langle b_{\mathbf{f}p} | b^\dagger_{\mathbf{g}p'} \rangle\rangle_E = {} & \sigma_p (2\pi)^{-1} \delta_{\mathbf{f}p,\mathbf{g}p'} + \Big(\sum_{p''} A^0_{pp''} + g\mu_B H_\parallel \Big) \langle\langle b_{\mathbf{f}p} | b^\dagger_{\mathbf{g}p'} \rangle\rangle_E \\ & - \sum_{\mathbf{h}p''} A_{\mathbf{f}p,\mathbf{h}p''} \langle\langle b_{\mathbf{h}p''} | b^\dagger_{\mathbf{g}p'} \rangle\rangle_E \\ & + 2 \sum_{\mathbf{h}p''} A_{\mathbf{f}p,\mathbf{h}p''} \langle\langle n_{\mathbf{f}p} b_{\mathbf{h}p''} | b^\dagger_{\mathbf{g}p'} \rangle\rangle_E \\ & - 2 \sum_{\mathbf{h}p''} A_{\mathbf{f}p,\mathbf{h}p''} \langle\langle n_{\mathbf{h}p''} b_{\mathbf{f}p} | b^\dagger_{\mathbf{g}p'} \rangle\rangle_E \end{aligned} \tag{9.10}$$

A certain decoupling procedure[274,275] has to be used in order to express

[273] Here the following definition of the Green function is used:

$$\langle\langle b_{\mathbf{f}p} \mid b^\dagger_{\mathbf{g}p'} \rangle\rangle^{(r)} = -i\theta(t - t')(\langle b_{\mathbf{f}p} b^\dagger_{\mathbf{g}p'} \rangle - \langle b^\dagger_{\mathbf{g}p'} b_{\mathbf{f}p} \rangle)$$

where $b_{\mathbf{f}p}$, $b_{\mathbf{g}p'}$ are time-dependent Heisenberg operators, and θ is the step function of Heaviside. The Fourier expansion of the Green function is

$$\langle\langle b_{\mathbf{f}p} \mid b^\dagger_{\mathbf{g}p'} \rangle\rangle^{(r)} = \int_{-\infty}^{\infty} dE \langle\langle b_{\mathbf{f}p} \mid b^\dagger_{\mathbf{g}p'} \rangle\rangle_E \exp[-iE(t - t')].$$

[274] S. V. Tyablikov, *Phys. Metal. Metaloved.* **16**, 321 (1963).
[275] H. B. Callen, *Phys. Rev.* **130**, 890 (1963).

the two last Green's functions by the former. Following Tyablikov we have

$$\langle\langle n_{\mathbf{f}p} b_{\mathbf{h}p''} | b^{\dagger}_{\mathbf{g}p'} \rangle\rangle_E \sim \langle n_{\mathbf{f}p} \rangle \langle\langle b_{\mathbf{h}p''} | b^{\dagger}_{\mathbf{g}p'} \rangle\rangle_E$$

$$= (1 - \sigma_p)/2 \langle\langle b_{\mathbf{h}p''} | b^{\dagger}_{\mathbf{g}p'} \rangle\rangle_E \qquad (9.11)$$

Using now the plane expansion of the type (8.10) of the Fourier component of the Green's function

$$\langle\langle b_{\mathbf{f}p} | b^{\dagger}_{\mathbf{g}p'} \rangle\rangle = N^{-1} \sum_{\boldsymbol{\nu}} G^{\boldsymbol{\nu}}_{pp'}(E) \exp[i(\mathbf{f} - \mathbf{g})\, \boldsymbol{\nu}], \qquad (9.12)$$

as suggested by the translational symmetry of the film, the problem is restricted to the solving of the system of equations

$$\left[E - \left(g\mu_{\mathrm{B}} H_{\parallel} + \sum_{p''} A^{0}_{pp''} \sigma_{p''}\right)\right] G^{\boldsymbol{\nu}}_{pp'}(E)$$

$$+ \sigma_p \sum_{p''} A_{pp''} G^{\boldsymbol{\nu}}_{pp''}(E) = \sigma_p (2\pi)^{-1} \delta_{pp'} \, . \qquad (9.13)$$

It is clear that this system occurs as a more complicated variant of the boundary condition problem, as reported in the molecular field and spin waves approximations.

In fact, the Green's function method was conceived as a renormalized spin wave approximation. Thanks to the specific magnetizations appearing in the Eqs. (9.13) the results on the spectrum of elementary excitations are improved. To solve approximately the system (9.13) one replaces on the left-hand side the specific monolayer magnetizations σ_p by their mean value σ. Using the notation

$$\epsilon = (E - g\mu_{\mathrm{B}} H_{\parallel})/(A\sigma) - \Gamma + \Gamma_0{}^{\boldsymbol{\nu}}, \qquad (9.14)$$

where Γ is the total number of nearest neighbours, the "uniformized system" is

$$\epsilon G^{\boldsymbol{\nu}}_{pp'}(E) = \sum_{t=1}^{D} \{\Gamma_t{}^{\boldsymbol{\nu}} G^{\boldsymbol{\nu}}_{p+t,p'}(E) + \Gamma^{\boldsymbol{\nu}}_{-t} G^{\boldsymbol{\nu}}_{p-t,p'}(E)\}$$

$$+ \sigma_p/(2\pi A\sigma)\, \delta_{pp'} , \qquad p, p' = 1, 2, \ldots, q. \qquad (9.15)$$

with the boundary conditions

$$\sum_{t=-D}^{D} (1 - \Delta(p,t)) \{\Gamma_t{}^{0} G^{\boldsymbol{\nu}}_{pp'}(E) - \Gamma_t{}^{\boldsymbol{\nu}} G^{\boldsymbol{\nu}}_{p+t,p'}(E)\} = 0,$$

$$1 \leqslant p \leqslant D, \qquad q - D + 1 \leqslant p \leqslant q. \qquad (9.16)$$

Here $2D$ is the maximum number of monolayers still containing first-order neighbors of an arbitrary inner atom. One sees that, except for the inhomogeneity term $\sigma_p/(2\pi A\sigma)$, the above problem is completely similar to that of finding the spin wave amplitudes and the dispersion law in the quasisaturation approximation, so that the Green's function method works indeed as a renormalized spin wave theory.

The total relative magnetization of a s.c. system with {100} surface is given by the following expression

$$\sigma^{-1} = (2\pi)^{-2} (Nq)^{-1} \mathscr{S} \sum_{\tau=0}^{q-1} \iint_{-\pi/a}^{\pi/a} d\nu_y \, d\nu_z \coth(2^{-1}\beta E_{\tau\nu}) \tag{9.17}$$

where $\mathscr{S}$ is the film surface,

$$E_{\tau\nu} = g\mu_B H_{\|} + \sigma[2A_{\|}(2 - \cos \nu_y a - \cos \nu_z a) + 2A_{\perp}(1 - \cos(\pi\tau/q))]. \tag{9.18}$$

$H_{\|}$ is related to an anisotropy term of unidirectional form $-g\mu_B H_{\|} \sum S^z_{fp}$ and $A_{\|}$ and $A_{\perp}$ are, respectively, the exchange integrals of the two atoms in the same layer and in different layers.

The quantity $E_{\tau\nu}$ may be considered as the energy of an elementary excitation in a ferromagnetic thin film, the "spin wave" energy. The expression for $E_{\tau\nu}$ is of the same form as the usual formula for the spin wave energy but differs from this through the dependence on σ (instead of S) and hence on the temperature. As a rule (see Wojtczak[244]) we can state that the poles of the retarded Green's function, $E_{\tau\nu}$, can be obtained from the usual dispersion relation of the spin wave theory, where for $\frac{1}{2}$ spin we have to replace $2S$ by σ.

As just mentioned, the anisotropy term introduced in the spin Hamiltonian is unidirectional. Corciovei[74] has pointed out the difficulties which arise when such a term is used. Corciovei and Costache[276,277] have supposed that

$$H_{\|} = K_{\|}\sigma, \tag{9.19}$$

i.e., in the plane of the film a two-ion anisotropy acts. This anisotropy is uniaxial and better describes the true physical conditions of the thin film.

Furthermore, the problem of the thickness dependence of the Curie temperature as well as that of the spontaneous magnetization was analytically studied.[223,277] The problem is closely connected with the transition from two-dimensional to three-dimensional systems; the way

[276] A. Corciovei and G. Costache, *Phys. Lett.* **25A**, 458 (1967).

[277] A. Corciovei and G. Costache, *Rev. Roum. Phys.* **12**, 687 (1967).

in which this transition actually occurs with increasing thickness was found.

Using the procedure due to Costache,[278] a direct comparison between the behavior of a ferromagnetic thin film and that of the corresponding bulk sample could be made. It was proved that for very thin films the film behavior predominates.

In the low temperature limit the spin wave result of Döring[230] are recovered:

$$\sigma^{-1} = 1 + 2(4\pi\beta A\sigma)^{-3/2} Z_{3/2}(\delta/\epsilon) - q^{-1}(2\beta\pi A\sigma)^{-1}\log(4q\delta/\sqrt{\epsilon}), \tag{9.20}$$

$$\delta = \mu_B K_\parallel/A, \qquad \epsilon = (\beta A\sigma)^{-1}, \qquad Z_l(\rho) = \sum_m e^{-m\rho}m^{-l},$$

provided the following assumptions are made

$$A_\parallel = A_\perp = A, \qquad T \geqslant 5°K, \qquad q \geqslant 4. \tag{9.21}$$

Actually it was proved that Eq. (9.20) can not be applied in the neighborhood of $T = 0°$K, thus avoiding the complications connected with the operation of the third principle of thermodynamics.

In the neighborhood of the transition temperature, the temperature dependence of the magnetization is the same as that predicted by the molecular field theory. However the Curie temperature obtained in this limit does not equal that of the Weiss field method. The dependence of the Curie temperature on the anisotropy qualitatively follows the conclusions given in Section 7 as expected.

For $\delta \ll 1$ and $q \geqslant 4$, the thickness dependence of the Curie temperature was found to be

$$T_c^{\delta=0}(\infty)/T_c^{\delta}(q) \simeq 1 - 8\sqrt{2\delta}/\pi - (4q)^{-1} - (\pi q)^{-1}\log(\delta q/\sqrt{2}). \tag{9.22}$$

The formal machinery involved in the theory of Green's functions certainly depends on the value of the spins constituting the system under study. Thus for $S = \frac{1}{2}$ the theory makes explicit use of the relation between spin operators and Pauli operators. Such a relation does not exist for larger spin values $S > \frac{1}{2}$ and then we have to use another type of Green's function.

The problem of the magnetization and Curie temperature with $S > \frac{1}{2}$

[278] G. Costache *Rev. Roum. Phys.* **14**, 1203 (1969).

was treated by Brodkorb and Haubenreisser.[68,279] The influence of the volume and shape anisotropies on the magnetic properties of the ferromagnetic thin films was investigated. In this case the Hamiltonian of the system contains three terms: the exchange term, the anisotropy term

$$\mathscr{H}_{\text{anis}} = -(K_{\parallel}{}^{s}/2) \sum_{\mathbf{f}p} (S_{\mathbf{f}p}^{z})^2, \tag{9.23}$$

and a dipolar term giving the shape anisotropy.

Following the same procedure with Green's functions defined by

$$\langle\langle S_{\mathbf{f}p}^{+}(t) | B_{\mathbf{g}p'}(t') \rangle\rangle = -i\theta(t-t') \operatorname{tr} e^{-\beta\mathscr{H}}[S_{\mathbf{f}p}^{+}(t), B_{\mathbf{g}p'}(t')], \tag{9.24}$$

and using a modified form of the Tyablikov decoupling procedure,

$$\langle\langle S_{\mathbf{g}p'}^{z} S_{\mathbf{f}p}^{\pm} | B_{\mathbf{h}0} \rangle\rangle \approx \sigma S \langle\langle S_{\mathbf{f}p}^{\pm} | B_{\mathbf{h}0} \rangle\rangle$$

the equations of motion of the retarded Green's functions were obtained. The dipole sums are calculated in the long wave approximation. Using Corciovei's method,[232] these equations were solved and the spin wave energies were found to be

$$E_{\tau\mathbf{v}} = \sigma S [E_{\tau\mathbf{v}}^{(1)} E_{\tau\mathbf{v}}^{(2)}]^{1/2}, \tag{9.25}$$

where

$$\begin{aligned} E_{\tau\mathbf{v}}^{(1)} &= K_{\parallel}{}^{s} + 2A(\Gamma_0{}^0 - \Gamma_0{}^{\mathbf{v}}) + (3/2)(n_1{}^x + 2n_2{}^x) \\ &\quad + 2A\Gamma_1{}^0(2 - \Gamma_1{}^{\mathbf{v}}/\Gamma_1{}^0 x_\tau{}^{\mathbf{v}}) - 2n_2{}^x(1 - 2^{-1}x_\tau{}^{\mathbf{v}}) - \gamma_{\tau\tau}^{\mathbf{v}} \\ E_{\tau\mathbf{v}}^{(2)} &= K_{\parallel}{}^{s} + 2A(\Gamma_0{}^0 - \Gamma_0{}^{\mathbf{v}}) \\ &\quad + 2A\Gamma_1{}^0[2 - (\Gamma_1{}^{\mathbf{v}}/\Gamma_1{}^0)\, x_\tau{}^{\mathbf{v}}] + n_2{}^x(1 - 2^{-1}x_\tau{}^{\mathbf{v}}) + \gamma_{\tau\tau}^{\mathbf{v}}. \end{aligned} \tag{9.26}$$

To obtain these expressions, the following notations were used

$$x_\tau{}^0 = 2\cos[(\pi/q)(\tau - 1)], \qquad x_\tau{}^{\mathbf{v}} \approx x_\tau{}^0, \qquad \tau = 1, 2, \ldots, q$$

$$\begin{aligned} \gamma_{\tau\tau'}^{\mathbf{v}} &= \tfrac{3}{4}n_2{}^x[2A\Gamma_1{}^0 + \tfrac{1}{2}n_2{}^x][2A\Gamma_1{}^{\mathbf{v}} \\ &\quad - \tfrac{1}{4}n_2{}^x](T_{1\tau}^{\mathbf{v}} T_{1\tau'}^{\mathbf{v}} + T_{q\tau}^{\mathbf{v}} T_{q\tau'}^{\mathbf{v}}) \end{aligned}$$

$n_1{}^x$ and $n_2{}^x$ are related the demagnetizing factors and $x_\tau{}^{\mathbf{v}}$ are the eigenvalues of the Corciovei eigenvalue problem[232] applied to this case.

[279] V. Brodkorb and W. Haubenreisser, *Phys. Status Solidi* **16**, 577 (1966).

The relative magnetization σ_p are given by[275]

$$\sigma_p = \{[S - \Phi_p^{(S)}][1 + \Phi_p^{(S)}]^{2S+1} + (S + 1 + \Phi_p^{(S)})(\Phi_p^{(S)})^{2S+1}\}$$
$$\times \{S[(1 + \Phi_p^{(S)})^{2S+1} - (\Phi_p^{(S)})^{2S+1}]\}^{-1}, \qquad (9.27)$$

where

$$\Phi_p^{(S)} = (2N)^{-1} \sum_{\tau=1}^{q} \sum_{\mathbf{v}} (T_{p\tau}^{\mathbf{v}})^2 (\mathscr{E}_{\tau\mathbf{v}}^{(1)}/E_{\tau\mathbf{v}}) \coth[(\beta/2) E_{\tau\mathbf{v}}] - \tfrac{1}{2},$$

$$\mathscr{E}_{\tau\mathbf{v}}^{(1)} = \sigma S\{K_{\parallel}{}^{s} + 2A(\Gamma_0{}^0 - \Gamma_0{}^{\mathbf{v}}) + (1/2)(n_1{}^x + 2n_2{}^x) + A\Gamma_1{}^0$$
$$+ (1/4)\, n_1{}^x - x_\tau{}^{\mathbf{v}}(2A\Gamma_1{}^{\mathbf{v}} - (1/4)\, n_2{}^x)\},$$

$$T_{p\tau}^{\mathbf{v}} \approx T_{p\tau}^{0}$$
$$= [(2 - \delta_{1\tau})/q]^{1/2} \cos[(\pi/(2q))(\tau - 1)(2p - 1)], \qquad p, \tau = 1, 2, \ldots, q.$$

The Curie temperature was found to be

$$T_c(q) = q^{-1}[2S(S + 1)/(3k_B)]$$
$$\times \sum_{p=1}^{q} \left\{(1/N) \sum_{\tau=1}^{q} \sum_{\mathbf{v}} (T_{p\tau}^{\mathbf{v}})^2 [(E_{\tau\mathbf{v}}^{(1)})^{-1} + (E_{\tau\mathbf{v}}^{(2)})^{-1}]\right\}. \qquad (9.28)$$

These results were applied to the study of the magnetic properties of iron thin films, with a {100} surface. The Curie temperature of this system was calculated and compared to that predicted by the molecular field theory and to experimental data as well. It was found that the Curie temperature obtained in the Green's function method is not equal to that predicted by the molecular field theory, but is much closer to that predicted in a *spherical* model of ferromagnetic thin film—a conclusion which was confirmed by the calculations of Corciovei and Costache.[277] The predicted values of Curie temperature were found to be in good agreement for films thicker than six layers with the experimental data of Lee *et al.*[280]

Starting with Eq. (9.27), a calculation of the spontaneous magnetization in the Valenta's approximation, of the entire film also may be made. Such a calculation has been made,[281] and the dependence of the relative spontaneous magnetization both on temperature and thickness was studied for an iron thin film with a {100} surface. Comparison with other

[280] E. L. Lee, P. E. Buldoc, and C. E. Violet, *Phys. Rev. Lett.* **13**, 800 (1964).

[281] W. Haubenreisser, W. Brodkorb, A. Corciovei, and G. Costache, *Phys. Status Solidi* **31**, 245 (1969).

theories was made, namely with the spin wave theory and molecular field theory, which are limiting cases of the Green's function method. The calculations seem to indicate a strong dependence of the spontaneous magnetization on the thickness; this dependence is stronger than that predicted by molecular field theory but more feeble than that given by spin wave theory.

Also some numerical calculations of the *distribution of the magnetization* starting from the expression (9.27) of the mean value σ_p were reported.[210,281] Once again comparison with other theories might be made The numerical calculations were made for the case of iron thin films with {100} surfaces. The spatial distribution of the magnetization depends on temperature. The decrease of the magnetization is negligible near the surface of the film at low temperatures. This decrease is more and more pronounced with increasing temperature but after a given temperature the distribution of magnetization becomes more uniform as the temperature approaches the Curie point when all σ_p vanish at the same time. As compared with the molecular field theory, the Green's function approach yields a less pronounced decrease of the magnetization near the surfaces, although both methods predict that the decrease in the magnetization is limited to a few atomic layers in the neighborhood of the surface.

V. Concluding Remarks

Much of the theoretical discussion in the present paper has been done on a phenomenological basis. To a great extent this has been necessitated by the lack of knowledge concerning a precise accurate theoretical model which would be able to give *ab initio* a quantitative understanding of the magnetic properties of thin films. Experimental facts, accumulated especially during the last ten years, revealed many new specific phenomena present in thin film magnetic configurations, which strongly depend on experimental factors like deposition, size, shape, structure, substrate, processing, contamination and so on. The present paper is limited to the topics which reveal the most striking features of the subject.

A large part of the paper was devoted to the discussion of the macroscopic description of various parts of the magnetic energy and to the corresponding experimental data. On this basis a full account was given to the microscopic description of the Hamiltonian of thin films. One part was devoted to magnetic configurations in thin films with special emphasis on domain structures, stripe domains, domain walls, and magnetization ripple. Some account on an ideal ellipsoidal micromagnet was also given. Some topics have been intentionally omitted such as magnetization curve, magnetization reversal, galvanomagnetic, and magnetooptic phenomena, and so on.

In the fourth part, attention was focused mainly on a typical specimen considered as a single domain thin film magnetized to saturation, preferably in its plane. Molecular field and high temperatures expansions were fully reviewed with special attention paid to various methods of calculation, phase transition, especially Curie temperatures, e.g., constant coupling approximation, Bethe–Peierls–Weiss method, Kirkwood's method. A special section was devoted to elementary excitations in thin ferromagnetic specimens, i.e. spin waves with careful consideration of natural surface defects, surface contamination and substrate influence, effect of pinning and boundary conditions in general. As a consequence magnetic properties at low temperatures and ferromagnetic resonances were examined. The paper ended with a brief account on calculations based on the new techniques of double time-temperature Green's functions.

In the paper several good possibilities for further explorations of the field are reviewed. In closing, it seems worth stressing the value of any studies that can increase our understanding of this very controversial subject of ferromagnetic thin film.

Acknowledgments

The authors are indebted to Mr. Gh. Adam for his kind support in preparing the manuscript, especially Part III.

Author Index

S

T

Subject Index

Cumulative Author Index, Volumes 1–27

E

F

G

H

S

T

V

W

Y

Z